ACS SYMPOSIUM SERIES **671**

Molecular Markers in Environmental Geochemistry

Robert P. Eganhouse, EDITOR

U.S. Geological Survey

Developed from a symposium sponsored by the
Divisions of Environmental Chemistry, Inc., and Geochemistry, Inc.

American Chemical Society, Washington, DC

UNIV. OF CALIFORNIA
WITHDRAWN

Library of Congress Cataloging-in-Publication Data

Molecular markers in environmental geochemistry / Robert P. Eganhouse, editor.

p.   cm.—(ACS symposium series, ISSN 0097–6156; 671)

"Developed from a symposium sponsored by the divisions of Environmental Chemistry, Inc., and Geochemistry, Inc., at the 212th National Meeting of the American Chemical Society, Orlando, Florida, August 25–29, 1996."

Includes bibliographical references and indexes.

ISBN 0–8412–3518–X

1. Environmental geochemistry—Congresses. 2. Biochemical markers—Congresses.

I. Eganhouse, Robert P. (Robert Paul), 1948–   . II. American Chemical Society. Division of Environmental Chemistry. III. American Chemical Society. Division of Geochemistry. IV. American Chemical Society. Meeting (212th: 1996: Orlando, Fla.) V. Series.

QE516.4.M65   1997
628—dc21   97–16625
                                                                      CIP

This book is printed on acid-free, recycled paper.

PRINTED IN THE UNITED STATES OF AMERICA

Foreword

THE ACS SYMPOSIUM SERIES was first published in 1974 to provide a mechanism for publishing symposia quickly in book form. The purpose of this series is to publish comprehensive books developed from symposia, which are usually "snapshots in time" of the current research being done on a topic, plus some review material on the topic. For this reason, it is necessary that the papers be published as quickly as possible.

Before a symposium-based book is put under contract, the proposed table of contents is reviewed for appropriateness to the topic and for comprehensiveness of the collection. Some papers are excluded at this point, and others are added to round out the scope of the volume. In addition, a draft of each paper is peer-reviewed prior to final acceptance or rejection. This anonymous review process is supervised by the organizer(s) of the symposium, who become the editor(s) of the book. The authors then revise their papers according to the recommendations of both the reviewers and the editors, prepare camera-ready copy, and submit the final papers to the editors, who check that all necessary revisions have been made.

As a rule, only original research papers and original review papers are included in the volumes. Verbatim reproductions of previously published papers are not accepted.

ACS BOOKS DEPARTMENT

Contents

CONTEMPORARY BIOGENIC MARKERS

FOSSIL BIOMARKERS

ANTHROPOGENIC MARKERS

INDEXES

Preface

MOLECULAR MARKERS ARE EXTREMELY POWERFUL TOOLS that can provide information about sources of organic matter and the physical, chemical, and biological processes that act on them. They become particularly valuable when there is no direct means of establishing the contribution of a nonspecific pollutant or when it is of interest to know whether the target compounds have undergone alteration following release to the environment. Unfortunately, molecular markers went largely unnoticed by environmental chemists for many years. At the same time, only a few individuals with formal training in organic geochemistry or related fields pursued research on environmental problems. Consequently, the growth of *environmental organic geochemistry*, a term I use here to characterize this field, has until recently been rather gradual.

In October 1993, almost exactly 20 years after first entering the nascent field of environmental organic geochemistry, I was fortunate to participate in a workshop sponsored by the International Atomic Energy Agency. The meeting brought together a handful of experts for the purpose of developing a working document for the coastal pollution monitoring program in the Mediterranean Sea. Participants were asked to provide detailed analytical procedures and programmatic guidelines for implementation of a variety of anthropogenic molecular markers in the monitoring program. Upon returning to my laboratory in the United States, I had discussions with a visiting Australian scientist, Mark Rayner, about the meeting. Mark was also interested in the use of molecular markers, in his case for delineation of coastal pollution problems off Sydney in relation to the presence of submarine wastewater outfalls (*cf.* Chapter 20). It was at this time that the idea for an international meeting, focused on molecular markers and their many applications in environmental geochemistry, crystallized in my mind. The time was right, it seemed to me, for such a meeting.

That idea culminated in a symposium presented at the 212th National Meeting of the American Chemical Society, titled "Application of Molecular Markers in Environmental Geochemistry", jointly sponsored by the ACS Divisions of Environmental Chemistry, Inc., and Geochemistry, Inc., in Orlando, Florida, August 25–29, 1996. The symposium consisted of 54 oral and 22 poster presentations delivered by scientists from many countries. My initial concept for the symposium was limited to one class of molecular markers, those derived from human sources (i.e., anthropogenic markers). However, the scope of the meeting quickly expanded as John Volkman and Peter Nichols, members of the "Australian connection", suggested that modern and ancient biological markers should also be included. The format for the symposium was set with the addition of contaminant assemblages, complex mixtures such as the polychlorinated biphenyls and polycyclic aromatic hydrocarbons, that can serve as powerful molecular process probes. The four theme areas were recent biogenic markers, anthropogenic markers, fossil biomarkers, and contaminant assemblages.

The symposium papers were presented by a diverse international cast of scientists representing a variety of disciplines, including organic geochemistry, environmental chemistry, atmospheric chemistry, microbial ecology, and marine biology. Because of this diversity and limitations imposed by traditional disciplinary boundaries, many of these scientists had not previously come in contact with each other. Thus, it became the primary objective of the meeting to facilitate interdisciplinary discussion and foster scientific collaboration. As judged by the many collaborations and communications initiated or enhanced during and after the symposium, it was a success. Emboldened by the raw enthusiasm of symposium participants, I decided that some follow-up and documentation were in order. Hence, this book.

To a large extent, the format of the symposium has been retained in the book. The reader will find contributions from the four previously mentioned theme areas. However, the book is organized into only three sections: contemporary biogenic markers, fossil biomarkers, and anthropogenic markers (including contaminant assemblages). The last section is the largest. The lead chapter of each section (i.e., chapters 2, 8, and 12) offers an overview of that theme area, whereby salient concepts are illustrated with examples from ongoing or recent research. These overview chapters are highly recommended for readers wishing to learn about the big picture in a given subject area before delving into the more specific subject matter found in subsequent chapters of that section. Chapters 8 and 12 also contain structural appendices that should prove useful. Although this book contains examples of many of the known applications of molecular markers, coverage is necessarily incomplete, and there are doubtless other applications yet to be discovered. However, the interested reader will find that the references provided in each chapter serve as a valuable guide to the literature.

In some respects, this book and the symposium upon which it is based represent a benchmark. They reflect the fact that the field of *environmental organic geochemistry* is beginning to show signs of maturity. It is my hope that this book will adequately convey some of the most exciting recent advances in this rapidly evolving field and thereby stimulate further research and interaction. If we have done our job correctly, the international flavor of the meeting and the excitement of the participants will be apparent to readers. The tremendous support and efforts of symposium participants and authors has made this an extremely enjoyable and deeply satisfying professional experience for me.

Acknowledgments

Many organizations and individuals deserve credit for the success of the symposium and the development of this book. I thank the symposium participants and authors, whose efforts made this book possible. For their important conceptual contributions and continuing encouragement, I extend special thanks to John Volkman, Peter Nichols, and Mark Rayner. Jim Pontolillo and Bonnie Hower provided day-to-day editorial services and valuable advice, and R. P. Eganhouse, Sr., contributed to the design of the cover. I sincerely appreciate the financial assistance given by the ACS Divisions of Environmental Chemistry, Inc., and Geochemistry, Inc., as well as the moral support of program chairs Martha Wells, Tim Eglinton, and George Luther. Acknowledgment is made to the donors of the Petroleum Research Fund, administered by the American Chemical Society, for partial support of this activity (ACS–PRF30870–SE). ACS–PRF funds were used to facilitate foreign travel for some of the non-U.S. speakers. I gratefully acknowledge the support of the U.S. Geological Survey during the planning of the symposium and the editing of this book. Finally, I pay tribute to Edward G. Wood, Jr., whose lifework and determination serve as an inspiration for the completion of big projects within very short periods of time.

ROBERT P. EGANHOUSE
U.S. Geological Survey
12201 Sunrise Valley Drive
Reston, VA 20192

March 12, 1997

Chapter 1

Molecular Markers and Environmental Organic Geochemistry: An Overview

Robert P. Eganhouse

U.S. Geological Survey, 12201 Sunrise Valley Drive, Reston, VA 20192

It is said that we live in an *Information Age*. Yet, throughout Man's tenure on Earth the ambient environment has been providing information on a continual basis. In the case of olfactory and taste senses, information is supplied and detected in the form of chemicals. So, in essence, we have been using chemical cues to discriminate among various substances for the entirety of our collective existence. [A good example is wine tasting.] In this book we will be interested in the detection of a particular category of chemicals, *molecular markers*, that can, under appropriate circumstances, provide information about sources of organic material. Some molecular markers are produced by humans either intentionally or inadvertently, whereas others are strictly biosynthetic. The characteristic shared by all molecular markers is the fact that their structures are linked to specific origins. Thus, the observation of one of these compounds in the environment necessarily signals the presence of a specific *source material*. In principle, when several markers are found, one can begin to unravel contributions made by multiple sources (source apportionment). This becomes particularly relevant when non-specific contaminants are also present because the markers can be used to indirectly estimate contributions of contaminants made by different sources.

The chapters to follow examine a variety of molecular markers having diverse origins. However, for purposes of organization we have placed markers into three general categories. *Contemporary Biogenic Markers* (Chapters 2-7) are compounds synthesized by microorganisms and/or vascular plants that are used as cellular constituents. These can be found in living organisms themselves, or, if persistent, as residues that have undergone little or no alteration in the modern environment. The second category, *Fossil Biomarkers* (Chapters 8-11), includes compounds that are present in fossil fuels and ancient sediments as a result of the burial of biogenic organic matter and its subsequent alteration. As such, these markers have clear structural relations to contemporary biogenic markers. The last category we have called *Anthropogenic Markers* (Chapters 12-26). These compounds fall into two general classes based on their real or perceived toxicity and source specificity. The first is populated by synthetic (or sometimes natural) organic compounds which are not considered to be pollutants *per se* but whose presence in the modern environment is directly related to human activity. Examples include the fecal sterol, coprostanol, and synthetic surfactants used in commercial detergents. The second class of anthropogenic markers is composed of *contaminant assemblages* which are known to be toxic and whose pervasive presence in the environment is entirely or primarily attributable to industrial activity. These include such compounds as the PCBs (polychlorinated biphenyls) and PAHs (polycyclic aromatic hydrocarbons), toxaphene and chlorinated pesticides.

In the chapters to follow the reader will see how these various types of markers can be used to gain a better understanding of our contemporary environment and to assess the impact of human activities on it. Many types of information can be derived from studies of these molecules, information that frequently cannot be obtained by conventional biogeochemical approaches or by direct examination of the contaminants themselves. In this sense they serve not only as source indicators but also as molecular 'process probes'.

Historical Perspective

Origin of the 'Molecular Marker Concept'. Our current appreciation of the *molecular marker concept* comes from the discipline of organic geochemistry. Early work by Alfred Treibs on the organic constituents of shales, bitumens and crude oils led to the discovery in 1934 of a geochemical class of compounds having the tetrapyrrole structure, the petroporphyrins (*1*). Based on similarities between their core molecular structures, Treibs proposed that the petroporphyrins were actually derived from chlorophylls synthesized by ancient plants (Figure 1). He hypothesized that following death, cellular constituents of plants such as the tetrapyrrole pigments undergo a series of biogeochemical transformations (*2*). [The processes responsible for these transformations are collectively referred to as *diagenesis* by geochemists.] Whereas the vast majority ($\approx$ 99.9%) of biologically produced organic matter is remineralized during diagenesis, the fundamental architecture of some of the more stable (or stabilized) molecules may be preserved (*3*). When this occurs over geologic time, a type of 'molecular' fossil (*4*) is formed, the presence of which offers strong evidence of a biological origin. The pathway proposed by Treibs between a recognizable **biological precursor** and a **diagenetic product** was highly significant at the time because it established the first unequivocal molecular evidence for the biological origin of petroleum and simultaneously ushered in the era of molecular organic geochemistry. In the context of the present book, his findings provide a compelling demonstration of the utility of the *molecular marker approach* in explaining complex natural phenomena.

Environmental Organic Geochemistry: Formative Years (1950-70). Between the time of Treibs' original reports and the early 1950s progress in the field of organic geochemistry continued to be limited by the low resolving power and poor sensitivity of available analytical technologies. It was only with great difficulty that individual compounds could be isolated from complex matrices and identified at trace levels.

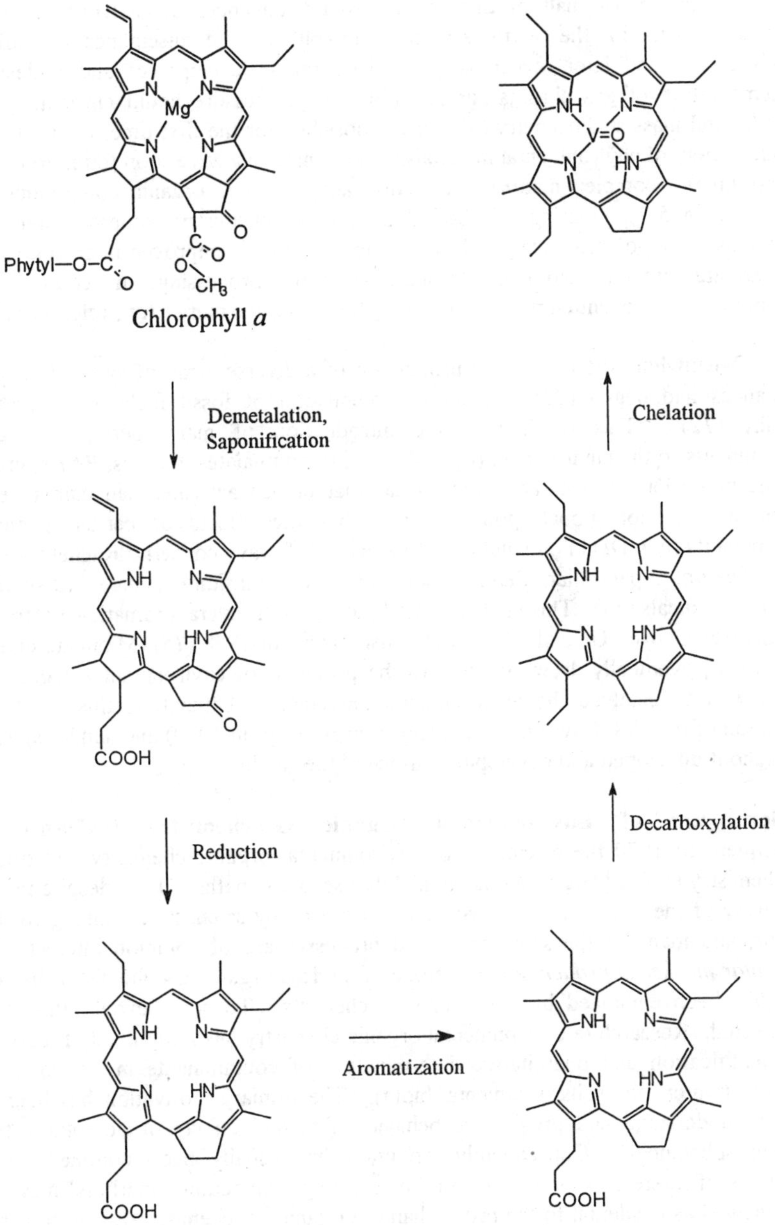

Figure 1. The *Treibs Scheme* for diagenetic conversion of chlorophyll *a* to petroporphyrins.

Environmental chemistry as we know it today simply did not exist. However, at the outset of the second half of this century steady advances in analytical chemistry spurred growth in the earth sciences, notably in the disciplines of organic geochemistry, and later, oceanography. The rapid development and widespread implementation of gas chromatography (GC), high pressure liquid chromatography (HPLC) and mass spectrometry (MS) made possible, for the first time, the analysis of a wide variety of environmental materials. These materials were soon found to contain extraordinarily complex mixtures of natural and synthetic organic compounds (cf., references in 5-7). During this period organic geochemistry witnessed significant growth as these powerful analytical tools enabled a greatly improved understanding of the architecture of biological molecules, their processing in contemporary environments and eventual transformation to fossil organic matter (cf., references in 8-10).

Meanwhile, the large-scale production of a diverse array of synthetic organic substances had begun (11), and global combustion of fossil fuels was increasing steadily (12). This resulted in the introduction of many persistent organic contaminants to the environment (e.g. DDT, PCBs, phthalates, dioxins, PAHs, etc...). By the mid-1960s, it was abundantly clear that human activities had culminated in widespread, if not global, pollution, and that the effects of certain persistent contaminants (e.g. DDT) on wildlife were a matter of grave concern. Rachel Carson's Silent Spring, in particular, identified some of the more insidious dangers of synthetic organic chemicals (13). This landmark book, along with several dramatic catastrophes [e.g. Torrey Canyon Oil spill-1967 (14), Yusho incident-1968 (15); Minimata disease-1953 (16)], eventually drew attention to the problems of environmental degradation and served to catalyze the environmental movement. Ultimately, this led to the formation of the U.S. Environmental Protection Agency (in 1970) and similar agencies throughout developed and developing nations of the world.

Modern Era of Environmental Organic Geochemistry (1970-present). Subsequent to 1970 the disciplines of environmental organic chemistry and organic geochemistry evolved largely along parallel, but separate, paths. This, despite the fact that many of the same analytical techniques are used by scientists belonging to these groups, and many of the same or related processes are of common interest. The molecular marker approach, a cornerstone of modern organic geochemistry, has only gradually been embraced by environmental chemists. To some extent, this is not unexpected. Research in environmental organic chemistry has traditionally focused on the identification and quantitative determination of contaminants in environmental matrices (water, air, soils, sediments, biota). The primary motivation has been the need to understand and predict the behavior, fate and effects of potentially toxic organic substances. Consequently, attention has usually been confined to the pollutants of interest, whereas the value of seemingly innocuous 'markers' may have been viewed as incidental to the task at hand. Organic geochemists, by contrast, have concentrated on establishing the dominant geochemical pathways and mechanisms by which biologically derived organic matter is transformed over geologic time into complex natural deposits such as coal, oil shales and petroleum. The approach is decidedly broad and holistic in character, incorporating as it does all available molecular and geologic information for purposes of providing a self-consistent and unified picture of the processes under study. A major product of this comprehensive effort over the last three decades has been the identification and structural elucidation of many classes of compounds that are commonly found in geological materials (*e.g.* steranes, triterpanes, acyclic isoprenoids, etc...; *cf.*, Chapter 8) as well as the biological precursors from which they were presumably derived (*cf.*, Chapters 2-7). These compounds have come to be known in the organic geochemistry community as '*biomarkers*', a term that bespeaks their biosynthetic origin (*17,18*). There has also been considerable growth in the closely related field of biogeochemistry which is concerned with processes responsible for the transformation and cycling of biogenic organic matter in the contemporary environment. Biogeochemistry plays a central role in studies of global cycling of the elements and the impact of the biosphere on climate.

The first applications of the *molecular marker approach* in environmental studies were reported in the 1960s. Early work with the fecal sterol, coprostanol, showed that this compound was produced by bacteria in the intestinal tracts of mammals *via* hydrogenation of cholesterol and could be detected in receiving water bodies and sediments impacted by human wastes (*19*). Thus, it was suggested that coprostanol might have potential as a molecular tracer of human fecal pollution (*20,21*). In subsequent years a large number of researchers conducted studies of the behavior and distribution of coprostanol in many aquatic environments. However, when considered individually, these investigations tended to be fragmented and limited in scope. For example, no systematic efforts have yet been undertaken to evaluate the stability of coprostanol under a broad spectrum of geochemical conditions or to determine its basic physical properties, and only limited information exists on the source specificity of this compound. In the late 1960s, the occurrence of several oil spills (Torrey Canyon-1967, Buzzard's Bay-1969, Santa Barbara Channel-1969) caught the attention of the public and led to cross-fertilization among the disciplines of organic geochemistry, oceanography and environmental chemistry. As a result, many field and laboratory studies of the fate and effects of oil in aquatic ecosystems were conducted during this period (*22*). Pioneering work by Max Blumer and co-workers demonstrated how the *molecular marker approach* could be used effectively to differentiate biogenic and petroleum hydrocarbons (*23,24*) as well as to track the fate of petroleum and other fossil hydrocarbons in the environment (*25,26*). To a large extent this early work established the basis for our current understanding of acute episodes such as the grounding of the *Exxon Valdez* (*27*). However, it also resulted in the recognition that chronic inputs of petroleum were far more important quantitatively than the highly publicized and readily observable spills (*22*). This, in turn, stimulated much research into the composition of organic matter in municipal wastes (*28-30*), urban runoff (*31,32*) and discrete industrial discharges (*33*).

During the late 1970s and early to mid-1980s, a new class of molecular markers emerged (*cf.*, Chapter 12). Several synthetic organic compounds, not ordinarily considered toxic and many of which are related to commercial detergents [*e.g.* vitamin E acetate (*34*), LCABs: long-chain alkylbenzenes (*30,35*), LAS: alkylbenzenesulfonates (*36,37*), alkylphenolpolyethoxylates (*38*), TAMs: trialkylamines (*39*)], were observed in municipal wastes and waste impacted environments. A flurry of research on the origins of these compounds ensued. As information concerning their properties and environmental occurrence accumulated, attention was increasingly directed at understanding their transport and ultimate fate. Research on these compounds continues today (*cf.*, Chapters 12-21) and is complemented by ongoing efforts to search for new markers such that there now is a greatly expanded list of molecules at the environmental organic geochemist's disposal (*40,41*). During this same period (late 1970s-1980s), the unique lipid signatures of many classes of microorganisms were elucidated (*42*). In time this yielded valuable information that is being used today to characterize the structure and activity of microbial communities and to assess their nutritional status (*43; cf.* Chapters 3-6).

During the last 10 years, the number of environmental applications in which molecular markers play a role has expanded, and there is growing evidence of an ongoing convergence among organic geochemistry, microbial ecology and environmental chemistry as disciplinary barriers show signs of breaking down. The talents and experience of classical organic geochemists have increasingly been brought to bear on environmental problems such as apportionment of natural and anthropogenic sources of organic matter in atmospheric aerosols (*e.g. 44; cf.*, Chapters 7,11) and studies of ground water (*45; cf.*, Chapter 26) and soil contamination (*46; cf.*, Chapter 14). At the same time, there appears to be more recognition among environmental chemists of the utility of these powerful tools (*47*). In some cases, the development of new analytical capabilities is reinforcing this evolutionary trend (*48-50*). The result is that the science of environmental organic geochemistry is steadily moving in the direction of a more comprehensive approach in which anthropogenic markers, contemporary biogenic markers and fossil biomarkers are used together to answer complex environmental questions (*51,52; e.g.* Chapter 16).

Atoms and Molecules as Information Carriers

Molecular markers can carry information in three principal ways: 1) the isotopic composition of constituent atoms, 2) three dimensional molecular structure, and 3) the composition of assemblages of contemporaneously formed (*i.e.* co-genetic) compounds (homologs and isomers). We will have a brief look at these manifestations of molecular information and attempt to illustrate their utility for the three categories of molecular markers.

Stable Isotopic Composition. With the exception of the PAHs (*49*) and more recently, chlorinated hydrocarbons (*53*), *anthropogenic markers* have not often been examined for their isotopic composition. Therefore, most of the discussion to follow will focus on the stable isotopic composition of contemporary and ancient residues of biogenic organic matter. Within the atomic nuclei of the light elements involved in life processes (C, N, H, O, S), differences in the number of neutrons give rise to stable and/or unstable isotopes. The isotopic composition of the atoms comprising a given biomolecule depend on the isotopic composition of the component building blocks, fractionations (mostly kinetic) that occur during uptake and at steps along the biosynthetic reaction pathway, and the structure of the molecule itself (*e.g.* chain length, functional groups). Within a single organism this gives rise to intermolecular and intramolecular variations in the isotopic composition of the constituent biochemicals. In addition, because of differences in the biosynthetic pathways used by various autotrophic organisms (*e.g.* C_3, C_4, CAM plants), the isotopic composition of organic matter produced during photosynthesis varies (*54*). Together, these variations and the large fractionation associated with photosynthesis and dissimilatory pathways provide isotopic signatures that can be exploited for purposes of determining the source of organic matter and/or the effects of diagenesis.

Until the last decade, the vast majority of measurements of stable isotopic composition of organic matter were made on bulk organic matter or somewhat less complex fractions isolated from it. Examples of such fractions include the humic acids, kerogen, extractable organic matter, chitin, hydrocarbons, fatty acids and so on. Although much has been learned about the major biogeochemical cycles and their variation over time from these analyses (*55*), the fractions nonetheless comprise complex mixtures of individual organic compounds and, thus, reflect multiple origins. In addition, the effects of diagenesis (and *catagenesis-* thermal breakdown of kerogen) serve to further obscure the isotopic signature of the originally deposited organic matter. Hence, the amount and/or quality of information that can be extracted from such analyses is necessarily limited and dependent upon the type of sample and the degree to which the fractions have be refined. With the advent of compound-specific isotope analyses including stable isotope ratio monitoring-GC/MS (*56*) and compound-specific radiocarbon analysis (*50*) some of these limitations have now been overcome. These techniques permit the chromatographic separation and isotopic analysis of individual compounds. Although environmental applications are just beginning to emerge, this combined molecular-isotopic approach is likely to yield new insights into the origins of selected natural (*e.g. n*-alkanes, perylene, unsaturated long chain alkenones, etc...; *57*) and anthropogenic compounds (*e.g.* PAH; *58*).

Molecular Structure. The second locus of information in a molecule is its three dimensional structure. Here, isomerism is often the structural information-bearing feature of paramount importance. There are two types of isomerism. *Structural isomerism* is characterized by compounds having the same molecular formula but differing in the positions of substituent atoms. Examples of structural isomerism are found among many complex molecular assemblages including those: 1) synthesized industrially (LCABs, PCBs), 2) produced as a result of human activities (*e.g.* PAHs, dioxins), 3) formed by living organisms (*e.g.* fatty acids and aliphatic hydrocarbons) and 4) generated diagenetically or during thermal alteration of kerogen (*e.g.* alkylbenzenes; *59*). *Stereoisomerism*, on the other hand, results from the presence of one or more chiral carbons in a molecule. Stereoisomers are compounds having the same molecular formula and substitution pattern but differing in the way the atoms are oriented in space. Stereoisomers are further classified as either *enantiomers* (mirror images of each other-one or more chiral carbon) or *diastereomers* (non-mirror images-more than one chiral carbon). A special subclass of diastereomers, *geometric isomers,* includes unsaturated compounds which have hindered rotation about the double bond(s).

Biosynthesis leads to the production of organic structures which are uniquely suited to a specific biological function. Examples include the phospholipids, amphipathic molecules that are key structural constituents of the lipid bilayer of cell membranes (*cf.*, Chapter 2), and bacteriohopanepolyols which are used as membrane rigidifiers by many prokaryotes (*cf.*, Chapter 8). As the number of atoms in a molecule increases, the number of structures that are theoretically possible grows exponentially. Because of the highly directed nature of biosynthesis, however, biological molecules having more than about 20 carbon atoms represent a vanishingly small proportion of the total number of possible structures for a given molecular formula. In short, although biomolecules show great diversity, they are structurally limited by design. Upon death of the host organism the molecular constituents of cells suffer a variety of fates (Figure 2). Diagenesis leads to nearly complete destruction of labile biochemicals, but some of the smaller molecules are preserved intact. Meanwhile, various reactions (*i.e.* condensation, reduction, decarboxylation, dehydration, aromatization, rearrangement, etc...) bring about transformation of biomolecules to many other species. Resistant biomacromolecules such as the *algaenans* are preserved as a component of the insoluble organic matter in sediments by virtue of their inherent stability (*60*), whereas (otherwise labile) functionalized lipids may rapidly become sequestered in the insoluble organic matrix during early stages of diagenesis (*61-63*). All of these diagenetic products, though more diverse than the parent biomolecules, still represent only a small fraction of the total number of theoretically possible compounds. Subsequent burial and continued heating (*i.e.* catagenesis) causes breakdown of the insoluble macromolecular organic matter (kerogen) and further alteration of persistent low molecular weight species resulting in the production of the complex mixtures we know as fossil fuels.

In addition to producing restricted 'types' of molecular structures, biosynthetic reactions are stereospecific. This results in molecules with unique stereochemistries that often do not have the thermodynamically favored configuration (*e.g.* sterols, acyclic isoprenoids; see references in *17*, Chapter 8). In these cases, diagenesis can lead to isomerization. Thus, what may have started off as a simple mixture of stereochemically unique biomolecules ends up as a complex assemblage of isomers due to defunctionalization followed by rearrangement and isomerization reactions. [This type of process has been demonstrated in laboratory simulations as shown in Figure 3.] The extent of isomerization and the chiral carbons affected during these reactions are variable and depend upon the structure of the compound in question as well as a variety of diagenetic parameters related to the environment of deposition, mechanism of isomerization, and burial history (time, temperature, depth, etc...) of the sediments. Petroleum geochemists have exploited these relationships in efforts to assess the maturity of source rocks, migration processes, biodegradation, oil maturation and to establish oil-oil and oil-source rock correlations (*17,18,65*). In summary, whereas similarities in structure between *contemporary biogenic markers* and *fossil biomarkers* can be used to infer precursor-product relations, structural differences reflect the effects of post-depositional diagenetic processes. In either case, molecular structure and in particular, isomerism, is the key to understanding functional relationships.

Compounds synthesized industrially may be thought of as an 'exogenous' manifestation of biosynthesis. Here, as in biosynthesis, the formation of specific molecular structures is directed by living organisms (*i.e.* humans) using controlled reactions with particular functions in mind. Nevertheless, there are important

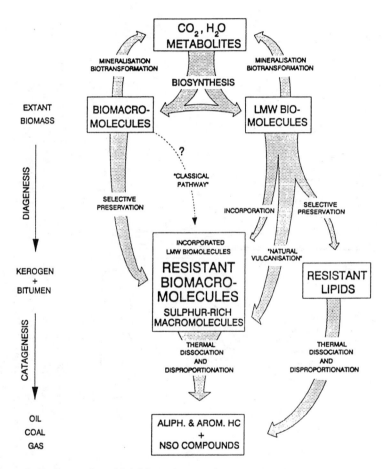

Figure 2. Pathways by which biogenic organic matter is converted to fossil fuels. (Reproduced with permission from reference 60. Copyright 1989 Elsevier Science Ltd.)

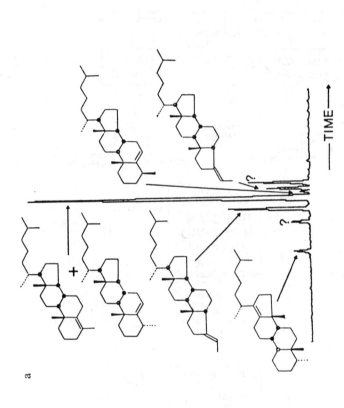

Figure 3. Model studies of the diagenetic transformation of 4-methyl sterenes: a) partial gas chromatogram of the products of acid treatment of 4-methylcholest-4-ene, b) proposed pathways leading from 4-methyl stanols to steranes and diasteranes. (Reproduced with permission from reference 64. Copyright 1986 Elsevier Science Ltd.)

Figure 3. *Continued.*

distinctions to be made between biogenic and synthetic molecules. First, in most cases synthetic chemicals are not also produced intracellularly by living organisms. Thus, the presence of a synthetic organic compound such as DDT in the environment is taken as unequivocal evidence of an anthropogenic source. Second, chemical synthetic processes involving optically inactive reactants that lead to molecules bearing one or more chiral carbons result in production of racemic (non-optically active) enantiomeric mixtures. Because of the presence of chiral carbon atoms in biomolecules and the stereoselectivity of enzyme-controlled reactions, the products of biosynthesis are usually optically active (*i.e.* composed of only one of two possible enantiomers). Therefore, even if biogenic and synthetic molecules were to have the same gross structures, they could, in principle, be distinguished on the basis of stereochemistry. [A more common application, however, is the use of enantiomeric composition to identify the actions of biological processes on synthetic compounds (*66; cf.*, Chapter 23).] Similarly, diagenetic/catagenetic molecular markers (*i.e.* biomarkers) are not synthesized industrially on a large scale. Consequently, there is usually little difficulty in distinguishing between diagenetic/catagenetic molecules and synthetic ones in modern environments. Molecular markers having no chiral carbons or having the same stereochemistry in both living organisms and fossil deposits are few (*9*). The most notable example is the normal alkanes. As discussed below, homolog distributions are often used to distinguish between biogenic sources of *n*-alkanes and those derived from thermal alteration of organic matter (*cf.*, Chapters 9-10,16).

Molecular Assemblages. In addition to the specific molecular structures that typically result from biosynthesis, diagenesis/catagenesis and chemical synthesis, characteristic distributions of homologous series of compounds are often produced (*e.g. n*-alkanes, *n*-fatty acids, sterols, alkylcyclohexanes, LCABs, etc...). As noted above, the complexity of these distributions is sometimes increased by the presence of isomers (either structural or stereochemical). This gives rise to *assemblages* of homologs and isomers (*e.g.* long chain alkenones, triacylglycerols, alkylbenzenes alkylphenolpolyethoxylates, LCABs) which can be viewed as both source indicators and process probes. Depending upon the pathway of formation, these assemblages may provide a unique signature that can be used to differentiate among sources of contributing materials (*cf.*, Chapters 7-11,16). Upon this basis, numerous compositional indices such as the CPI (carbon preference index; *67*) and $5\beta/5\alpha + 5\beta$ (epimeric stanol ratio; *68*) have been proposed as source indicators. Others, such as the I/E ratio (internal/external LAB isomers; *69; cf.*, Chapters 12,16,18), DAR (deethylatrazine/atrazine; *70*) ratio and U^k_{37} (alkenone paleothermometer; *71*), are considered to be process or (paleo)environmental indices. More recently, multivariate statistical techniques (*51,72,73; e.g.* Chapter 16) and other numerical approaches (*74*) have been used to establish sources and explore patterns of temporal and spatial distribution of various marker assemblages.

Because these assemblages are made up of homologs which have systematically different physical properties, and isomers which either have similar (*i.e.* structural isomers, diastereomers) or identical (*i.e.* enantiomers) properties, their compositions can be used effectively as process probes (*e.g.* Chapters 22-26).

Changes in isomeric composition during transport through the environment can provide valuable insights about the influence of physical, chemical and biological processes. For example, deviation of the enantiomeric ratio of hexachlorocyclohexane and chlordane isomers from racemic (*i.e.* 1:1) to non-racemic compositions signals metabolic activity because of the stereoselectivity of such processes *vis à vis* strictly chemical or physical transformations (*75,76;* Chapter 23). Similarly, variations in the rate of attenuation of isomeric alkylbenzenes in oil-contaminated ground water has been used to demonstrate the predominance of microbial degradation over physical removal processes (*77*). On the other hand, differences in the phase distribution of refractory substances reflect the effects of physical partitioning (*78,79;* Chapter 25). These are but a few examples of the ways in which assemblages of markers can be and have been used to probe natural processes.

Criteria for the Ideal Marker

There are two *principal* criteria for an ideal molecular marker. These are:

• Source specificity
• Conservative behavior

Source Specificity. Source specificity refers to the link between a molecular marker (or marker assemblage) and a given source. Under ideal conditions, this link would be direct and unique. In practice, this criterion is met to varying degrees depending upon how the compound in question is formed/produced and where and how it is used (if synthetic). For example, the fecal sterol, coprostanol, is produced *via* biohydrogenation of cholesterol in the intestinal tracts of mammals. Yet, it has also been found as a constituent (albeit a minor one) of the feces of birds (*80*), and under reducing conditions in sediments it can be generated *in situ* by reduction of the double bond in cholesterol (*81*), an ubiquitous sterol (*cf.*, Chapters 20,21). Thus, in reducing environments not heavily impacted by mammalian feces or where marine or other non-human mammalian fecal inputs are dominant, the observation of coprostanol could mistakenly be taken as an indication of the presence of human wastes. This example highlights an important aspect of markers that are not uniquely source specific, namely, that the abundance of the marker in the primary source relative to its abundance in other potential minor sources determines the degree to which a marker can be used for purposes of quantitative source apportionment (*3*).

At the other end of the spectrum are compounds synthesized industrially or otherwise generated as a result of human activities which have no natural counterpart. They have become (to some degree) pervasive because they are used in a wide range of applications (PCBs, phthalates), applied to vast areas of the Earth's surface (pesticides), and/or are effectively mobilized in the atmosphere-hydrosphere-biosphere system (PCBs, PAHs, dioxin, toxaphene). These types of compounds are usually considered to be general, albeit definitive, indicators of human activity/impact. In certain circumstances, however, they can also prove useful as source-specific markers. For example, where point source contamination has been extreme (*e.g.* Hudson River, NY-PCBs; Palos Verdes, CA-DDT; Seveso, Italy-dioxin) and/or the character of the assemblages are unique (*82*) these compounds can be used effectively as source indicators. Where the history of production or use is known they can serve as geochronological tools in sediments (*11*). More often, these conditions do not obtain. This is when other markers with more specific origins (*e.g.* LCABs) and similar physical properties can be used to infer the contribution of a non-specific contaminant (*e.g.* PCBs) to a contaminated site (*83; cf.*, Chapter 12).

Conservative Behavior. The ideal molecular marker should be conservative. In absolute terms this implies a molecule that is refractory over time scales relevant to the process(es) under study. Once introduced to the environment, however, any substance comes immediately under the influence of a multitude of natural processes which can affect its spatial distribution and fate (*84*). Physical processes such as particle transport (advection, sedimentation, mixing, resuspension, aggregation) and phase transfer (volatilization, dissolution, adsorption, desorption) will affect each compound differently depending on its physico-chemical properties and phase association(s). Some compounds may be subject to photolytic or chemical reactions in the atmosphere or in solution. Most molecules will be subject to microbially-mediated transformation and/or complete mineralization in the water column or in sediments. However, the rates are highly dependent on a variety of environmental factors including the dominant terminal electron accepting process, the availability of labile organic matter, temperature and so on. Hydrophobic substances tend to accumulate in tissues of living organisms or may be excreted following metabolism (*cf.*, Chapter 18). In order to properly assess the viability of a potential marker as a quantitative tool, one must know about all of these aspects of their behavior in the environment. Currently, we have a smattering of information on the physical properties of some compounds, and a limited number of laboratory tests have been conducted on the biodegradation of certain markers, but our knowledge in these areas is far from complete. Finally, it is important to remember that a marker serves as an indicator or tracer of something. Consequently, conservative behavior may not always be a necessary condition for its use. The argument can be made that as long as a marker follows a biogeochemical pathway that adequately reflects that of the source material of interest, it is a valid tracer. Hedges and Prahl (*3*) have presented a cogent discussion of the limitations and assumptions inherent in using molecular markers to estimate the abundance of corresponding source materials.

Future Directions and Research Needs

As awareness of molecular markers becomes more widespread, use of these powerful tools will undoubtedly proliferate. One goal of this book is to assist this process by highlighting the many ways in which markers are presently being exploited by environmental researchers in a variety of fields. However, increased awareness and use must be accompanied by refinement of the tools themselves. We are presently at the point where a large number of molecular markers are known, but they are employed primarily as **qualitative** indicators. This limits their utility in the regulatory arena where questions related to sources of contaminants (*i.e.* culpability), the efficacy of remediation efforts, and processes affecting contaminant transport and fate are all-important. Therefore, a critical next step, already beginning to take shape in some studies (*e.g.* Chapters 2,6,12,13,20,26), is the extension of the *molecular marker approach* to **quantitative** applications. Examples include: source apportionment, sediment geochronology, and process partitioning. It is important to recognize that this evolutionary step will require not only an improved understanding of markers that are already identified (*e.g.* Chapters 17,21), but also continuing efforts to discover new ones (*cf.*, Chapter 19).

Generalizing about the data gaps for existing markers is difficult because each marker (or assemblage of markers) is unique. We might start, however, by considering one of the two principal criteria for evaluating the utility/validity of a potential molecular marker, namely, *source-specificity*. In some cases, it is not known with confidence whether a marker has additional sources (*e.g. 85*) or if it might be produced in the environment from other, related precursors (*e.g.* the fecal stanols; Chapter 21). This is particularly important for biological markers which are common to many organisms (*cf.*, Chapters 2,3,7) because some environmentally significant species (of biota) may not yet have been tested. In other instances (*i.e.* synthetic compounds), information on the history of production (*e.g.* optical brighteners; Chapter 15) or the range of applications in which the compound of interest is used (*e.g.* trialkylamines; Chapter 17) may be lacking or unavailable. These issues of source-specificity become especially critical when complex systems receiving multiple inputs are under study because uncertainties in the uniqueness of the source-marker relationship are propagated during source apportionment. Here, the availability of more than one marker for a given source can provide a powerful test of concordance (*3*; Chapters 8,12) or reveal unforeseen differences in behavior (*e.g.* Chapter 17). Researchers should, therefore, strive to adopt a more comprehensive approach in which a variety of markers is used together (*e.g. 51*, Chapter 16). In this context, compound-specific isotopic analyses (^{13}C, ^{14}C, ^{15}N) will become important ancillary tools, particularly when source-specific markers are not readily available (*48-50*).

But, what of the other criterion, **conservative behavior**? Unfortunately, our knowledge of the behavior of various markers in the environment is still often inadequate for quantitative applications. As a first step, it seems important to gather more data on the basic physico-chemical properties of these compounds (aqueous solubility, octanol-water partition coefficient, vapor pressure, etc...) using a variety of analytical and estimation techniques (*84*). The existing database is spotty, at best (*e.g. 41*), and this limits the use of numerical modeling to predict the transport and fate of these compounds (*86,87*). Another important consideration is the nature of phase associations established during synthesis/production and how these affect exchange between solid and aqueous phases (*e.g.* Chapters 3,9,19,22,24) or susceptibility to biodegradation (*e.g.* Chapter 26).

We clearly need more data on the rates at which various markers are biodegraded. These should preferably be collected *in situ* and/or in the laboratory under conditions that closely mimic those found in the natural environment. While a large amount of data on biodegradation of natural and synthetic organic markers has been amassed over the last few decades, the data have never been synthesized and systematically compared. This should be done if only to see what data exist. More importantly, biodegradation studies have typically involved individual markers, not the complex mixtures one encounters in the environment. [A notable exception is the degradation of petroleum and hydrocarbon fuels (*cf.*, Chapter 12).] This makes comparison of biodegradation rates among markers and between markers and related contaminants of interest questionable, at best. Future research efforts in this area should be designed with intercomparability of data in mind, preferably through the simultaneous use of multiple tracers/contaminants. One of the gravest limitations of the present database is a lack of knowledge concerning the effects of critical environmental variables (*e.g.* dominant terminal electron accepting processes, temperature, availability of labile organic matter, etc...) on rates of biodegradation and the metabolic intermediates that are formed. Too often in the past, biodegradation experiments were simply conducted under operationally defined (aerobic or anaerobic) conditions with no knowledge of the fundamental biogeochemical processes driving the observed changes in marker concentration. Moreover, degradation pathways have often been assumed by analogy with other related compounds. In the future these types of studies would best be carried out under well controlled conditions paying due attention to all potentially important master variables and transformation products. Finally, as in the case of biologically mediated degradation processes, there is a general need for more information on the chemical and photochemical transformations these compounds undergo in the atmosphere, surface waters and soils as well as their potential for bioaccumulation in terrestrial and aquatic organisms.

The most pressing need in the area of new markers is the identification of hydrophilic compounds that can be used to trace the movement of water in aquatic systems (*e.g.* ground water) and the soluble organic contaminants that are often found in them. There are, however, two potential obstacles. First, many water-soluble substances are also readily biodegraded. This is especially true of biogenic molecules for which effective dissimilatory enzyme systems have evolved. Second, the analytical determination of highly water soluble organic compounds is frequently difficult. Here, recent advances in techniques such as solid phase extraction and HPLC/MS should offer more opportunities for the isolation and structural elucidation of many uncharacterized components of the dissolved organic matter pool. In any event, the discovery of new hydrophilic markers will require a sustained effort in both the laboratory and the field.

One final frontier that may prove fertile ground for expansion is the numerical analysis of field data. It seems inevitable that as source-marker relations are further defined and analytical methods continue to improve, we will accumulate data sets of increasing size and complexity. The environmental organic geochemist faced with the sheer magnitude of this information will necessarily come to rely on computer-assisted numerical tools such as artificial neural networks and multivariate statistics for the reduction and analysis of data. Although driven by pragmatic needs, this approach should lead to new and exciting insights about Man's impact on the natural environment. In this context, molecular markers can reasonably be expected to play a prominent, if not critical, role.

Literature Cited

1. Treibs, A. *Ann. Chem.* **1934**, *510*, 42-62.

2.   Treibs, A. *Angew. Chem.* **1936**, *49*, 682-686
3.   Hedges, J.I.; Prahl, F.G. In *Organic Geochemistry: Principles and Applications*; Engel, M.H., Macko, S.A., Eds; Plenum Press: New York, NY, 1993, 237-252.
4.   Calvin, M. *Chemical Evolution*; Oxford University Press: New York, NY, 1969; 278p.
5.   *Identification & Analysis of Organic Pollutants in Water*; Keith, L.H., Ed.; Ann Arbor Science: Ann Arbor, MI, 1976; 718p. And references cited therein.
6.   *Environmental Chemistry*; Eglinton, G., Ed.; The Chemical Society Burlington House: London, UK, 1975; Vol. 1; 199p. And references cited therein.
7.   Graedel, T.E. *Chemical Compounds in the Atmosphere*; Academic Press: New York, NY, 1978; 440p.
8.   *Organic Geochemistry: Methods and Results*; Eglinton, G.; Murphy, M.T.J., Eds.; Springer-Verlag: New York, NY, 1969; 828p. And references cited therein.
9.   Tissot, B.P.; Welte, D.H. *Petroleum Formation and Occurrence*; 2nd ed., Springer-Verlag: Berlin, 1984; 699p. And references cited therein.
10. Hunt, J.M. *Petroleum Geochemistry and Geology*, W.H. Freeman and Co.: San Francisco, 1995, 743p. And references cited therein.
11. Eisenreich, S.J.; Capel, P.D., Robbins, J.A.; Bourbonniere, R. *Environ. Sci. Technol.* **1989**, *23*, 1116-1126.
12. Marland, G.; Andres, R.J.; Boden, T.A. In *Trends '93: A Compendium of Data on Global Change*; Boden, T.A., Kaiser, D.P., Sepanski., R., Stoss, F.W., Eds.; Carbon Dioxide Information Analysis Center: Oak Ridge National Laboratory Publication No. ORNL/CDIAC-65, Oak Ridge, TN, 1994; pp. 505-584.
13. Carson, R. *Silent Spring*; Houghton Mifflin: Boston, MA, 1962; 366p.
14. *'Torrey Canyon' Pollution and Marine Life*; Smith, J.E., Ed.; Cambridge University Press: Cambridge, U.K., 1968; 196p.
15. Kuratsune, M. In *Halogenated Biphenyls, Terphenyls, Naphthalenes, Dibenzodioxins and Related Products*; Kimbrough, R.D., Ed.; Elsevier/North-Holland Biomedical Press: New York, NY, 1980; pp. 282-302.
16. Kurland, L.T.; Fara, S.N.; Siedler, H.S. *World Neurol.* **1960**, *1*, 320.
17. *Biological Markers in the Sedimentary Record*; Johns, R.B., Ed.; Methods in Geochemistry and Geophysics, 24; Elsevier: Amsterdam, 1986; 364p. And references cited therein.
18. Peters, K.E.; Moldowan, J.M. *The Biomarker Guide: Interpreting Molecular Fossils in Petroleum and Ancient Sediments*; Prentice Hall: Englewood Cliffs, NJ, 1993; 363p. And references cited therein.
19. Walker, R.W.; Wun, C.K.; Litsky, W. *CRC Critical Reviews in Environmental Control*, **1982**, *12*, 91-112. And references cited therein.
20. Goodfellow, R.M.; Cardoso, J.; Eglinton, G.; Dawson, J.P.; Best, G.A. *Mar. Pollut. Bull.* **1977**, *8*, 272-276.
21. Hatcher, P.G.; Keister, L.E.; McGillivary, P.A. *Bull. Environ. Contam. Toxicol.* **1977**, *17*, 491-498.
22. *Oil in the Sea: Inputs, Fates, and Effects*; Ocean Sciences Board, National Academy of Sciences; National Academy Press: Washington, D.C., 1985; 601pp. And references cited therein.

23. Blumer, M.; Blokker, P.C.; Cowell, E.B. In *A Guide to Marine Pollution*; Goldberg, E.D., Ed.; Gordon and Breach: New York, NY, 1972; pp. 19-40.

24. Farrington, J.W.; Meyers, P.A. In *Environmental Chemistry*; Eglinton, G., Ed.; The Chemical Society: London, UK, 1975, Vol. 1; pp. 109-136.

25. Blumer, M.; Sass, J. *Science* 1972, *176*, 1120-1122.

26. Teal, J.M.; Farrington, J.W.; Burns, K.A.; Stegeman, J.J.; Tripp, B.W.; Woodin, B.; Phinney, C. *Mar. Pollut. Bull.* 1992, *24*, pp. 607-614.

27. Bence, A.E.; Kvenvolden, K.A.; Kennicutt, M.C. *Org. Geochem.* 1996, *24*, 7-42.

28. Van Vleet, E.S.; Quinn, J.G. *Environ. Sci. Technol.* 1977, *11*, 1086-1092.

29. Barrick, R.C. *Environ. Sci. Technol.* 1982, *16*, 682-692.

30. Eganhouse, R.P.; Kaplan, I.R. *Environ. Sci. Technol.* 1982, *16*, 541-551.

31. Hoffman, E.J.; Mills, G.L.; Latimer, J.S.; Quinn, J.G. *Environ. Sci. Technol.* 1984, *18*, 580-587.

32. Eganhouse, R.P.; Simoneit, B.R.T.; Kaplan, I.R. *Environ. Sci. Technol.* 1981, *15*, 35-326.

33. Jungclaus, G.A.; Lopez-Avila, V.; Hites, R.A. *Environ. Sci. Technol.* 1978, *12*, 88-96.

34. Eganhouse, R.P.; Kaplan, I.R. *Environ. Sci. Technol.* 1985, *19*, 282-285.

35. Ishiwatari, R.; Takada, H.; Yun, S-J. *Nature* 1983, *301*, 599-600.

36. McEvoy, J.; Giger, W. *Naturwissenschaften* 1985, *72*, 429-431.

37. De Henau, H.; Mathijs, E.; Hopping, W.D. *Intern. J. Environ. Anal. Chem.* 1986, *26*, 279-293.

38. Giger, W.; Stephanou, E.; Schaffner, C. *Chemosphere* 1981, *10*, 1253-1263.

39. Valls, M.; Bayona, J.M.; Albaigés, J. *Nature* 1989, *337*, 722-724.

40. Vivian, C.M.G *Sci. Total Environ.* 1986, *53*, 5-40. And references cited therein.

41. Takada, H.; Eganhouse, R.P. In *Encyclopedia of Environmental Analysis and Remediation*, John Wiley & Sons: (in review). And references cited therein.

42. Gillan, F.T.; and Johns, R.B. In *Biological Markers in the Sedimentary Record*; Johns, R.B., Ed.; Methods in Geochemistry and Geophysics, 24; Elsevier: Amsterdam, The Netherlands, 1986; pp. 291-309. And references cited therein.

43. White, D.C.; Ringelberg, D.B.; Macnaughton, S.J.; Alugupalli, S.; Schram, D. (Chapter 2, this volume).

44. Mazurek, M.A.; Cass, G.R.; Simoneit, B.R.T. *Environ. Sci. Technol.* 1991, *25*, 684-694. And references cited therein.

45. Eganhouse, R.P.; Baedecker, M.J.; Cozzarelli, I.M.; Aiken, G.R.; Thorn, K.A.; Dorsey, T.F. *Appl. Geoch.* 1993, *8*, 551-567.

46. Richnow, H.H.; Seifert, R.; Hefter, J.; Kästner, M.; Mahro, B.; Michaeles, W. *Org. Geochem.* 1993, *22*, 671-681.

47. Douglas, G.S.; Bence, A.E.; Prince, R.C.; McMillen, S.J.; Butler, E.L. *Environ. Sci. Technol.* 1996, *30*, 2332-2339.

48. Lichtfouse, E.; Eglinton, T.I. *Org. Geochem.* 1995, *23*, 969-973.

49. O'Malley, V.P.; Abrajana, T.A.; Hellou, J. *Environ. Sci. Technol.* 1996, *30*, 634-639.

50. Eglinton, T.I.; Aluwihare, L.I.; Bauer, J.E.; Druffel, E.R.M.; McNichol, A.P. *Anal. Chem.* 1996, *68*, 904-912.

51. Readman, J.W.; Mantoura, R.F.C.; Llewellyn, C.A.; Preston, M.R.; Reeves, A.D. *Intern. J. Environ. Anal. Chem.* **1986**, *27*, 29-54.
52. Bouloubassi, I.; Saliot, A. *Oceanol. Acta* **1993**, *16*, 145-161.
53. Aravena, R.; Frape, S.K.; Van Warmerdam, E.M.; Drimmie, R.J.; Moore, B.J. *Isotopes in Water Resources Management*; Symposium Proceedings, IAEA/UNESCO: Vienna, Austria, March 20-24, 1995,1996; Vol. 1, pp. 31-42.
54. O'Leary, M.H. *Phytochem.* **1981**, *20*, 553-567.
55. *Earth's Earliest Biosphere*; Schopf, J.W., Ed.; Princeton University Press: Princeton, NJ, 1983; 543pp.
56. Hayes, J.M.; Freeman, K.H.; Popp, B.N.; Hoham, C.H. *Org. Geochem.* **1990**, *16*, 1115-1128.
57. Bird, M.I.; Summons, R.E.; Gagan, M.K.; Roksandic, Z.; Dowling, L.; Head, J.; Fifield, L.K.; Cresswell, R.G.; Johnson, D.P. *Geochim. Cosmochim. Acta* **1995**, *59*, 2853-2857.
58. Sheffield, A.E.; Gordon, G.E.; Currie, L.A.; Riederer, G.E. *Atmosph. Env.* **1994**, *28*, 1371-1384.
59. Ellis, L.; Singh, R.K.; Alexander, R.; Kagi, R.I. *Geochim. Cosmochim. Acta* **1995**, *59*, 5133-5140. And references cited therein.
60. Tegelaar, E.W.; de Leeuw, J.W.; Derenne, S.; Largeau, C. *Geochim. Cosmochim. Acta* **1989**, *53*, 3103-3106.
61. Sinninghe, Damsté, J.S.; Eglinton, T.I.; de Leeuw, J.W.; Schenck, P.A. *Geochim. Cosmochim. Acta* **1989**, *53*, 873-889.
62. Sinninghe Damsté, J.S.; van Duin, A.C.T.; Hollander, D.; Kohnen, M.E.L.; de Leeuw, J.W. *Geochim. Cosmochim. Acta* **1995**, *59*, 5141-5147.
63. Mycke, F.; Marjes, F.; Michaelis, W. *Nature* **1987**, *326*, 179-181.
64. Wolff, G.A.; Lamb, N.A.; Maxwell, J.R. *Geochim. Cosmochim. Acta* **1986**, *50*, 335-342.
65. Philp, R.P.; *Mass Spec. Rev.* **1985**, *4*, 1-54.
66. Faller, J.; Hühnerfuss, H.; König, W.A.; Krebber, R.; Ludwig, P. *Env. Sci. Technol.* **1991**, *25*, 676-678.
67. Bray, E.E.; Evans, E.D. *Geochim. Cosmochim. Acta* **1961**,*22*,2-15.
68. Grimalt, J.O.; Fernàndez, P.; Bayona, J.M.; Albaigés, J. *Env. Sci. Technol.* **1990**, *24*, 357-363.
69. Takada, H.; Ishiwatari, R.; *Env. Sci. Technol.* **1987**, *21*, 875-883.
70. Thurman, E.M., Meyer, M.T., Mills, M.S., Zimmerman, L.R., Perry, C.A. *Env. Sci. Technol.*, **1994**, *28*, 2267-2277.
71. Brassel, S.C.; Eglinton, G.; Marlowe, I.T.; Pflaumann, U.; Sarnthein, M. *Nature* **1986**, *320*, 1-5.
72. Yunker, M.B.;Snowdon, L.R.; MacDonald, R.W.; Smith, J.N.; Fowler, M.G.; Skibo, D.N.; McLaughlin, F.A.; Danyushevskaya, A.I.; Petrova, V.I.; Ivanov, G.I. *Env. Sci. Technol.* **1996**, *30*, 1310-1320.
73. Reemstma,T.; Ittekot, V.; *Org. Geochem.* **1992**, *18*, 121-129.
74. Almeida, J.S.; Sonesson, A.; Ringelberg, D.B.; White, D.C. *Bin. Comput. Microb.* **1995**, *7*, 53-59.

75. Falconer, R.L.; Bidleman, T.F.; Gregor, D.J.; Semkin, R.; Teixeira, C. *Env. Sci. Technol.* **1995**, *29*, 1297-1302.
76. Buser, H-R.; Müller, M.D.; Rappe, C. *Env. Sci. Technol.* **1992**, *26*, 1533-1540.
77. Eganhouse, R.P.; Dorsey, T.F.; Phinney, C.S.; Westcott, A.M. *Env. Sci. Technol.* **1996**, *30*, 3304-3312.
78. Burkhard, L.P.; Armstrong, D.E.; Andren, A.W. *Chemosphere* **1985**, *14*, 1703-1716.
79. Baker, J.E.; Capel, P.D.; Eisenreich, S.J. *Env. Sci. Technol.* **1986**, *20*, 1136-1143.
80. Leeming, R.; Latham, V.; Rayner, M.; Nichols, P. (Chapter 20, this volume).
81. Gaskell, S.J.; Eglinton, G. *Nature* **1975**, *254*, 209-211.
82. Merrill, E.G.; Wade, T.L. *Env. Sci. Technol.* **1985**, *19*, 597-603.
83. Takada, H.; Farrington, J.W.; Bothner, M.H.; Johnson, C.G.; Tripp, B.W. *Env. Sci. Technol.* **1994**, *28*, 1062-1072.
84. Schwarzenbach, R.P.; Gschwend, P.M.; Imboden, D.M. *Environmental Organic Geochemistry*; John Wiley & Sons, Inc.: New York, NY, 1993; 681p.
85. Ellis, L., Langworthy, T.A., Winans, R. *Org. Geochem.* **1996**, *24*, 57-69.
86. Holysh, M., Paterson, S., Mackay, D. *Chemosphere* **1986**, *15*, 3-20.
87. Gledhill, W.E., Saeger, V.W., Trehy, M.L. *Env. Toxic. Chem.* **1991**, *10*, 169-178.

CONTEMPORARY BIOGENIC MARKERS

Chapter 2

Signature Lipid Biomarker Analysis for Quantitative Assessment In Situ of Environmental Microbial Ecology

David C. White[1,2], David B. Ringelberg[1], Sarah J. Macnaughton[3],
Srinivas Alugupalli[3], and David Schram[3]

[1]Center for Environmental Biotechnology, University of Tennessee,
10515 Research Drive, Knoxville, TN 37932–2575
[2]Environmental Sciences Division, Oak Ridge National Laboratory,
Oak Ridge, TN 37831
[3]Microbial Insights, Inc., Knoxville, TN 37953–3044

Examination of the lipid components of microbes in recent sediments has provided a convenient, quantitative, and comprehensive method to define the viable biomass, community composition, and nutritional/physiological activities of the biological communities in the sediments. The lipid extraction provides both a concentration and purification of the lipids from the soils and sediments. The subsequent fractionation, purification, and derivatization, sets up the definitive separation and structural identification by capillary gas chromatography with mass spectral identification of each component. As a part of this signature lipid biomarker (SLB) analysis, the lipid extraction also lyses the cells and allows for recovery of purified nucleic acids for subsequent gene probing with and without enzymatic amplification. This polyphasic analysis adds powerful specificity to the analysis of community microbial ecology. Since the SLB analysis involves detection by mass spectrometry, rates of incorporation of non-radioactive ^{13}C and ^{15}N mass-labeled precursors into signature biomarkers can be utilized to gain insight into specific metabolic activities. Application of electrospray and other external ionization sources to ion-trap mass spectrometry will greatly increase the specificity and sensitivity of the SLB analysis.

The assessment of the microbes and their *in situ* interactions in various environments has proven to be a major problem as it has become increasingly apparent that communities of microbes act differently in geochemical cycles than the sum of the isolated individuals. This has required the application of non-traditional methodology. Classical microbiological methods, that were so successful with infectious disease, have severe limitations for the analysis of environmental samples. Pure-culture isolation, biochemical testing, and/or enumeration by direct microscopic counting or most probable number (MPN) destroy most of the interactions between the various components within the environment. These disruptive methods requiring isolation are not well suited for the estimation of total biomass or the assessment of community composition within environmental samples. Moreover, these classical methods provide little insight into the *in situ* phenotypic activity of the extant microbiota because several of the techniques depend on microbial growth and, thus, select against many microorganisms which are non-culturable under a wide range of conditions. It has been repeatedly documented in the literature that viable counts or direct counts of bacteria attached to sediment grains are difficult to quantify and may grossly underestimate the biomass and community composition of the existing community [1-5]. In addition, the traditional tests including the new molecular biomarker technologies [6] provide little indication of the *in situ* nutritional status or evidence of toxicity within the microbial community.

A step towards the comprehensive analysis of extant microbial communities is known as the MIDI, Microbial Identification System (Microbial ID, Inc., Newark, DE). MIDI analysis permits measurement of free and ester-linked fatty acids from microorganisms which have been **isolated** and subsequently **cultured** from the microbial community. This has been commercialized, and the MIDI system sells a comprehensive database. Bacterial isolates are identified by comparing their fatty acid profiles to the MIDI database which contains over 8000 entries. The utilization of this system for identification of clinical isolates has been remarkably successful. However, application of the MIDI system to the analysis of environmental samples has significant drawbacks. The MIDI system was developed to identify clinical microorganisms and requires their isolation and culture on trypticase soy agar at 27°C. Since many environmental isolates are unable to grow under these restrictive growth conditions, the system does not lend itself to identification of some environmental organisms. Other culture conditions that more closely correspond to environmental conditions in soils and sediments allow culture of greater numbers of organisms, but they often require the generation of new databases. In the MIDI system, the identification of the specific ester-linked components as methyl esters is by their gas chromatographic mobility without mass spectral confirmation. Components can be misidentified, and some specificity can be lost.

A more holistic community-based analysis that has proven to be more applicable methodology for biomarker analysis of sediments is based on the one-phase liquid extraction, fractionation of the lipids, sequential hydrolysis of lipid components, and derivatization followed by quantitative analysis/identification using gas chromatography/ mass spectrometry (GC/MS). Research thus far has concentrated on several unique classes of lipids, including steroids, diglycerides (DG), triglycerides (TG), respiratory quinones (RQ), mycocerosic acids and phenol waxes in the neutral lipid fraction. The glycolipid fraction contains the poly β-hydroxyalkanoate (PHA) and the complex glucosyl glycerides. The polar lipid fraction contains the phospholipid lipid fatty acids (PLFA), lipo-amino acids, plasmalogens, acyl ethers, and sphingolipids. The lipid-extracted residue also contains covalently linked lipids that can be released (after acid methanolysis) and extracted and derivatized. Of these the lipopolysaccharide hydroxy fatty acids (LPS-OHFA) can be analyzed by GC/MS. The combination of these SLB analyses can be used to characterize microorganisms or communities of microorganisms in sediments.

All living cells are surrounded by a membrane containing polar lipids. Lysis of the cellular membrane results in cell death. Since the major polar lipids in sediments are phospholipids, the component fatty acids in phospholipids are one of the most important SLB classes. PLFA are essential membrane components of living cells. Unlike most other biomarkers, phospholipids are typically degraded within hours following cell death [7]. This rapid degradation of the phospholipids establishes the PLFA as ideal biomarkers for viable cells. Thus, the quantification of total PLFA is an accurate measurement of living biomass.

PLFA are particularly useful biomarkers because microbes contain a wide variety of structures that are readily determined by GC/MS, and this structural insight can be utilized in taxonomic identification. Different groups of microorganisms synthesize a variety of PLFA through various biochemical pathways, and PLFA are effective taxonomic markers useful for defining community composition. There is great overlap in PLFA composition amongst many species, so defining each species with a unique pattern of PLFA is impossible in a community analysis. Only about 1% of the organisms in the total sedimentary community has actually been isolated and had their PLFA analyzed. However, the patterns of PLFA when elucidated by careful structural identification of each component by mass spectrometry, and in combination with a more comprehensive analysis of the other lipid components, have proven to be very useful as a community analysis tool. The PLFA pattern of a sediment is much like the IR spectrum or a TOF-SIMS (time of flight secondary ion mass spectrometry) analysis of a complex macromolecule--all the information is there, but it is often not specifically interpretable as to which component is responsible for each spectral line. What is readily interpretable are changes in over-all patterns. Thus, similarities and differences between **communities** can be quantitatively defined. PLFA and other biomarkers have been successfully extracted from environmental matrices such as soils and sediments providing a means for direct *in situ* measurements of the microbiota. PLFA patterns recovered from different microbial communities (Figures 1, 2) illustrate the differences in the PLFA component of the SLB analysis of a subsurface soil and ground water membrane filter retentate from a different site.

Generation of electron withdrawing derivatives such as pentafluorobenzyl esters for GC/MS analysis utilizing negative ion detection gives femtomolar sensitivities that are equivalent to hundreds of bacteria the size of *E. coli* [8]. Microbes contain fatty acids with methyl branching at the *omega* (alkyl, ω) end, the iso-branched fatty acids, or 2 carbons from the ω end, the anteiso branched fatty acids, or branches at various mid chain positions. Multiple methyl branches characterize the micocerosic acids that are particularly useful in defining the pathogenic mycobactria [9]. Double bonds are useful as microbes contain monoenoic PLFA with the double bond usually at the ω 7 or ω 9 position which can be in the *cis* or *trans* configuration. The ω9 monoenoic PLFA are formed from the aerobic desaturase pathway common to all cells, whereas the ω7 PLFA are formed from the anaerobic desaturase pathway that is most often a prokaryotic biochemical pathway [4]. Other bacteria can form monoenoic PLFA with the unsaturation in unusual positions that are often characteristic of the distinct physiologic

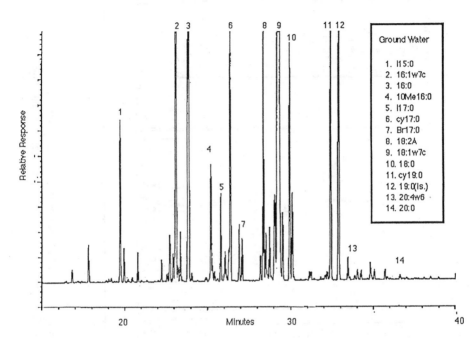

Figure 1.   PLFA patterns from a subsurface soil sample. Lipids were extracted from soil, fractionated, the polar lipid transmethylated and analyzed by GC/MS. PLFA are designated as the number of carbon atoms: number of double bonds and position of the first double bond counting from the alkyl or omega (ω) end of the molecule followed by the c for *cis* or t for *trans* conformation. Prefixes indicate the position counting from the number of carbon atoms from the carboxyl end of the molecule with OH for hydroxyl, br for methyl branching, and cy for cyclopropane ring. Prefixes i represents iso or a for anteiso methyl branching in the PLFA. Responses indicate relative proportions based on flame ionization detection with 19:0 as internal standard.

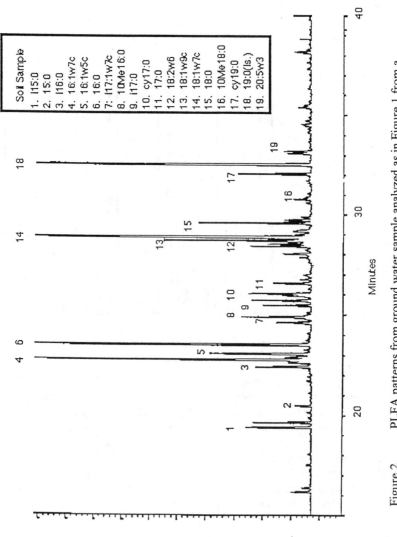

Figure 2   PLFA patterns from ground water sample analyzed as in Figure 1 from a membrane filter retentate (0.2 μm pore diameter) from a different site than Figure 1.

groups of bacteria, plants, or fungi [10]. The position of the unsaturation can be readily determined by fragmentation of dimethyldisulfide adducts of fatty acid methyl esters during GC/MS [11]. Multiple double bonds containing (polyenoic) PLFA are rare in bacteria. PLFA may also have cyclopropane rings that form from monoenoic PLFA in the intact lipids. These are best assessed after mild alkaline transesterification because they are sensitive to acid conditions, and artifacts (o-methoxy esters) may result [12]. True o-methoxy PLFA are known in some bacteria as are PLFA with multiple methyl or polar substituents on the carbon chain [13]. Hydroxy fatty acids with substitution most often at the 2 or 3 position, from the carboxyl, are important biomarkers. LPS-OHFA from bacteria are most often 3-OH whereas sphingolipid derived OHFA are often 2-0H fatty acids.

Comparisons of PLFA patterns by hierarchical cluster analysis shows relationships between individual species of methane-oxidizing bacteria that exactly parallel relationships defined by biochemical activities and the phylogenetic relationships based on the sequence homology of 16S ribosomal RNA [14]. Sequence homology of 16S ribosomal RNA indicates close evolutionary relationships. With time these highly conserved gene sequences diverge, so differences can be related to evolutionary events. A similar study of the PLFA patterns of sulfate-reducing bacteria showed that relationships based on PLFA patterns paralleled the evolutionary phylogeny based on the sequence homology of 16S ribosomal RNA with the exception of one species [15]. Study of that species could possibly establish that some substantial part of its genome came from horizontal gene transfer. Horizontal gene transfer is often detected as gene transfer between different species. Hierarchical cluster analysis of the micocerosic acids and secondary alcohols of 19 isolated and culturally defined species of mycobacteria showed relationships that parallel the physiological properties and genetic analyses of these species [9].

Further the knowledge of specific lipid biosynthetic pathways can provide insight into the physiological or nutritional status of the microbial community. Certain fatty acids, such as *trans* and cyclopropyl PLFA, provide an indications of environmental stress (see below).

Lipids are extracted with organic solvents and then separated into compound classes using column chromatography. The effectiveness of the lipid extraction can be assessed by comparing the extraction before and after sequential alkaline and acid hydrolysis which will liberate all the ester and amide linked fatty components of the cells. In some cases increasing the temperature and pressure of the extraction can substantially increase the yield of lipid and, thus, more effectively indicate the true biomass. Fractions containing SLBs may be further separated by thin layer or liquid chromatography, and individual SLB can be determined after sequential mild alkaline methanolysis, mild acid methanolysis and strong acid methanolysis [10]. Utilizing mild alkaline transesterification for the ester-linked lipid components avoids confusing artifacts such as free fatty acids or loss of acid-sensitive lipid components [4,7]. This analytical scheme (Figure 3) has been successfully applied for quantitatively assessing microbial communities (bacteria, fungi, protozoa, and metazoa) in slimes, muds, soils, filter retentates, bioreactors, and sediments [1-5,10, 16]. SLB methodology provides a

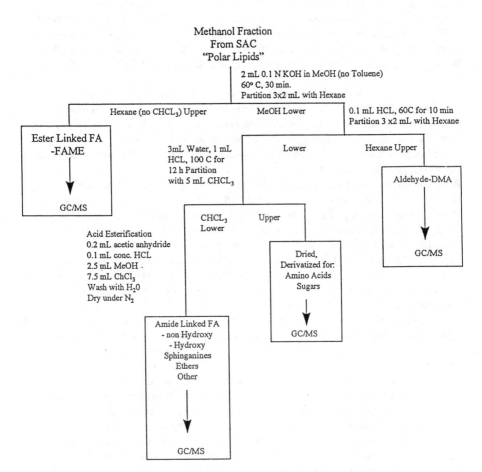

Figure 3.   Sequential hydrolysis of lipid samples to define ester, vinyl ether, alkyl ether, and amide linkages in components of the lipids.

quantitative means to measure: 1) viable microbial biomass, 2) microbial community composition, and 3) community nutritional status.

1). **Viable Biomass** The determination of the total phospholipid ester-linked fatty acids (PLFA) provides a quantitative measure of the viable or potentially viable biomass. The viable microbes have an intact membrane which contains phospholipids (and PLFA). Cellular enzymes hydrolyze and release the phosphate group within minutes to hours following cell death *[7]*. The lipid component that remains is diglyceride (DG). The resulting diglycerides contain the same signature fatty acids as the original phospholipid, at least for days to years in the subsurface sediments. Consequently, a comparison of the ratio of phospholipid fatty acid profiles to diglyceride fatty acid profiles provides a measure of the viable to non-viable microbial abundance and composition. The DG to PLFA ratio increases in many subsurface sediments from <0.2 at the surface to over 2.0 at > 200 m *[17]*. A study of subsurface sediment showed that viable biomass as determined by PLFA was equivalent (but with a much smaller standard deviation) to that estimated by intracellular ATP, cell wall muramic acid, and very carefully conducted acridine orange direct counts (AODC) *[18]*. Problems of converting viable biomass in chemical terms to numbers of microbes has been discussed comprehensively *[19]*, and the number of cells/gm dry weight can vary over more than an order of magnitude in different environments and for different microbes.

2). **Community Composition** The presence of certain groups of microorganisms can be inferred by the detection of unique lipids that originate from specific biosynthetic pathways *[1-5]*. Consequently, the analysis of SLB classes provides a quantitative definition of the microbial community. For example, specific PLFA are prominent in *Desulfovibrio* sulfate-reducing bacteria, whereas the *Desulfobacter* type of sulfate-reducing bacteria contain distinctly different PLFA *[20,21]*. Despite the fact that that only about 1% of the sedimentary microbes in a given sample can be isolated and the SLB characterized, SLB analysis has proven very useful in the characterization of the extant microbial community. The SLB analysis of sediments provides community composition with signature biomarkers for groups of organisms that are reasonably expected to be in that specific community. The induction of microbial community compositional shifts by altering the microenvironment results in changes that are often predictable, based on past experience with microbial communities. For example, biofouling communities incubated in seawater at altered pH in the presence of antibiotics and specific nutrients resulted in a community dominated by fungi, while other conditions resulted in a community dominated almost exclusively by bacteria *[22]*. In these experiments, the morphology of the biofilm seen in scanning electron microscopy correlated with the chemical analysis *[22]*. Similar experiments showed that light-induced shifts which occurred within microbial biofilms induced the expected changes in morphology and shifts in signature lipid biomarkers. Isolation of a specific organism or groups of organisms with the subsequent detection of the same organisms by signature lipid analysis in consortia under conditions where their growth was induced has been done in microcosms *[3]*. It is possible to induce a "crash" in methanogenesis in a bioreactor by inducing the growth of sulfate-reducing bacteria, or by adding traces of chloroform or oxygen. These crashes are accompanied by shifts in the signature lipid biomarkers that are correlated to the changes in the microbial communities. Specific microbes whose signature biomarkers have been determined in isolates can be detected in biofilm communities, clinical specimens, or subsurface soils. Specific sulfate-reducing bacterial groups can be "induced" in estuarine muds as can methane-oxidizing populations, or propane-oxidizing actinomycetes through the addition of appropriate substrates. Again, all of these community shifts are evidenced by measurable changes in lipid signatures and in lipid patterns. An additional validation is the detection of specific shifts in microbial communities as a result of specific grazing by predators. Results of the signature lipid biomarker analysis agreed with the cellular morphologies present as shown by scanning electron microscopy. These validations for the SLB have been reviewed [3,10].

The analysis of other lipids such as the sterols (for the microeukaryotes-- nematodes, algae, protozoa) [22], glycolipids (phototrophs, gram-positive bacteria), or the hydroxy fatty acids from the lipid A component of lipopolysaccharide of gram-negative bacteria [23,24], sphinganines from sphingolipids [25], fatty dimethyl acetals derived from vinyl ether containing plasmalogens [4], and alkyl ether polar lipids derived from the Archae_[26] can provide a more detailed community composition analysis.

3). **Nutritional/Phsiological Status** Growth without cell division can occur in microbes when carbon source(s) and terminal electron acceptors are present but some essential nutrient is missing. Under these conditions of "unbalanced growth" bacterial can accumulate poly β-hydroxyalkanoic acid (PHA) [27-29] and microeucaryotea accumulate triglyceride [30]. The relative amounts of these endogenous storage lipids, as compared to the PLFA, provides a measure of the nutritional status. Specific patterns of PLFA can indicate physiological stress [31,32]. For example exposure to toxic environments can lead to "minicell" formation and a relative increase in specific PLFA. Increased conversion of cis to trans PLFA occurs in Pseudomonas species with exposure to higher concentrations of phenol or organic solvents in the absence of bacterial growth [33]. Increasing proportions of trans-monoenoic PLFA is not the only critical feature of solvent resistance in Pseudomonas putida. Comparison of a solvent sensitive strain of P. pudita to P. pudita (Idaho) strain which is resistant to saturating concentrations of solvents and surfactants, showed that both exhibited increases in trans-monoenoic PLFA. The solvent resistant strain also shifts its lipid composition by increasing the proportion of monoenoic PLFA to saturated PLFA, increasing the lipopolysaccharide hydroxy fatty acids, and decreasing permeability to the hydrophobic antibiotic difloxacin [34]. These changes were not detected in the solvent sensitive strain [34]. Prolonged exposure to conditions inducing stationary growth phase such as limiting concentrations of carbon substrates can induce the formation of cyclopropane PLFA [4, 31,32]. Respiratory quinone composition can be utilized to indicate the proportion of microbial metabolism that is aerobic [35]. Environments with high potential terminal electron acceptors (oxygen, nitrate) induce formation of benzoquinones in Gram-negative bacteria in contrast to microbes respiring on organic substrates form naphthoquinones [35]. Some specific but useful insights come from analysis of organisms like the Pseudomonas species which form acyl-ornithine lipids when growing with limited bioavailable phosphate *[36]* while some gram-positive bacteria form increased levels of acylamino acid phosphatidylglycerols when grown at sub-optimal acid pH levels *[37]*.

Recently it has been shown that the solvent extraction utilized in the SLB lyses cells in the environmental matrix which facilitates the subsequent extraction of nucleic acids from the lipid-extracted residue *[38]*. These cellular nucleic acids can be used for enzymatic amplification and gene probing *[38]*. Over 50% of the gene *nahA* (an enzyme of naphthalene catabolism) present in intact *Pseudomonas fluorescens* cells added to soil was recovered using the lipid extraction protocol as compared to recovery by the standard techniques *[39]*. The DNA recovered from the lipid extraction was of high quality, and is suitable for enzymatic amplification. The combined lipid extraction and recovery of nucleic acids can be very useful in biomass and community composition determinations. The DNA probe analysis offers powerful insights because of the exquisite specificity in the detection of genes in the extant microbial community. The concomitant DNA/lipid analysis readily provides quantitative recoveries independent of the ability to isolate or culture the microbes. The lipid analysis gives indications of the phenotypic properties of the community that indicates extant microbial activity by providing *in situ* indications of starvation, growth rate, exposure to toxicity, unbalanced growth, deficiencies of specific nutrients, and the aerobic/anaerobic metabolic balance *[40]* whereas DNA probes define the physiologic potential and highly specific community composition of the microbial community. The combined DNA/lipid analysis overcomes some deficiencies in microbial ecology studies involving only nucleic acid analysis *[6]*.

Current SLB methodology requires procedures that are time consuming and labor intensive and extraction procedures require extensive attention to detail in the purification of solvents, reagents and glassware. The interpretation of the SLB analysis requires an extensive understanding of a widely dispersed database. New methods to potentially automate the SLB procedures and combine them with nucleic acid analysis are currently under active development so this quantitative analysis can be much more readily applied.

Acknowledgments

This work has been supported in part by grant DE-FG05-90ER60988 from the Subsurface Science Program, administered by F. W. Wobber, grant 94UOT001S and from the National Institute for Global Environmental Change, South East Regional Center, from the U. S. Department of Energy and grants N00014-93-1-1317, N00014-94-1-0441, and N00014-94-1-0765 from the office of Naval Research.

References

1.   White, D.C. Analysis of microorganisms in terms of quantity and activity in natural environments. In *Microbes in their natural environments*; J.H. Slater, R. Whittenbury and J.W.T. Wimpenny Eds.; *Society for General Microbiology Symposium.* **1983**, *34*, 37-66.

2.   White, D.C. Environmental effects testing with quantitative microbial analysis: Chemical signatures correlated with *in situ* biofilm analysis by FT/IR. *Toxicity Assessment.* **1986,** *1,* 315-338.

3.   White, D.C. Validation of quantitative analysis for microbial biomass, community structure, and metabolic activity. *Advances in Limnology.* **1988,** *31,* 1-18.

4.   Tunlid, A. and D.C. White. Biochemical analysis of biomass, community structure, nutritional status, and metabolic activity of the microbial communities in soil. In *Soil Biochemistry* J-M. Bollag, G. Stotzky, Eds; **1991,** *7,* 229-262.

5.   Colwell, R.R., P.R. Brayton, D.J. Grimes, D.B. Rozak, S.A. Huq and L.M. Palmer. Viable but non-culturable *Vibrio cholerae* and related pathogens in the environment: Implications for the release of genetically engineered microorganisms. *Biotechnology.* **1985,** *3,* 817-820.

6.   White, D.C. Is there anything else you need to understand about the microbiota that cannot be derived from analysis of nucleic acids? *Microb. Ecol.* **1994,** *28,* 163-166.

7.   White, D.C., W.M. Davis, J.S. Nickels, J.D. King and R.J. Bobbie. Determination of the sedimentary microbial biomass by extractable lipid phosphate. *Oecologia..* **1979,** *40,* 51-62.

8.   Odham, G., A. Tunlid, G. Westerdahl, L. Larsson, J.B. Guckert and D.C. White. Determination of microbial fatty acid profiles at femtomolar levels in human urine and the initial marine microfouling community by capillary gas chromatography-chemical ionization mass spectrometry with negative ion detection. *J. Microbiol. Methods.* **1985,** *3,* 331-344.

9.   Almeida, J.S., A. Sonesson, D.B. Ringelberg and D.C. White. Application of artificial neural networks (ANN) to the detection of *Mycobacterium tuberculosis,* its antibiotic resistance and prediction of pathogenicity amongst *Mycobacterium spp.* based on signature lipid biomarkers. *Binary Computing in Microbiology.* **1995,** *7,* 53-59.

10. White, D.C., J.O . Stair and D.B. Ringelberg. Quantitative Comparisons of *in situ* Microbial Biodiversity by Signature Biomarker Analysis. **1996,** *J. Indust. Microbiol.* In press.

11. Nichols, P.D., J.B. Guckert and D.C. White. Determination of monounsaturated fatty acid double-bond position and geometry for microbial monocultures and complex consortia by capillary GC-MS of their dimethyl disulphide adducts. *J. Microbiol. Methods.* **1986,.** *5,* 49-55.

12. Maybery W.M. and J.R. Lane. Sequential alkaline saponification/ acidhydrolysis/esterification: A one-tube method with enhanced recovery of both cyclopropane and hydroxylated fatty acids. *J. Microbial Methods.* **1993,** *18,* 21-32.

13. Kerger, B.D., P.D. Nichols, C.P. Antworth, W. Sand, E. Bock, J.C. Cox, T.A. Langworthy and D.C. White. Signature fatty acids in the polar lipids of acid producing *Thiobacilli*: methoxy, cyclopropyl, alpha-hydroxy-cyclopropyl and branched and normal monoenoic fatty acids. *FEMS Microbiol. Ecology* **1986,** *38,* 67-77.

14. Guckert, J.B., D.B. Ringelberg, D.C. White, R.S. Henson and B.J. Bratina. Membrane fatty acids as phenotypic markers in the polyphasic taxonomy of methylotrophs within the proteobacteria. *J. Gen. Microbiol.* **1991,** *137,* 2631-2641.
15. Kohring, L.L., D.B. Ringelberg, R. Devereux, D. Stahl, M.W. Mittelman and D.C. White. Comparison of phylogenetic relationships based on phospholipid fatty acid profiles and ribosomal RNA sequence similarities among dissimilatory sulfate-reducing bacteria. *FEMS Microbiol. Letters.* **1994,** *119,* 303-308.
16. Federle, T.W., M.A. Hullar, R.J. Livingston, D.A. Meter and D.C. White. Spatial distribution of bochemical parameters indicating biomass and community composition of microbial assemblies in estuarine mud flat sediments. *Appl. Environ. Microbiol.* **1983,** *45,* 58-63.
17. Ringelberg, D.B., S. Sutton, ad D.C. White. Microbial ecology of the deep subsurface: Analysis of ester-linked fatty acids, *Microbial Ecology,* **1997,** in press.
18. Balkwill, D.L., F.R. Leach, J.T. Wilson, J.F. McNabb and D.C. White. Equivalence of microbial biomass measures based on membrane lipid and cell wall components, adenosine triphosphate, and direct counts in subsurface sediments. *Microbial Ecology.* **1988,** *16,* 73-84.
19. White, D.C. H.C. Pinkart and D.B. Ringelberg. Biomass Measurements: Biochemical Approaches. In *Manual of Environmental Microbiology,* 1st Edition, *American Society for Microbiology Press,* Washington, DC., C. H. Hurst, G. Knudsen, M. McInerney, L. D. Stetzenbach, and M. Walter, Eds; **1996,** American Society for Microbiology Press, Washington, DC. pp. 91-101.
20. Edlund, A., P.D. Nichols, R. Roffey and D.C. White. Extractable and lipopolysaccharide fatty acid and hydroxy acid profiles from *Desulfovibrio* species. *J. Lipid Res.* **1985,** *26,* 982-988.
21. Dowling, N.J.E., F. Widdel and D.C. White. Phospholipid ester-linked fatty acid biomarkers of acetate-oxidizing sulfate reducers and other sulfide forming bacteria. *J. Gen. Microbiol.* **1986,** *132,* 1815-1825.
22. White, D.C., R.J. Bobbie, J.S. Nickels, S.D. Fazio and W.M. Davis. Nonselective biochemical methods for the determination of fungal mass and community structure in estuarine detrital microflora. *Botanica Marina.* **1980,** *23,* 239-250.
23. Parker, J.H., G.A. Smith, H.L. Fredrickson, J.R. Vestal and D.C. White. Sensitive assay, based on hydroxy-fatty acids from lipopolysaccharide lipid A for gram negative bacteria in sediments. *Appl. Environ. Microbiol.* **1982,** *44,* 1170-1177.
24. Bhat, R.U. and R.W. Carlson. A new method for the analysis of amide-linked hydroxy fatty acids in lipid-A from gram-negative bacteria. *Glycobiology.* **1992,** *2,*: 535-539.
25. Fredrickson, J.K., D.L. Balkwill, G.R. Drake, M.F. Romine, D.B. Ringelberg and D.C. White. Aromatic-degrading *Sphingomonas* isolates from the deep subsurface. *Appl. Envron. Micro.* **1995,** *61,* 1917-1922.

34   MOLECULAR MARKERS IN ENVIRONMENTAL GEOCHEMISTRY

26.   Hedrick, D.B., J.B. Guckert and D.C. White. Archaebacterial ether lipid
       diversity analyzed by supercritical fluid chromatography: Integration with a
       bacterial lipid protocol. *J. Lipid Res.* **1991**, *32*, 659-666.
27.   Nickels, J.S., J.D. King and D.C. White. Poly-beta-hydroxybutyrate
       accumulation as a measure of unbalanced growth of the estuarine detrital
       microbiota. *Appl. Environ. Microbiol.* **1979**, *7*, 459-465.
28.   Findlay, R.H. and D.C. White. Polymeric beta-hydroxyalkanoates from
       environmental samples and *Bacillus megaterium*. *Appl. Environ.
       Microbiol.* **1983**, *45*, 71-78.
29.   Doi, Y.; Microbial Polyesters, VCH Publishers Inc., New York, NY. 1990;
       pp. 1-8.
30.   Gehron, M.J. and D.C. White. Quantitative determination of the nutritional
       status of detrital microbiota and the grazing fauna by triglyceride glycerol
       analysis. *J. Exp. Mar. Biol.* **1982**, *64*, 145-158.
31.   Guckert, J.B., M.A. Hood and D.C. White. Phospholipid, ester-linked fatty
       acid profile changes during nutrient deprivation of *Vibrio cholerae*:
       increases in the *trans/cis* ratio and proportions of cyclopropyl fatty acids.
       Appl. Environ. Microbiol. **1986**, *52*, 794-801.
32.   Guckert, J.B., C.P. Antworth, P.D. Nichols and D.C. White. Phospholipid
       ester-linked fatty acid profiles as reproducible assays for changes in
       prokaryotic community structure of estuarine sediment. *FEMS Microbiol.
       Ecology.* **1985**, *31*, 147-158.23.
33.   Heipieper, H-J., R. Diffenbach and H. Keweloh. Conversion of *cis*
       unsaturated fatty acids to *trans*, a possible mechanism for the protection of
       phenol degrading *Pseudomonas putida* P8 from substrate toxicity. *Appl.
       Environ. Microbiol.* **1992**, *58*, 1847-1852.
34.   Pinkart, H.C., J.W. Wolfram, R. Rogers and D.C. White. Cell envelope changes in
       Solvent-tolerant and solvent sensitive *Pseudomonas putida* strains following
       exposure to O-xylene. *Appl. Env. Microbiol.* **1996**, *62*, 1127-1131.
35.   Hedrick, D.B. and D.C. White. Microbial respiratory quinones in the
       environment I. A sensitive liquid chromatographic method. *J. Microbiol.
       Methods.* **1986**, *5*, 243-254.
36.   Minnikin D.E., and Abdolrahimzadeh H. The replacement of
       phosphatidylethanlolamine and acidic phospholipids by ornithine-amide lipid
       and a minor phosphorus-free lipid in *Pseudomonas fluorescens* NCMB129.
       FEBS Letters. **1974**, *43*, 257-260.
37.   Lennarz, W.J. Bacterial lipids. *In*: Lipid Metabolism. Academic Press, New
       York, NY, Wakil, S. Ed.; 1970; pp 155-183.
38.   Kehrmeyer, S.R., B.M. Appelgate, H. Pinkert, D.B. Hedrick, D.C. White
       and G.S. Sayler. Combined lipid/DNA extraction method for environmental
       samples. *J. Microbiological Methods.* **1996**, *25*, 153-163.
39.   Ogram, A., G.S. Sayler and T. Barkay. The extraction and purification of
       microbial DNA from sediments. *J. Microbiol. Methods.* **1987**, *7*, 57-66.
40.   White, D.C. Chemical ecology: Possible linkage between macro-and
       microbial ecology. *Oikos.* **1995**, *74*, 174-181.

Chapter 3

Diether and Tetraether Phospholipids and Glycolipids as Molecular Markers for Archaebacteria (Archaea)

Morris Kates

Department of Biochemistry, University of Ottawa,
Ottawa, Ontario K1N 6N5, Canada

Archaebacteria (Archaea) inhabit extreme environments globally: extreme halophiles in nearly saturated salty locales; methanogens in anoxic environments; and extreme thermophiles or thermoacidophiles at high temperatures or at both high temperatures and low pH. A new group of archaebacteria has recently been found to be highly concentrated and widely distributed in marine bacterioplankton. Archaea are clearly delineated from all other organisms by their unique ribosomal RNA (rRNA) sequences, cell wall structures and membrane lipids. The latter are derived from diphytanylglycerol diether (archaeol) and its dimer, dibiphytanyldiglycerol tetraether (caldarchaeol), as lipid cores (hydrophobic moieties) instead of from diacylglycerol diesters. Archaeal phospholipids, glycolipids and archaeol or caldarchaeol lipid cores should be useful molecular markers as aids for identification of archaeal subgroups and in some cases of individual genera, and for determination of their biomass in mixed microbial communities. These molecular markers should be useful as an adjunct to rRNA sequencing for detecting and characterizing new archaeal genera in the environment, such as those recently discovered in oceanic bacterioplankton.

Archaea (Archaebacteria) are prokaryotic (non-nucleated) microorganisms that have been assigned separate Domain status, distinct from the two previous Domains, Bacteria and Eukarya (nucleated), but more closely related to the Eukarya than to the Bacteria (1). Archaea, which presumably arose in the Archaean era, are generally known to inhabit extreme environments globally: a) extreme halophiles in saturated or nearly-saturated salty locales (salt flats, salt lakes such as Great Salt Lake, alkaline lakes, and inland seas, such as the Dead Sea, etc.); b) methanogens in anaerobic environments (marshes, rice paddies, wetlands, rumen of herbivores, sewage, marine sediments, deep-ocean hydrothermal vents, etc.); and c) extreme thermophiles and thermoacidophiles in high temperature locales and at both high temperature and low pH, respectively (hot springs such as in Yellowstone National Park, volcanic soils and solfatara, etc.)(2). Recently, however, Archaea have been detected, in high concentration, in marine bacterioplankton samples from the Pacific Ocean, Antarctica and Alaska (3,4), showing that Archaea are not confined to extreme environmental niches, but are widespread and abundant in the global biota (3-5).

Archaea are clearly delineated from all other organisms by their ribosomal RNA (rRNA) sequences, their glycosylated polypeptide and other non-murein (non-Bacterial) cell wall structures, and their unique glycerol ether-type membrane lipids (6). The latter (7,8) are derived from the lipid core diphytanylglycerol diether (1)(termed $C_{20}C_{20}$-archaeol (9)) and variants such as $C_{20}C_{25}$- (1A), $C_{25}C_{25}$- (1B), C_{40}-macrocyclic (1C), and hydroxylated (1D,1E) archaeols (Figure 1) and/or its dimer, dibiphytanyldiglycerol tetraether (2) (caldarchaeol (9)) and variants such as calditoglycerocaldarchaeol (2A)(9,10) with or without cyclopentane rings (2B-2E)(Figure 2), instead of from diacylglycerol diesters as in all other organisms. Note that the term "lipid core" refers to the hydrophobic or lipid portion of membrane lipids. Within the Archaea, the compositions (profiles) of lipid cores (Figures 1 & 2) and phospholipids and glycolipids (Figures 3 to 6) can clearly distinguish between the three subgroups of Archaea: extreme halophiles, methanogens and extreme thermophiles or thermoacidophiles, and also to some extent between genera within each subgroup (Tables I and II) (see below).

The present paper, apart from presenting the specific structures of these archaeal polar lipid molecular markers (Figures 1 to 6) and their distribution and taxonomic correlations among the Archaea (Tables I & II) (see below), includes a review of procedures for extraction, isolation and structure determination of polar lipids from halophiles, methanogens and thermophiles and for preparation and analysis of their lipid cores and other derivatives. Methods for extraction of lipids from marine sediments and soil samples, for analysis of both fatty acids and glycerol diethers/tetraethers and for biomass estimation in such samples of mixed microbial communities will also be reviewed. The significance of these archaeal lipid markers to environmental geochemistry and to ongoing environmental research will be discussed, in particular, the recent reports of the identification of archaea in microbial communities in marine bacterioplankton (3,4).

Extraction and Fractionation of Polar Lipids

Procedures for extraction and fractionation of lipids from Archaea cells, soil samples, etc., and preparation of lipid core and other lipid markers are summarized in Figure 7.

a) From Halophilic Archaea. Total lipids of extreme halophiles with easily broken cell walls (e.g., *Halobacterium, Haloferax, Haloarcula* species) can be extracted quantitatively by a modified neutral Bligh-Dyer extraction procedure (11,12). For halophile cells with rigid cell walls (e.g. *Halococcus* species), cell suspensions are subjected to sonication before extraction by the neutral Bligh/Dyer procedure (extraction with chloroform/methanol/water, 1:2:0.8, v/v; then formation of a biphasic system, chloroform/methano/ water, 1:1:0.9, v/v) (11). The total lipids in the lower chloroform layer are then separated into polar and non-polar lipids by acetone precipitation (11,12). The precipitated polar lipids are analyzed for phospholipids and glycolipids by quantitation of lipid phosphate (lipid-P) and sugars (12), respectively, and are separated into individual components by preparative silicic acid thin-layer chromatography (TLC)(11,12).

Alternatively, the total lipids may be fractionated by silicic acid column chromatography into a neutral lipid fraction (elution with chloroform), a glycolipid fraction (elution with chloroform/acetone, 1:1, v/v, then acetone) and a phospholipid fraction (elution with methanol)(12). The glycolipid and phospholipid fractions can then be separated into individual lipids by preparative TLC (11,12).

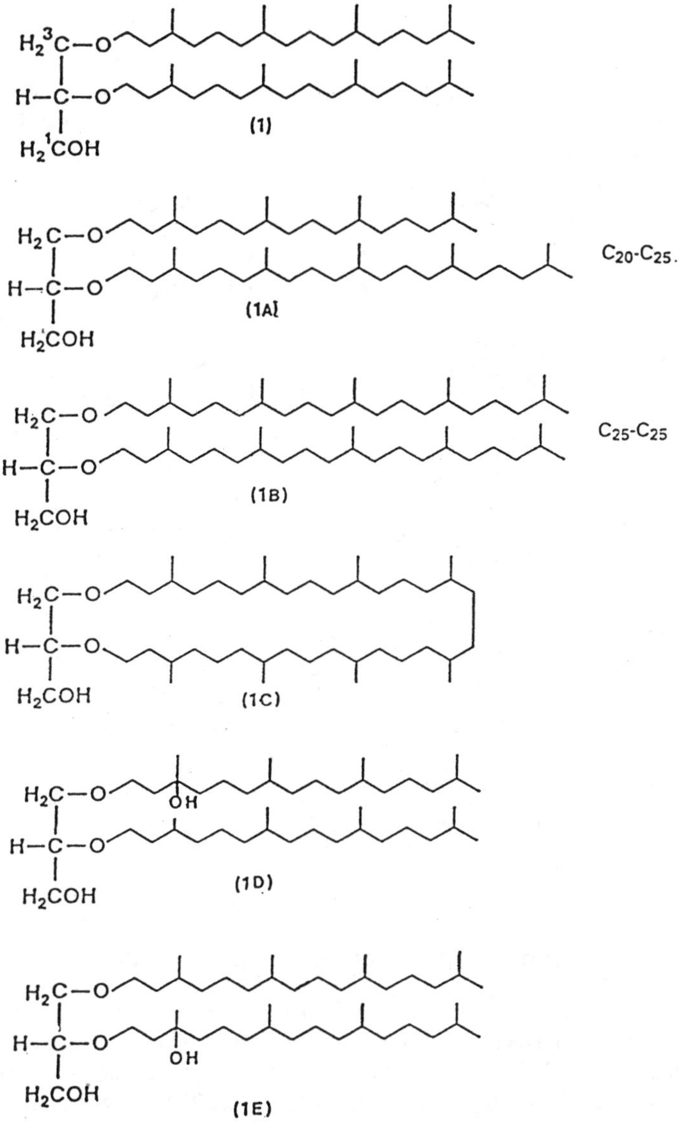

Figure 1. Structures of archaeol and variant lipid cores (*7,8,35-37*):
(1) diphytanylglycerol (C_{20}-C_{20}-archaeol) in extreme halophiles and methanogens;
(1A) C_{20}-C_{25}-archaeol and (1B) C_{25}-C_{25}-archaeol in haloalkaliphiles and some
halococci; (1C) C_{40}-macrocyclic diether in *M. jannaschii*; (1D & 1E)
3-hydroxyarchaeols in *Methanosaeta (Methanothrix)* and *Methanosarcina* species.

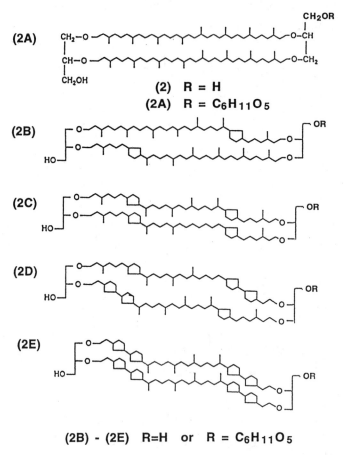

(2B) - (2E)   R=H   or   R = C₆H₁₁O₅

Figure 2. Structures of caldarchaeol and some variant lipid cores (*7,8,10,35-37*):
(**2**) dibiphytanyldiglycerol tetraether (caldarchaeol) in methanogens and
thermophiles; (**2A**) dibiphytanyl glycerol calditol tetraether
(calditoglycerocaldarchaeol) in thermoacidophiles; (**2B-2E**) cyclized caldarchaeols
and calditoglycerocaldarchaeols (1 to 4 cyclopentane rings per chain) in extreme
thermophiles and thermoacidophiles, respectively.

CH2-O-PO-(OH)2
|
R-O-C-H
|
R-O-CH2

Phosphatidic acid
(PA)

CH2-O-PO-O-CH2
| OH |
R-O-C-H   H-C-OH
| |
R-O-CH2   H2C-OH

Phosphatidylglycerol
(PG)

CH2-O-PO-O-CH2
| OH |
R-O-C-H   H-C-OH
| |
R-O-CH2   H2C-O-SO3
               OH

Phosphatidylglycerosulfate
(PGS)

CH2-O-PO-O-CH2
| OH |
R-O-C-H   H-C-OH
| |
R-O-CH2   H2C-O-PO-OCH3
                OH

Phosphatidylglycerolmethylphosphate
(PGP-Me)

R = Phytanyl group = CH2[CH(CH2)3]3CH(CH2)2—
                    CH3   CH3

Figure 3. Structures of archaeol phospholipids in extreme halophiles (*7,8*).

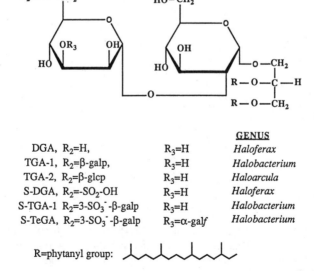

		GENUS
DGA, R_2=H,	R_3=H	*Haloferax*
TGA-1, R_2=β-galp,	R_3=H	*Halobacterium*
TGA-2, R_2=β-glcp	R_3=H	*Haloarcula*
S-DGA, R_2=-SO$_2$-OH	R_3=H	*Haloferax*
S-TGA-1 R_2=3-SO$_3^-$ -β-galp	R_3=H	*Halobacterium*
S-TeGA, R_2=3-SO$_3^-$ -β-galp	R_3=α-gal*f*	*Halobacterium*

R=phytanyl group:

Figure 4. Structures of some archaeol glycolipids in extreme halophiles (*7,8*).

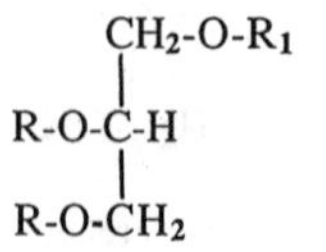

Figure 5. Structures of some typical archaeol complex lipids (**3**) and caldarchaeol complex lipids (**4**) in methanogens (*7,8,35-37*); R_1 = phosphoamino (P-serine, P-ethanolamine, P-aminopentantetrol) and/or P-inositol in archaeol phospholipids, or glycosyl (Glc*p*, Gal*f*, Man*p*) groups in archaeol glycolipids; R_2 = H or phosphoamino groups; and R_3 = glycosyl groups or H, in caldarchaeol phospholipids or glycolipids, respectively.

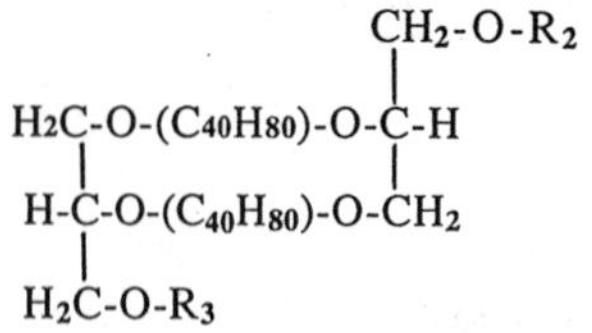

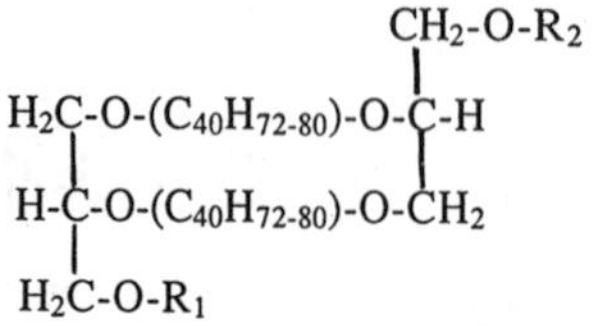

Figure 6. Structures of typical complex lipids (**5**) of caldarchaeol (with 0-4 rings per chain) and (**6**) of calditoglycerocaldarchaeol (with 0-4 rings per chain) in thermophiles/thermoacidophiles (*7,8,10,35,36*): R_1,R_3 = glycosyl(Glc*p*,Gal*p*) groups; R_2,R_4 = H or P-inositol.

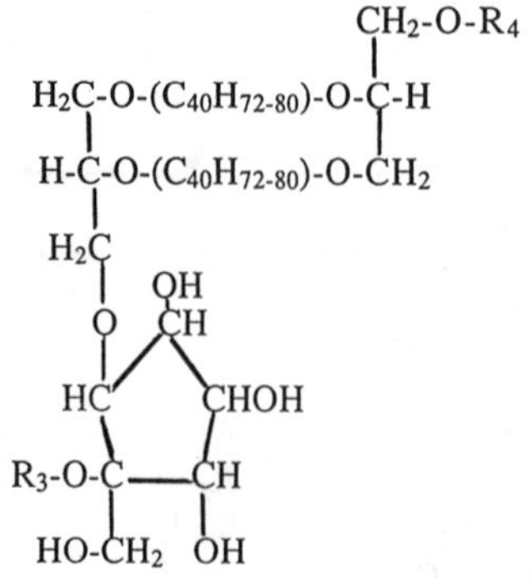

Table I. Distribution of Archaeol, Caldarchaeol and Variants Among Archaea[a]

Genus	Archaeols(A) C20C20	C20C25	C25C25	Caldarchaeols(CA) 0 rings	2-8 rings	CalditoglyceroCA 0-8 rings
Halophiles						
Halobacterium	+					
Haloferax	+					
Haloarcula	+					
Halococci	+	+				
Haloalkaliphiles	+	+	+			
Methanogens						
Methanococcus [b]	+			trace		
Methanosarcina[b]	+			trace		
Methanobacterium	++			+		
Methanothermus	+			++		
Thermophiles						
Desulfurococcus	+			+		
Thermoplasma	+			+	+	
Sulfolobus	trace			+	+	+
Thermoproteus	+			+	+	+

[a]Data from references *7,8,20,24,35-37.*
[b]These genera also contain hydroxyarchaeol lipid cores.

Table II. Comparison of Archaeal and Bacterial Polar Lipids[a]

Organism	Lipid Core	Polar Head-Groups
Archaea		
Halophiles	Archaeols	phosphoglycerols di-,tri- glycosyl sulfates
Methanogens	Archeols	phosphoamino phosphoinositol glycosyl groups
	+ Caldarchaeols	phosphoamino + glycosyl
Thermophiles	Caldarchaeols + Calditoglycerocaldarchaeol (0 - 8 cyclopentane rings)	phosphoinositol phosphoglycerol + glycosyl
Bacteria		
Gram positive	Diacylglycerols	phosphoamino phosphoglycerols di-, tri-glycosyls
Gram negative	Diacylglycerols	phosphoamino phosphoglycerol monoglycosyls

[a]Data from references *7,8,20,24,35-37.*

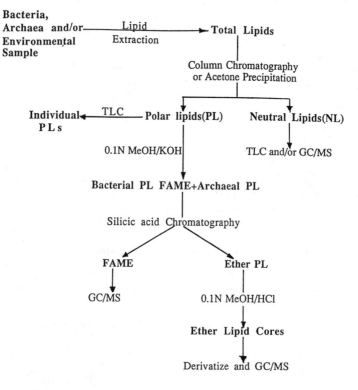

FAME = Fatty acid methyl esters;
GC/MS = Gas Chromatography/Mass Spectrometry;
TLC = Thin-layer Chromatography.

Figure 7. Procedures for preparation of total lipids, individual phospholipids and glycolipids, and lipid cores from Bacteria, Archaea and environmental samples.

b) From Methanogenic Archaea. Methanogens that contain largely neutral polar lipids (e.g. *Methanospirillum hungatei* and *Methanosarcina mazei*) may be extracted virtually quantitatively by a modified neutral Bligh/Dyer procedure (*13*). Methanogens containing highly polar and acidic lipids, including aceticlastic (acetic acid utilizing) methanogens (e.g., *Methanosaeta, Methanosarcina*, etc.) which contain acid-hydrolyzable hydroxyarchaeols, may be extracted first with the neutral Bligh/Dyer one-phase solvent and then rapidly with a mildly acidic Bligh/Dyer one-phase solvent, chloroform/methanol/5% aqueous trichloracetic acid (TCA) (1:2:0.8, v/v), as described by Koga et al.(*14, 15*). This method can also be applied to environmental samples containing methanogens (*15*). The total lipids in the chloroform phase should be washed with upper phase (methanol/water, 10:9, v/v) to remove any traces of acid, and then can be fractionated into individual phospholipids, glycolipids and phosphoglycolipids by silicic acid column chromatography and preparative TLC (*13,14,16*). Core lipids of individual polar lipids or of the total polar lipids may then be prepared (*13-16*) as described below.

c) From Thermophilic Archaea and Environmental Mixed Microbial Cultures. Lipids of thermophiles/thermoacidophiles, as well as other Archaea, Bacteria and Eukarya in pure or mixed culture or in environmental samples are extracted by a modified Bligh/Dyer procedure using a neutral one-phase solvent, chloroform/methanol/phosphate buffer pH 7.5 (1:2:0.8, v/v)(*17,18*). After formation of the Bligh/Dyer biphasic system, the total lipids are obtained from the lower chloroform phase, while the upper methanol/water (10:9, v/v) phase is used for isolation of DNA (*18*). In the presence of soil, however, most of the DNA is associated with the insoluble soil phase and can be recovered by the procedure described by D.C. White and colleagues (*18*).

The total lipids are separated into neutral, glycolipid and polar lipid (phospholipid) fractions by silicic acid column chromatography (*12,17,18*). The glycolipid and polar lipid (phospholipid) fractions are then subjected to mild alkaline methanolysis in 0.1 N methanolic KOH to convert any ester-linked fatty acids to their methyl esters (FAMEs) which are separated from the intact isopranoid/isoprenoid ether glycolipids, phospholipids or phosphoglycolipids by silicic acid column chromatography (*17*,18). The ether lipids can then be fractionated by preparative TLC and converted to their lipid cores (diethers and/or tetraethers) by methanolysis in methanolic-HCl (*12,15,17*), as described below.

Identification of Archaeal Phospholipid and Glycolipid Components

Diether phospholipids and glycolipids in the total polar lipid fraction, or in the isolated individual lipids, can be identified in halophilic (*11,19,20*) or methanogenic (*16,21-23*) Archaea by their relative mobilities (R_f values) on TLC in acidic or basic solvent systems, by spectral analysis, such as nuclear magnetic resonance (NMR) spectrometry (*21,22,24,25*), fast atom bombardment mass spectrometry (FAB-MS)(*22-25,28*) and infrared (IR) spectrometry (*23,25,26*) and by chemical degradation and analysis of degradation products (*12,16,19-23,25,28*). Anomeric configurations, linkage positions and the position of substituents of glycosyl groups in glycolipids can be determined by ^{3}H- and ^{13}C-NMR (*19,20*), but this should be confirmed, if sufficient material is available, by measurement of optical rotation and by permethylation analysis (*20,23*), respectively. Similar procedures are used for the identification of tetraether phospholipids, glycolipids and phosphoglycolipids (Figure 5) in methanogens *(10,16,21,22)* and the corresponding tetraether and calditoglycerotetraether analogs (Figure 6) in thermophilic or sulfur dependent Archaea *(7,8,20,24)*. Identification of the various archaeal lipid cores is described below.

Preparation and Identification of Archaeal Ether Lipid Cores

For archaeal lipids derived from saturated isopranoid diethers (Figure 1), tetraethers and calditoglycerotetraethers (Figure 2), which have acid-stable ether linkages, a "strong acid" methanolysis procedure using 0.6N to 2N methanolic-HCl at 80-100°C for 1-3 hr gives essentially quantitative yields of the corresponding lipid cores *(12,14,17)*. Archaeol phospholipids lacking a free hydroxyl group adjacent to the phosphodiester group, such as phosphatidylserine (Figure 5), are not readily degraded by acid methanolysis to the glycerol diether and the polar head group, but the diether lipid core may be completely released by acetolysis *(14,25)*.

Lipid cores containing acid-labile ether linkages, such as isoprenoid or 3-hydroxy-isopranoid diethers or isoprenoid tetraethers, are degraded during strong acid methanolysis, to glycerol monoethers *(22,26,27)*, but may be obtained without degradation and in good yield by mild acid methanolysis in 0.18% HCl (0.05N) in methanol *(22,26,27)* or in 5% methanolic HCl/chloroform (1:27, v/v) *(15)* at 50°C for 24 hr. Hydrolysis at 4°C in 48% HF for 16 hr *(25,26)* or 30 hr *(15)* will also yield the hydroxy diether lipid cores. However, phosphatidylserine (Figure 3) or other phospholipids lacking a hydroxyl adjacent to the phosphodiester bond form the diether phosphatidic acid (Figure 3) rather than the diether *(25)*.

The diether or tetraether (with or without cyclopentane rings) lipid cores may be separated by TLC, high performance liquid chromatography (HPLC), supercritical fluid chromatography or high temperature gas chromatography (GC) and identified by IR, NMR and mass spectrometry (for reviews see refs. *17,20,24,37*). Mixtures of 3-hydroxydiethers and other diethers can be resolved by TLC *(21,22,26,27)*, HPLC of 9-anthroyl derivatives *(15)* or GC of acetylated derivatives *(22)* and identified by NMR *(22,26)* and chemical ionization mass spectrometry (CI-MS) *(22,27)*. A simple TLC procedure may be used to resolve mixtures of tetraether lipid cores with up to 8 cyclopentane rings *(29)*.

Final identification of hydrocarbon type and chain length of diether and tetraether lipid cores is done by cleavage of the ether linkages with BCl_3, BBr_3 or HI and analysis of the respective alkyl halides or derived acetates or hydrocarbons by GC or GC-MS (for reviews see refs. *17,20,24,37*). Changes in the proportions of diether (including the macrocyclic diether (**1C**)) and tetraether type lipids in *M. thermoautotrophicum* have been determined as a function of growth phase by GC analysis of hydrocarbons derived by direct action of HI on total polar lipids, followed by $LiAlH_4$ reduction *(30)*. Proportions of diether (including macrocyclic diether) and tetraethers of *M. jannaschii* have been determined as a function of growth temperature by TLC separation of [^{14}C]-labeled lipid cores and counting of the radioactivity in each labeled lipid core *(31)*.

Mixtures of microbial phospholipid fatty acids and diether lipid cores (termed "signature lipid biomarkers" by D.C. White and colleagues) may be analyzed quantitatively by supercritical fluid chromatography *(32)* or more conveniently by capillary GC *(33)* or gas chromatography/mass spectrometry (GC/MS)*(34)*; fatty acids are analyzed as methyl esters (FAMEs) and diethers as trimethylsilyl (TMS) derivatives. Absolute quantitation as mg or mmol signature lipids is achieved by addition of a known amount of internal standard (C_{19}-FAME for fatty acids and dihexadecylglycerol for diethers) to the sample before chromatographic analysis; for estimation of mg or mmol lipid per cell, cell density is determined by measurement of carbon content (*15*) or by DNA extraction and quantitation (18). The data obtained can be used to estimate the biomass (as cell number or mg cell carbon/g sediment) of the total microbial community and of the archaeal components. Examples are biomass quantitation of aceticlastic methanogens in mixed cultures or in environmental samples such as anaerobic digestor sludges and sediment from a river bed in Japan (*15*), and of bacteria and methanogens in surface sediments from Ryans Billabong in Australia (*33*) (see refs. *18* and *37* for other examples).

Distribution of Archaeol and Caldarchaeol Lipid Cores Among Archaea (*7,8,20,24,35-37*)

The distribution of diether and tetraether lipid cores among halophilic, methanogenic and thermophilic archaea is given in Table I. It is clear that extreme halophile lipids are derived only from archaeol and archaeol variants; methanogen lipids may be derived from both archaeol and caldarchaeol (with 0 cyclopentane rings), although some species of *Methanococcus* and *Methanosarcina* have little or no caldarchaeol-lipids; and lipids of extreme thermophiles are derived largely from caldarchaeol or variants (containing 2-8 cyclopentane rings) or from calditoglycerocaldarchaeol (with 0-8 rings), with at most 1-10% archaeol-lipids. It should be noted that calditoglycerocaldarchaeol lipids are characteristic of *Thermoplasma* and *Sulfolobus* species and that the number of rings in the tetraether lipid cores increases with increasing growth temperature (*8,24,35,36*). Also, *M. jannaschii*, an isolate from a submarine hydrothermal vent growing over the temperature range 45 to 75°C, can alter its lipid core proportions of diether, macrocyclic diether and tetraether (Figures 1 & 2) as a function of temperature, diethers predominating at the lower temperatures and macrocyclic diether and tetraethers predominating at the higher temperatures (*31*).

Distribution of Phospholipids, Glycolipids and Phosphoglycolipids Among Archaea (*7,20,35-37*)

Phospholipids of extreme halophiles contain only archaeol-linked phosphoglycerol (PG), phosphoglycerosulfate (PGS) and phosphoglycerol methylphosphate (PGP-Me) headgroups (Figure 3) and completely lack any phosphoamino headgroups such as phosphoserine or ethanolamine. Glycolipids of extreme halophiles contain archaeol-linked sulfated or unsulfated glycosyl headgroups, with 1 to 4 sugars, the structures of which are characteristic of the genus (Figure 4). Glycolipid structures have been useful in establishing the generic classification of new or unclassified extreme halophile species (*7,38*).

In contrast, methanogen phospholipids contain both archaeol- and caldarchaeol-linked phosphoamino headgroups, such as the bacterial-type phosphoserine, phosphoethanolamine, dimethylamino- and trimethylamino-pentanetetrol, as well as the eukaryal-type phosphoinositol (*7,35,37*)(Figure 5). Glycolipids in methanogens contain 1 to 3 unsulfated glycosyl groups (commonly the diglycosyl group, gentiobiose) linked to archaeol or caldarchaeol without rings (Figure 5). In the complex phosphoglycolipids of methanogens the above phosphoamino and glycosyl polar headgroups are located on opposite ends of caldarchaeol (Figure 5 and Table II).

Lipids of extreme thermophiles consist largely of glycolipids and phosphoglycolipids with only small to trace amounts of phospholipids. The glycolipids contain 1 or 2 glycosyl groups (usually glucose) linked to caldarchaeol or calditoglycerocaldarchaeol, both having 0 to 8 rings (Figure 6). In the phosphoglycolipids, phosphoinositol (the major phosphorus-containing headgroup) and the above glycosyl groups are linked to opposite ends of caldarchaeol or variants with 2 to 8 cyclopentane rings or, in *Sulfolobus* and *Thermoproteus* species, to calditoglycerocaldarchaeol (with 0-8 rings)(Figure 6 and Table II).

Discussion

The lipid core profiles and the phospholipid and glycolipid head group compositions of the three sub-groups of Archaea and of gram-positive and gram-negative Bacteria are summarized in Table II. It is clear that these lipid profiles can be very useful in distinguishing between Archaea and Bacteria in microbial communities (see refs. *15,17,18,33,37*) and to identify the archaeal subgroup (halophile, methanogen or thermophile) and in some cases, particularly the extreme halophiles (see Figure 4), the genus of the archaebacterium (*7*). Furthermore, the fatty acid and glycolipid profiles of eubacteria are generally well correlated with their taxonomic classification on the level of the genus (*20*), and these profiles can be very useful in identifying the bacterial species (*18,33*). However, it is well to remember that the structural data for archaebacterial lipid cores, phospholipids, glycolipids and phosphoglycolipids, is not complete and new structures will most likely be found as more archaeal species are discovered.

In connection with the question of the presence and stability of archaeal lipids in geological samples, it has been known for some time that some of these molecular markers are present in geological samples and in petroleum (*39*). An early study of the lipids of relatively recent (<1000 y) sediments of the Dead Sea (*40*) showed that intact archaeol phospholipids and glycolipids similar to those in *Halobacteria* were present, along with archaeol lipid breakdown products (diether phosphatidic acid, Figure 3, free diphytanylglycerol diether, phytanol and phytanic acid, derived from phytanol or the phytanyldiether), in addition to fatty acids of plant origin. These results indicate that some archaeal lipid markers are sufficiently stable to survive up to 1000 y in sediments, and that even the breakdown products can be indicative of the presence of archaebacteria. More recent studies have shown that significant proportions of isoprenoid hydrocarbons derived from archaebacterial lipid cores are present in sediments as old as the Precambrian (*39*), in petroleum (*41*), in marine (*42*) and swamp (*43*) sediments, and in kerogen (*44*); and intact diether and tetraether lipid cores, with or without attached hydrophilic groups, have been found in sediments and petroleum of recent and ancient ages and of various origins (*45*). These findings are consistent with the concept of a widespread occurrence of Archaea throughout geological time and with the stability of the archaeal isopranoid chains and ether linkages under geological conditions.

Conclusions

The recently described procedures for simultaneous extraction of DNA and lipids from soil samples (*18*) by D.C. White and his group and for simultaneous analysis of microbial fatty acids and diether lipid cores ("signature lipid biomarkers") by supercritical fluid chromatography (*32*), by capillary gas chromatography (*33*) and by GC-MS (*34*) should greatly facilitate the identification of these biomarkers in Bacteria and Archaea. In conjunction with rRNA sequences, lipid biomarkers should help to characterize, identify and quantitate individual members of mixed microbial communities. Research in this area should thus be greatly stimulated and the results should be applicable to problems of environmental pollution, such as in landfill leachates, contaminated marine or lacustrine sediments, etc.

The recent discovery of novel groups of archaea that are distinct from the known archaeal groups and are widely distributed in oceanic environments (*3,4*), and probably also in terrestial environments, has opened up a new and exciting area in the field of environmental geochemistry. The lipid molecules discussed above, in conjunction with rRNA sequences, should prove to be reliable microbial molecular markers for identification of these novel Archaea and for their biomass determination and other characteristics in marine plankton, sediments, soil samples, etc. Archaea may well prove to be major components of mixed microbial communities globally and it is therefore important to be able to identify them in the environment.

Literature Cited

(*1*)   Woese, C.R.; Kandler, O.; Wheelis, M.L., *Proc. Natl. Acad. Sci. USA* . **1990,** *87*, 4576-4579.

(*2*)   *Archaebacteria*; Woese, C.R. and Wolfe, R.S., Eds.; The Bacteria, Vol.VIII, Academic Press, New York, NY, 1985.

(*3*)   Fuhrman, J.A.; McCallum, K; Davis, A.A., *Nature* **1992**, *356*, 148-149.

(*4*)   DeLong, E.F.; Wu, K.Y.; Prezelin, B.B.; Jovine, R.V.M., *Nature* **1994**, *371*, 695-697.

(*5*)   Olsen, G. J. *Nature* **1994**, *371*, 657-658.

(*6*)   *The Biochemistry of Archaea (Archaebacteria);* Kates, M., Kushner, D.J. and Matheson, A.T., Eds.; New Comprehensive Biochemistry Vol. 26; Elsevier, Amsterdam, The Netherlands, 1993 .

(*7*)   Kates, M. In *The Biochemistry of Archaea (Archaebacteria);* Kates, M., Kushner, D.J. and Matheson, A.T., Eds.; New Comprehensive Biochemistry Vol. 26; Elsevier, Amsterdam, The Netherlands, 1993, pp 261-295.

(*8*)   Langworthy, T.A. In *Archaebacteria*; Woese, C.R. and Wolfe, R.S. Eds.; The Bacteria, Vol. VIII; Academic Press, New York NY, 1985, pp 459-497.

(*9*)   Nishihara, M.; Morii, H.; Koga, Y. *J. Biochem.* **1987**, 101, 1007-1015.

(*10*) Sugai,A.; Sakuma, R.; Fukudo, I.; Kurosawa, N.; Itoh, Y.H.; Kon, K.; Ando, S.; Itoh, T. *Lipids*, **1995**, *30*, 339-344.

(*11*) Kates, M.; Kushwaha, S.C. In *Archaea: A Laboratory Manual, Halophiles;* DasSarma, S.; Fleischman, E.M.,Eds.; Cold Spring Harbor Laboratory Press, New York, NY, 1995, pp 35-54.

(*12*) Kates, *M. Techniques of Lipidology*; Laboratory Techniques in Biochemistry and Molecular Biology, Vol.3, Part 2; Elsevier; Amsterdam, The Netherlands; New York, NY, 1986; 2nd revised edition; pp 106-107, 113-115, 191-194.

(*13*) Sprott, G.D.; Choque, C.G.; Patel,G.B. In *Archaea: A Laboratory Manual, Methanogens*; Sowers, K.R.; Schreier, H.J., Eds.; Cold Spring Harbor Laboratory Press, New York, NY, 1995, pp 329-340.

(*14*) Nishihara, M.; Koga, Y. *J. Biochem.* **1987**, 101, 997-1005.

(*15*) Ohtsubo, S.; Kanno, M.; Niyahara, H.; Kohno, S.; Koga, Y.; Miura, I. *FEMS Microbiol. Ecol. 1993, 12, 39-50.*

(*16*) Kushwaha, S.C.; Kates, M.; Sprott, G.D.; and Smith, I.C.P. *Biochim. Biophys. Acta*, **1981**, *664*, 156-173.

(*17*) Hedrick, D.B.; White, D.C. In *Archaea: A Laboratory Manual, Thermophiles*; Robb, F.T.; Place, A.R., Eds.; Cold Spring Harbor Laboratory Press, New York, NY, 1995, pp 73-80.

(18) Kehrmeyer, S.R.; Applegate, B.M.; Pinkart, H.C.; Hedrick, D.B.; White, D.C.; Sayler, G.S., J. Microbiol. Meth. **1996**, *25*, 153-163.

(19) Kates, M. J. Microbiol. Meth. **1996**, *25*, 113-128.

(20) Kates, M. In *Glycolipids, Phosphoglycolipids and Sulfoglycolipids;* Kates,M., Ed; Handbook of Lipid Research, Vol. 6; Plenum Press, New York, NY, **1990**; pp 1-122.

(21) Sprott, G.D.; Ferrante, G.; Ekiel, I. *Biochim. Biophys Acta*, **1994**, *1214*, 234-242.

(22) Nishihara, M.; Koga, Y. *Biochim. Biophys Acta*, **1991**,*1082*, 211-217.

(23) Nishihara, M.; Morii, H.; Koga, Y. *J. Biochem.* **1987**, *101*, 1007-1015.

(24) De Rosa, M.; Gambacorta, A. *Prog. Lipid Res.* **1988**, *27*, 153-175.

(25) Morii, M.; Nishihara, M.; Ohga, M.; Koga, Y. *J. Lipid Res.* **1986**, *27*, 724-730.

(26) Ferrante, G.; Ekiel, I.; Patel, G.B.;Sprott, G.D. *Biochim. Biophys Acta*, **1988**, *963*, 173-182.

(27) Ekiel, I.; Sprott, G.D.*Can. J. Microbiol.* **1992**, *38, 764-768.*

(28) Sprott, G.D.; Dicaire, C.J.; Patel, G.B. *Can. J. Microbiol.* **1994**,*40, 837-843.*

(29) Trincone, A; De Rosa, M.; Gambacorta, A.; Lanzotti, V.; Nicolaus, B.; Harris, J.E.; Grant, W.D. *J. Gen. Microbiol.* **1988**, *134*, 3159-3163.

(30) Morii, H.; Koga, Y. *FEMS Microbiol. Lett.* **1993**, *109*, 283-288.

(31) Sprott, G.D.; Meloche, M.; Richards, J.C. *J. Bacteriol.* **1991**, *173*, 3907-3910.

(32) Hedrick, D.B.; Guckert, J.B.; White, D.C. *J. Lipid Res.* **1991**, *32, 659-666.*

(33) Virtue, P.; Nichols, P.D.; Boon, P.I. *J. Microbiol. Meth.* **1996**, *25*, 177-185.

(34) Teixidor, P; Grimalt, J.O. *J. Chromatogr.* **1992**, *607*, 253-259.

(35) Sprott, G.D. *J. Bioenerg. Biomembr.* **1992**, *24*, 555-566.

(36) De Rosa, M.; Gambacorta,A.; Gliozzi, A. *Microbiol. Rev.* **1986**, *50*, 70-80.

(37) Koga, Y.; Nishihara, M.; Morii, H; Akagawa-Matsushita, M. *Microbiol. Rev.* **1993**, *57*, 164-182.

(38) Kamekura, M; Dyall-Smith, M. *J. Gen. Appl. Microbiol.* **1995**, *41*, 333-350.

(39) Hahn, J.; Haug, P. In *Archaebacteria*; Woese, C.R. and Wolfe, R.S. Eds.; The Bacteria, Vol. VIII; Academic Press, New York NY, 1985, pp 215-253.

(40) Anderson, R.; Kates, M.; Baedecker, M.J.; Kaplan, J.R.; Ackman, R.G. *Geochim. Cosmochim. Acta*, **1977**, *41*, 1381-1390.

(41) Albaigés, J.; Borbon, J.; Walker, W. *Org. Geochem.* **1985**, *8*, 293-297.

(42) Brassell, S.C.; Wardroper, A.M.K.; Thomson, I.D.; Maxwell, J.R.; Eglinton, G. *Nature*, **1981**, *290*, 693-4.

(43) Pauly, G.G.; Van Vleet, E.S. *Geochim. Cosmochim. Acta*, **1986**, *50*, 1117-1125.

(44) Michaelis, W.; Albrecht, P. *Naturwissenschaften*, **1979**, *66*, 420-22.

(45) Chappe, B.; Albrecht,P.; Michaelis, W. *Science*, **1982**, *217*, 65-66.

Chapter 4

Seasonal Variation in Sedimentary Microbial Community Structure as a Backdrop for the Detection of Anthropogenic Stress

Robert H. Findlay[1] and Les Watling[2]

[1]Department of Microbiology, Miami University, Oxford, OH 45056
[2]Darling Marine Center, University of Maine, Walpole, ME 04573

Benthic microbial communities exhibit natural seasonal variations in community structure. Analysis of microbial community structure by phospholipid fatty acid profiles and principal component analysis revealed similar seasonal patterns of change for two marine sites. One site was considered pristine while the other experienced anthropogenic organic enrichment (the waste food and feces from a salmon net-pen facility). Seasonal patterns of change were found to dominate at both sites, but the changes induced by organic enrichment could be detected after variation due to the seasonal patterns were removed. The seasonal patterns of change were best described as increased importance of microeukaryotes and aerobes ($20:5\omega3$, $16:4\omega1$, $18:1\omega7$) during cold-water months and increased importance of bacteria and anaerobes ($i16:0$, $a15:0$, $16:1\omega7t$) during warm-water months. Organic enrichment induced increased importance of a chemolithotrophic community within the sediments that is best characterized by the microbial assemblage found in *Beggiatoa*-type mats. These results once again demonstrate the utility of the phospholipid fatty acid profiles for the quantitative description of microbial community structure and suggest that care must be taken to include seasonal factors into experimental designs attempting to determine the effects of anthropogenic stress.

Introduction

All natural biological systems are characterized by two features; they are spatially heterogeneous, and they are dynamic in that community structure changes with time. Hence, the greatest challenge to determining ecosystem health by investigating changes in populations is to be able to discriminate between changes induced by natural versus anthropogenic causes.

In marine and estuarine environments the most common pollutants are organic. Duursma and Marchand (*1*) list sewage, pulp mill effluent, detergents, agriculture runoff, hydrocarbons, and xenobiotics as common marine and estuarine organic pollutants. All, with exception of xenobiotics, cause, to a lesser or greater extent, a similar pattern of change in marine and estuarine benthic communities. Most commonly utilized for monitoring purposes is the pattern of change observed in the macrobenthic community (2). Changes in the benthic microbial community, however, can also serve effectively as an indicator of organic enrichment in estuarine environments and allow for the early detection of possible ecosystem change due to anthropogenic stress. The microbial component of the benthos offers many advantages over using patterns of change in the macrobenthic community for monitoring purposes. These are: 1. As living members of the sediment community they must adapt to environmental stress or perish. 2. Their response to sediment contamination facilitates the spatial definition of impacts. 3. Benthic microorganisms are effective indicators of impacts at higher levels of organization because of their central role in overall ecosystem structure and function. 4. Not only do bacteria have the potential to mediate transfer of toxic substances to higher trophic levels, they can transform toxic substances sometimes increasing their toxicity or their mobility within the food web. 5. Benthic microorganisms mediate nutrient recycling from the sediments into the water column. 6. The biomass of benthic microorganisms is generally controlled by sediment grain size and organic content and, as such, is sensitive to organic enrichment. 7. They are numerous, allowing small sample sizes and frequent replication. 8. They have short generation times, thus, offering the opportunity for rapid response. Coupled with the fact that bacteria and bacterial products comprise between 2-10% of sedimentary organic matter, the above points form a strong rationale for the need to understand bacterial biomass and community structure in environmental geochemical studies.

Most of the bacteria in sediments are heterotrophic and, therefore, dependent on the decomposition of organic matter (3). Catabolic processes in sediments generally proceed following a definite succession of oxidizing agents. The reason why this succession occurs is generally explained in terms of the metabolic free energy yield of the reactions (4). It is assumed that the greater the theoretical energy yield of a bacterial reaction, the greater the probability that the reaction will predominate over other competing reactions. Hence, diagenetic reactions within sediments can be modeled in terms of the availability of organic carbon and terminal electron acceptors. The prevalence of oxygen and sulfate in marine waters leads to their importance as electron acceptors within sediment systems, although nitrate, iron, manganese and carbon can also play a role. This model of early diagenesis is most often observed in the zonation of prevalent metabolic types within a sediment (Figure 1). Microbial degradation will proceed aerobically as long as molecular oxygen is available. If the supply of oxygen exceeds the demand generated by the availability of carbon, the sediments remain aerobic. When demand exceeds availability, oxygen is depleted and anaerobic processes will predominate, including the reduction of sulfate to hydrogen sulfide. The hydrogen sulfide produced can then, in turn, serve as an energy source for chemoautotrophic bacteria (e.g. *Beggiatoa*-type organisms) that reside at the sediment horizon where molecular oxygen and hydrogen sulfide coexist. If the amount of metabolizable organic matter is extreme, most of the available O_2 is consumed as an oxidant for sulfides and not

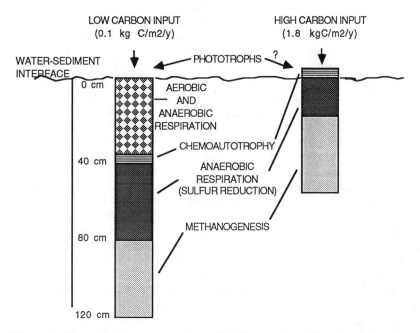

Figure 1. Diagrammatic representation of the distribution of functional groups of microorganisms within sediments and its relationship to organic matter input. (Drawn, in part, using data published in ref. *28-29*).

for organic carbon (5). Under these conditions visible *Beggiatoa* mats form at the sediment-water interface. If sufficient organic material is present, sulfate will also be depleted in the sediments and diagenesis will proceed *via* production of biogenic methane. Findlay et al. (6) have demonstrated that organic enrichment does indeed induce such changes in community structure and Findlay and Watling (in press, *Mar. Ecol. Prog. Ser.*) have shown that these changes are not only dependent on benthic carbon flux but also dependent on delivery of O_2 to the benthos. Combined, these two studies demonstrate that changes caused by variation in the balance of electron acceptors and metabolizable organic matter induce characteristic changes in microbial community structure detectable using phospholipid fatty acids (PLFA) profiles. What remains is to separate changes in microbial community structure induced by anthropogenic stresses from those that occur due to natural processes.

It is simplistic to assume that carbon content and terminal electron acceptor availability are the only factors to affect microbial community structure. Factors likely to influence the structure of microbial communities in sediments are: 1) season (Findlay and Watling submitted, *Micro. Ecol.*), 2) sediment grain size and shape (7-8), 3) organic carbon content, 4) fluid flux over the bed (Findlay and Watling in press, *Mar. Ecol. Prog. Ser.*), 5) disturbance or disruption of the sediments (9-12) and 6) animal-microbe interactions (13-14).

In this paper, we compare and contrast the patterns of change found in microbial community structure for two shallow-water sites from coastal Maine. Both sites were sampled approximately monthly during 1991, and microbial community structure was determined by phospholipid fatty acid profiles (15). The sites differ in that one is considered pristine (Damariscotta River estuary) while the other is impacted by organic waste (Toothacher Cove). Data from both sites were generated independently and were used to examine two separate research problems. The Toothacher Cove site was sampled to determine the effects of increased organic flux to the benthos generated by salmon aquaculture on the benthos. These results have been previously published (6). The Damariscotta River site was used to explore the natural seasonal variation in sedimentary microbial community structure and the manuscript relating these changes is currently in review (Findlay and Watling submitted, *Micro. Ecol.*). Because both sites were sampled approximately monthly for a year and microbial community structure was determined by the same method (PLFA and principal component analysis), a comparison of the findings of these two independent studies affords a unique opportunity to evaluate the relative strength of two important structuring processes; seasonality and organic enrichment. This paper evaluates interactions between seasonal changes and organic enrichment with the goal of separating those changes in microbial community structure induced by anthropogenic stresses from those that occur due to natural processes.

Materials and Methods.

Experimental Design. Permanent benthic study sites were established within two Maine coastal embayments. These study sites differed in that one embayment was pristine and the other anthropogenically impacted. Sampling stations were marked with moored buoys. Three replicate benthic sediment samples were collected monthly (or as closely as possible given the rigors of diving in coastal waters) by SCUBA-assisted divers using push cores from each sampling station and subsamples of approximately 1.1 cc were removed using 5 cc plastic syringes with the canula end removed. This effectively sampled the 0-1 cm sediment horizon with minimal disturbance to the sediment. Benthic microbial community structure was determined by phospholipid fatty acid (PLFA) analysis (*15*). Patterns of seasonal variation in benthic microbial community structure were determined using Principal Components Analysis (PCA).

Study Sites. The first study site was located approximately 100 m north of the Darling Marine Center dock in the Damariscotta River estuary system and is approximately 7 river miles from the mouth of the estuary. A complete description of this site and the observed seasonal changes in microbial community structure can be found in Findlay and Watling (submitted, *Microb. Ecol.*). This site is in approximately 10 m of water (Mean Low Water). Currents range from 0-20 cm s^{-1} and flow parallel to the shore (Findlay and Watling submitted, *Micro. Ecol.*). The bottom consists of a fine muddy sand. Water temperatures annually range from 0 to 20 $^{\circ}$C, and salinities typically range from 25 to 33 o/$^{\circ\circ}$ (*16*). Total fecal coliform counts at mid channel are typically less than 2 Colony Forming Units/100 ml. This is a pristine site with no evidence of anthropogenic stress. A single sampling station established at this site was sampled monthly from April 1991 - June 1992.

The second site was a working salmon aquaculture farm in Toothacher Cove, Swans Island, Maine. Swans Island is approximately 15 miles offshore, and the cove is exposed to ocean waves to the south-southwest. Water depth at this site is 15.7 m at mean low water, annual temperatures ranged from -1.2 to 15.5 $^{\circ}$C, the directions of prevailing currents were west-southwest and east-northeast and ranged from 0 to 10 cm s^{-1} at 1 m above the sediment surface (*6*). Sediments consisted of a muddy sand over glacial till. The study was conducted between February 1991 and June 1992 with the majority of sampling effort (monthly samplings -- again as closely as possible given the challenges of working with small boats and diving fifteen miles offshore in cold-temperate waters) concentrated during the summer and fall of 1991. We intensively studied sediments beneath and adjacent to a single salmon net-pen and sediments located approximately 100 m to the north-northwest of the pen (located perpendicular to the predominant flow to minimize the chance of a pen influence). A total of six sampling stations were established at this site; three ambient stations 100 m from the net-pen and spaced 10 m apart and three pen stations placed 1, 10 and 20 m from the edge of the net-pen. The rationale for this placement was that the ambient stations would be unimpacted by the aquaculture activity while the pen stations would experience a gradient of increasing benthic carbon flux. During this time the pen contained approximately 8,000 Atlantic salmon which were hand-fed approximately 0.012 kg semi-moist feed kg^{-1} fish biomass daily (2 daily feedings). During the approximately 150 days that sediment traps were successfully deployed during the summer of 1991 ambient sediments received ca. 250 g C m^{-2} (including sediment resuspension), sediments below the net-pen received 500-750 g C m^{-2} and sediment 10 m downcurrent of the net-pen received ca. 325 g C m^{-2} (*6*). Detailed descriptions of this site and the nature and extent of the benthic impact (from geochemical to epibenthic) associated with the increased carbon flux can be found in Findlay *et al.* (*6*).

Determination of Microbial Community Structure. Microbial community structure was determined using PLFA analysis following the protocols of Findlay (*15*). Briefly, total lipids were extracted using a modified Bligh and Dyer extraction (*17*), and phospholipid fatty acids were purified using silicic acid column chromatography. Fatty acid methyl esters were formed by transmethylation (0.2 M methanolic KOH) and purified by C$_{18}$ reverse-phase column chromatography. Fatty acid methyl esters were identified and quantified by gas chromatography. Individual fatty acids were expressed as weight percent, that is, (grams individual fatty acid x grams total fatty acids^{-1}) x 100. A total of 42 individual PLFAs were identified and quantified. Fatty acids are named using the following convention -- chain length, :, number of double bonds, ω, the position of first double bond from the *omega* or aliphatic end of the molecule. For example the fatty acid, 16:1ω7, is 16 carbons long, has one double bond, and it occurs at the seventh carbon from the *omega* end of the molecule. All bonds are assumed to be of the *cis* configuration unless noted by a *t*. Terminal branching patterns are designated by *i* (*ios*) and *a* (*anteiso*); internal branching patterns are designated by the number of the carbon (with respect to the acyl end) at which the branch occurs and the abbreviation Me.

Statistical Analysis. Patterns of seasonal variation in microbial community structure were determined by PCA using Systat (Systat Inc., Chicago). For the Damariscotta site 46 samples were analyzed, and for the Toothacher Cove site 107 samples were analyzed resulting in two matrices (45 x 42 and 107 x 42) from which the major components of variation were determined. Data were transformed [ln(x=1)] prior to statistical analysis. PLFA patterns were interpreted using the algorithms outlined in Findlay (*15*). Functional group designations for the PLFAs and the citations supporting the designations are given in Table I.

Safety. Several safety issues were addressed during the course of this work. They were SCUBA-assisted diving, the use of hydrogen as a carrier and flame gas during gas chromatography and use of organic solvents. Each of these carries a risk that is minimized by proper training and adherence to established safety procedures.

Results and Discussion

Seasonal Variation at a Pristine Site. The pattern of seasonal variation at the Damariscotta River site as determined by PLFA profiles and PCA was one of the predictable changes in PLFA profiles that correlated with months of the year (Figure 2). Samples taken from April and May had the most negative component scores for principal component 1 while samples from August, September and November (no October samples were collected) had the highest positive component scores for principal component 1. In general, sediments from cold-water months had negative component scores and sediments from warm-water months had

Table I. Phospholipid fatty acids that defined major components of variation at the Damariscota River and Toothacher Cove study sites

Fatty acid	Functional group assignment
1. 14:0	Gram-positive and some Gram-negative anaerobic bacteria (*18*)
2. *i*15:0	Gram-positive and some Gram-negative anaerobic bacteria (*18*)
3. *a*15:0	Gram-positive and some Gram-negative anaerobic bacteria (*18*)
4. 15:0	Gram-positive and some Gram-negative anaerobic bacteria (*18*)
5. 16:4ω1	Microeukaryotes (*18*)
6. 16:3	Microeukaryotes (*18*); Diatoms (*19*)
7. 16:1ω7t	Gram-negative bacteria experiencing stress (*20*)
8. 16:1ω13t	Eukaryotic phototrophs (*18*); Diatoms (*19*)
9. *i*16:0	Gram-positive and some Gram-negative anaerobic bacteria (*18*)
10. 10Me16:0	Sulfate-reducing bacteria and other anaerobes (*18*); *Desulfobacter* (*19*)
11. 17:1ω6	Aerobic bacteria and eukaryotes (*18*); *Beggiatoa*-type mat if 17:1ω6, 22:6ω3, 22:5ω3, 22:1ω11 are present (*6*).
12. *a*17:0	Sulfate-reducing bacteria and other anaerobes (*18*)
13. 18:2ω6	Aerobic bacteria and eukaryotes (*18*); *Beggiatoa*-type mat if 17:1ω6, 22:6ω3, 22:5ω3, 22:1ω11 are present (*6*); Fungi (*19*)
14. 18:1ω7	Aerobic bacteria and eukaryotes (*18*)
15. 20:5ω3	Microeukaryotes (*18*); Diatoms or higher plants (*19*)
16. 20:4ω6	Heterotrophic microeukaryotes (*18*); Protozoa (*19*)
17. 22:6ω3	Microeukaryotes (*18*); *Beggiatoa*-type mat if 17:1ω6, 22:6ω3, 22:5ω3, 22:1ω11 are present (*6*)
18. 22:5ω3	Microeukaryotes (*18*); *Beggiatoa*-type mat if 17:1ω6, 22:6ω3, 22:5ω3, 22:1ω11 are present (*6*)
19. 22:1ω11	*Beggiatoa*-type mat if 17:1ω6, 22:6ω3, 22:5ω3, 22:1ω11 are present (*6*); *Francisella tularensis* (*19*)

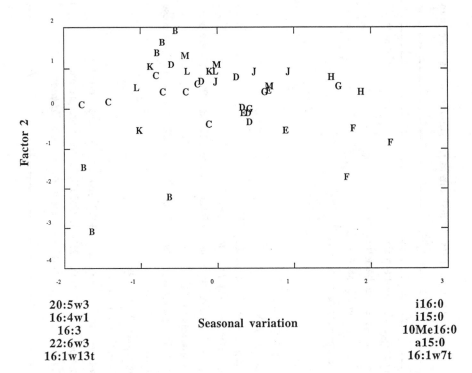

Figure 2. Plot of factor scores resulting from principal component analysis of
weight percent PLFA data from Damariscotta River estuary sediments.
Component 1 was designated seasonal variability. Scores are plotted by sample
date (April 9, 1991 through June 15, 1992). Symbols are: B = April , C = May,
D = June, E = July, F = August, G = September, H = November, J = December,
K = January, L = February, M = March. Note that there were no samples taken
during October. (Adapted from ref. Findlay and Watling submitted, *Microb.
Ecol.*)

positive component scores for principal component 1. Samples from periods of temperature transition (December and June) had component scores that bracketed zero. Four functional groups of microorganisms were represented by 10 fatty acids carrying the highest eigenvalues for principle component 1. These were: 1) phototrophic microeukaryotes [20:5ω3, 16:4ω1, 16:3, 22:6ω3 and 16:1ω13*t*], 2) Gram-positive bacteria and some Gram-negative anaerobic bacteria [*i*16:0, *i*15:0 and *a*15:0], 3) sulfate-reducing and other anaerobic bacteria [10me16:0] and 4) aerobic Gram-negative bacteria exposed to environmental stress [16:1ω7*t*] (known stressor include reduced O_2 tension, starvation and exposure to phenol (*18,20*)). The annual variation in microbial community structure at this site appeared to be driven, at least in part, by changes in biomass of phototrophic eukaryotes (Findlay and Watling submitted, *Micro. Ecol.*). Phototrophic eukaryotic biomass varied 6- to 11-fold while total biomass and biomass of the two functional groups of anaerobic bacteria varied 2- to 3-fold. Periods of peak total biomass and phototrophic eukaryotic biomass occurred during transition or cold-water months. During these periods, phototrophic eukaryotes accounted for 70% to 85% of total microbial biomass expressed as carbon (total biomass determined as total PLFA; PLFA concentration converted to carbon using the algorithms given in Findlay and Dobbs (*18*)). During warm-water months phototrophic eukaryotic biomass decreased and accounted for 35% to 55% of total microbial biomass. The biomass of anaerobic bacteria also increased during periods of highest microbial biomass but was relatively constant during other months. As total biomass was lowest during the warm-water months, the contribution of anaerobic bacteria to the community increased during these periods. The increase in the fatty acid 16:1ω7*t* indicated that aerobic bacteria were experiencing some form of environmental stress -- in this case it was most likely decreased O_2 concentrations in the sediments (Findlay and Watling in press, *Mar. Ecol. Prog. Ser.*). These results indicated that the observed pattern in microbial community structure was related to both the balance between eukaryotic and prokaryotic organisms and the balance between aerobic and anaerobic microorganisms within the sediments. As total microbial biomass and the abundance of eukaryotic phototrophs were strongly linked, the changes in the abundance of the latter strongly influenced total sedimentary microbial biomass in these sediments. The relatively small range in abundance of the marker fatty acids for the two anaerobic functional groups of bacteria indicated that their biomass was relatively constant compared to that of phototrophic microeukaryotes and that their increased importance within the sedimentary microbial community during the warm-water months was in part due to the decrease in eukaryotic biomass. Again, the pattern exhibited strong seasonal trends with maximum contributions to microbial biomass by eukaryotes and aerobes occurring during cold-water months and maximum contributions by bacteria and anaerobes occurring during warm-water months.

Seasonal Variation at an Organically Enriched Site. The pattern of seasonal variation at the Toothacher Cove site as determined by PLFA profiles and PCA was a predictable change in PLFA profiles that correlated with months of the year (Figure 3). Samples taken from April and May had the most negative component

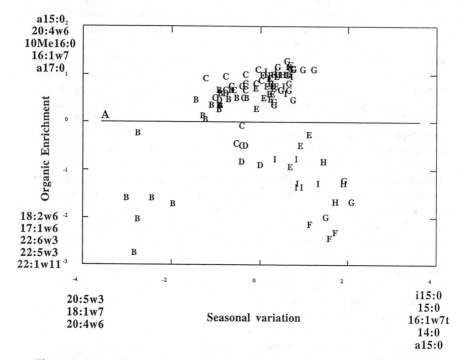

Figure 3. Plot of factor scores (component 1 x component 2; designated seasonal variability and organic enrichment, respectively) resulting from principal component analysis of weight percent PLFA data from Toothacher Cove sediments. Scores are plotted by sample date and symbols as in Figure 2, except that H = samples taken November 6, 1991 . (Adapted from ref. 6)

scores for principal component 1 while samples from August, September and October had the highest positive component scores for principal component 1. In general, sediments from cold-water months had negative component scores, and sediments from warm-water months had positive component scores for principal component 1. Samples from June and July had component scores that bracketed zero. Four functional groups of microorganisms were represented by eight fatty acids carrying the highest eigenvalues for principle component 1 (only three fatty acids had negative eigenvalues). These were: 1) phototrophic microeukaryotes [20:5ω3], 2) heterotrophic microeukaryotes [20:4ω6], 3) Gram-positive bacteria and some Gram-negative anaerobic bacteria [14:0, i16:0, i15:0 and a15:0] and 4) aerobic Gram-negative bacteria [18:1ω7, 16:1ω7t]. Fatty acid 16:1ω7t, as before, indicates exposure to stress (*20*). The annual variation in microbial community structure at this site appeared to be driven, at least in part, by changes in biomass of phototrophic and heterotrophic eukaryotes. Phototrophic and heterotrophic eukaryotic biomass varied 7- and 5-fold, respectively, while total biomass and biomass of anaerobic bacteria varied 2- to 3-fold (*6*). Periods of peak total biomass and eukaryotic biomass occurred during transition or cold-water months. Further geochemical analysis (comparison of sterol/fatty acid ratios, sediment protein, and sediment carbon and nitrogen content) indicated that living biomass was greatest during April and declined throughout the rest of the sampling period (*6*). Total sedimentary organic matter, however, peaked at the transition between cold- and warm-water conditions (early July). At the two periods of maximum eukaryotic biomass, eukaryotes contributed 29% of the total living microbial biomass expressed as carbon (*6*). The first period of maximum eukaryotic biomass occurred during April and phototrophic eukaryotes dominated the community. The second peak in eukaryotic biomass coincided with the peak in sedimentary organic matter as eukaryotic biomass was dominated by heterotrophs. The relatively small range in abundance for the fatty acids indicative of anaerobic bacteria suggests that their biomass was relatively constant compared to that of phototrophic microeukaryotes and that their increased importance within the sedimentary microbial community during the warm-water months was in part due to the loss of eukaryotic biomass. Again, the pattern exhibited strong seasonal trends with maximum contribution to microbial biomass by eukaryotes and aerobes occurring during cold-water months and maximum contribution by bacteria and anaerobes occurring during warm-water months.

Variation in Benthic Microbial Community Structure Associated with Organic Enrichment. Samples taken from Toothacher Cove also showed a strong impact that was associated with increased organic input to the sediments. Sediments proximal to the net-pen experienced a 2- to 3-fold increase in benthic carbon flux. This material was readily biodegradable and, in general, increased benthic respiration (*6*). For coastal Maine sediments, increased carbon flux to the sediments increased benthic O_2 demand in a 1:1 molar ratio (Findlay and Watling in press, *Mar. Ecol. Prog. Ser.*). PLFA profiling coupled with PCA was also able to detect changes in microbial community structure caused by the increased benthic carbon flux. Component 2 mapped all samples taken 1 m from the net-pen and six samples taken 10 m from the net-pen (those taken on 3 April 1991 and 20 November 1991) distinct from all other samples (Fig. 2). Samples below line A (Fig. 2) showed high weight-percent values for fatty acids 18:2ω6, 17:1ω6, 22:6ω3, 22:5ω3 and 22:1ω11. Samples above line A showed high weight-percent values for fatty acids a15:0, 20:4ω6, 10Me16:0, 16:1ω7 and a17:0. A comparison of the fatty acids found in *Beggiatoa*-type mats to ambient sediments demonstrated that fatty acids 18:2ω6, 17:1ω6, 22:6ω3, 22:5ω3 and 22:1ω11 were more abundant in *Beggiatoa*-type mats as compared to ambient sediments and that fatty acids a15:0, 20:4ω6, 10Me16:0, 16:1ω7 and a17:0 showed the opposite trend, that is, these fatty acids were more abundant in ambient sediments (*6*). Hence, principal components analysis indicated that sediments taken within 1 m of the net-pen, regardless of sample date, were enriched in PLFAs common in *Beggiatoa*-type mats. These organisms are aerobic, chemolithotrophic, and live at the interface where molecular oxygen and hydrogen sulfide are both present. In marine sedimentary environments the presence of these organisms is generally interpreted as an indication of local organic enrichment, and the biochemical detection of this community in the sediments proximal to the pen was the first consequence of the increased organic flux to be observed independent of sampling date. As such, it may prove useful as a monitoring tool for use in determining the nature and extent of environmental impact associated with organic enrichment. In addition to increasing the importance of the chemolithotrophs within the sedimentary microbial community, greater benthic carbon flux also increased the magnitude of the seasonal variation observed. By definition, PCA describes the major components or axis of variation within a data set. Principal component 1 maps the major component of variation. The predominance of samples within 1 m of the net-pen at the extremes of this mapping indicates that the areas proximal (samples plotted below line A) to the pen showed the greatest variation in microbial community structure. We interpret this variation as a shift in the relative importance of anaerobic and aerobic microorganisms, both procaryotic and eucaryotic. In other words, the microbial community structure of the sediments beneath the pen had a greater percentage of aerobes than ambient sediments during April and a greater percentage of anaerobes than ambient sediments during August, September and October.

Comparison of the Pristine and Anthropogenically Impacted Sites. There are at least three significant differences (other than anthropogenic effects) between the two sampling sites. The Damariscotta River site is shallower, 2-4°C warmer and far more protected than the Toothacher Cove site. Total microbial biomass was similar at the two sites, but the PLFA patterns indicated that Toothacher Cove sediments contain more bacteria and fewer eucaryotic microorganisms than Damariscotta River estuary sediments. This probably resulted from greater light penetration and less frequent sediment resuspension events at the Damariscotta River site. Greater light intensity will aid in phototrophic growth. Sediment resuspension events were common in Toothacher Cove (*6*). These will decrease light penetration and also cause disturbance induced shifts in community structure. In short-term studies, microeukaryotes show the greatest loss of biomass following a disturbance, as well as the slowest recovery rates. Examples of mixing and translocation disturbances are sieving (*11*), ray feeding activity (*14*) and commercial dragging, (*12*). In longer-term studies, disruption of the sediments also caused increases in the importance of bacteria (relative to eucaryotic microorganisms) within the community. Examples include sediments from the burrows of macrofauna (*13*) and sediments where the abundance of epibenthic predators have been experimentally increased (*9*). The differences between the microbial community structure in Toothacher Cove and Damariscotta River sediments are consistent with the Damariscotta River sediments being shallower and Toothacher Cove sediments undergoing episodic resuspension. In spite of these differences, the major variations (as determined by PCA analysis) in the structure of the sedimentary microbial community at two sites correlated with season. At both sites, seasonal variation was described in terms of shifts in the relative importance of aerobes and anaerobes, and microeukaryotes and bacteria. The maximum contribution to microbial biomass by eukaryotes and aerobes occurred during cold-water months while the maximum contribution by bacteria and anaerobes occurred during warm-water months. The cooler water temperatures at the Toothacher Cove site appeared to have little effect on the observed seasonal patterns of change in sedimentary microbial community structure except that transition periods were 20 to 30 days later in Spring and earlier in Fall.

Implications of Seasonal Variations in Benthic Microbial Community Structure for Environmental Geochemistry. Seasonal patterns of change in microbial community structure were observed regardless of the presence or absence of anthropogenic organic enrichment. The patterns were similar and many of the observed differences in microbial community structure can be attributed to the physical differences between the sites and not to organic enrichment. This is significant because the organic enrichment proximal to the net-pen was sufficient to induce an azoic sediment/*Beggiatoa*-type mat endpoint at the Toothacher Cove site (*6*). This, in turn, suggests that the environmental conditions associated with seasonal changes (temperature, light intensity, predation, and granulometry and its control on sedimentary carbon reserves -- see *21*) exert significant controls on sedimentary microbial community structure. Temperatures below $5^{\circ}C$ are likely to decrease the activity of benthic predators that remove microbial biomass and will also inhibit the activity of sulfate-reducing bacteria (*22-23*). It is unlikely that benthic phototrophs and diatoms, in particular, would be so affected. Increased light intensity will stimulate photosynthesis. Together low temperatures and high light levels explain the increased phototrophic biomass early in the calendar year, as has been demonstrated for phytoplankton communities in the Gulf of Maine (*24-25*). Progressively lower biomass of phototrophs as temperatures warm suggests that predation does eventually remove some of the biomass built up during the cold-water months of February through May or June (depending on the site). Coupled with increasing heterotrophic bacterial activity during the period of increasing temperatures, these processes may be sufficient to account for the observed variations in community structure. The robust nature of this pattern in the presence of organic enrichment has several consequences for researchers interested in determining the nature and extent of the effects of anthropogenic stress in environmental geochemistry.

Conclusions. The detection of biological effects of anthropogenic stresses typically involves detection of changes in the natural state of some community through a program of monitoring. Due to its central role in sedimentary biogeochemistry the microbial community is a natural candidate to include in any monitoring program. Because microbes are capable of much faster response to environmental change compared to macrofauna, we suggest that geochemical studies of this kind will be of greater use in detecting habitat degradation due to organic enrichment. With the advent of quantitative techniques to determine microbial community structure (PLFA analysis) it is now practical to carry out such studies. Preliminary results at a riverine site highly impacted with polycyclic aromatic hydrocarbons also show that seasonal variations dominate anthropogenic induced changes in community structure (Langworthy and Findlay, unpublished data) and extend the above findings to shallow freshwater systems and other types of anthropogenic stresses.

The implications of strong seasonal patterns in microbial community structure, at least in temperate waters, are that to be successful, experimental designs must account for and document seasonal variations as well as anthropogenic effects. The strength of the patterns observed in this study also suggest that statistical techniques such as PCA, that can separate multiple components of variation, must be included into experimental designs. Any study that does not include separate, non-impacted ambient sites in its design may not differentiate seasonal and anthropogenic effects. Common designs, such as the ABC method, where samples taken before the onset of stress are used as baseline or control data and compared to samples after the onset of stress are doomed to equivocal results. Note that these warnings are also being generated in studies of temporal variation in the distribution and abundance of macrobenthic organisms (*26-27*).

Finally, it should be noted that PLFA techniques produce data capable of detecting anthropogenic stress in benthic environments at reasonable cost. Many sediments can be characterized with as few as three replicates and at a cost of $100-200 per replicate. This compares quite favorably with more classical methods such as macrobenthos analysis where five or more replicates are typically needed at a cost of $400-600 per replicate or determinations of BOD, which while commonly used in water column studies, are rarely successfully applied in benthic environments.

Acknowledgments

This research was supported in part by grants from EPA Office of Exploratory Research (Assistance ID Number R-817196-01-0), EPA District 1 (Assistance ID Number X001837-01-0), National Marine Fisheries Service (U.S. Department of Commerce, National Oceanic and Atmospheric Administration, award number NA16FL0067-01), the UM/UNH Sea Grant College Program (U.S. Department of Commerce, National Oceanic and Atmospheric Administration, project numbers

R/FMD-215 and R/FMD-223), the Maine Aquaculture Innovation Center (project numbers 90-6, 91-13 and 93-08) and the PADI Foundation. We would like to gratefully acknowledge the assistance of a great number of people whose efforts helped bring this project to fruition. These include E. Harrison, R. Doering, T. Sawyer, J.B. Pelletier, L. Klippel, L. McCann, M. Tarrentino, W. Tripp, J. Anderson, J. Fay, D. Meisenheimer, and C. Heinig and B. Tarbox of Intertide Corp. for assistance with diving and field operations. M. Tarrentino and J.B. Pelletier ably served as dive safety officers and T. Sawyer directed field operations in the absence of RHF. S. Sutton, J. Smoot, the editor and two anonymous reviewers provide helpful comments on earlier versions of the manuscript.

Literature Cited

1. Duursma, E. K.; Marchand, M. *Oceanogr. Mar. Biol. Ann. Rev.* **1974** *12*, 315-431.
2. Pearson, T. H.; Rosenberg, R. *Oceanogr. and Mar. Biol. Ann. Rev.* **1978**, *16*, 229-311.
3. Fletcher, M. *Acrh. Microbiol.* **1979**, *122*, 271-274.
4. Berner, R. A. *Early Diagenesis.* Princeton University Press, Princeton, NJ, 1980; pp. 241.
5. Jørgensen, B.B. *Nature* **1982**, *296*, 643-645.
6. Findlay, R.H.; Watling, L.; Mayer, L. M. *Estuaries* **1995**, *18*, 145-179.
7. Nickels, J. S.; Bobbie, R. J;. Martz, R. F; Smith, G. A.; White, D. C.; Richards, N. L. *Appl. Environ. Microbiol.* **1981**, *41*, 1261-1268.
8. Findlay, R.H.; Kim, S.L.; Butman, C.A. *Mar. Ecol. Prog. Ser.* **1992**, *90*, 73-88.
9. Federle, T. W.; Livingston, R. J.; Metter, D. A.; White, D. C. *J. Exp. Mar. Biol. Ecol.* **1983** *73*, 81-94.
10. Wainright, S. *Science* **1987**, *238*, 1710-1712.
11. Findlay, R. H.; Trexler, M. B.; Guckert, J. B.; White, D. C. *Mar. Ecol. Prog. Ser.* **1990**, *61*, 121-133.
12. Mayer, L. M.; Schick, D. F.; Findlay, R. H.; Rice, D. L. *Mar. Environ. Res.* **1991**, *31*, 249-261.
13. Dobbs, F. C.; Guckert, J. B. *Mar. Ecol. Prog. Ser.* **1988** *45*, 69-79.
14. Findlay, R. H.; Trexler, M. B.; White, D. C. *Mar. Ecol. Prog. Ser.* **1990**, *61*, 135-148.
15. Findlay, R.H.; In: *Molecular Microbial Ecology Manual*, Akkermans, A.D.L.; van Elsas, J.D.; de Bruijn, F.J.; Eds.; Kluwer Academic Publishers: the Netherlands. 1996, Chapter 4.1.4, p. 1-17.
16. McAlice, B. J. *Maine Sea Grant Technical Report 43*, 1979.
17. Bligh, E. G.; Dyer, W. J. *Can. J. Biochem. Physiol.* **1959** *31*, 911-917.
18. Findlay, R.H.; Dobbs, F.C. In: *Current Methods in Aquatic Microbial Ecology*, Kemp, P. F.; Sherr, B. F.; Sherr, E. B.; Cole, J. J. Eds.; Lewis Publishers: Boca Raton, FL 1993, p. 271-284.
19. White, D. C.; Pinkart, H. C.; Ringelberg, A. B. In: *Manual of Environmental Microbiology*, Hurst, C. J.; Knudsen, G. R.; McInerney, M. J.; Stetzenback, L. D.; Walter, M. V.; Eds.; ASM Press: Washington, D. C. 1997, p. 91-101.

20. Guckert, J. B.; Hood, M. A.; White, D. C. *Appl. Environ. Microbiol.* **1986**, *52*, 794-801.
21. Mayer, L. M.; Rahaim, P.; Guerin, W.; Macko, S. A.; Watling, L.; Anderson, F. E. *Estuar. Coast. Shelf Sci.* **1986**, *20*, 491-503.
22. Isaksen, M. F.; Bak, F.; Jørgensen, B. B. *FEMS Microbiol. Ecol.* **1994**, *14*, 1-8.
23. Holmer, M.; Kristensen, E. *Biogeochemistry* **1996**, *32*, 15-39.
24. Townsend, D. W.; Keller, M. D.; Sieracki, M. E.; Ackleson, S. G. *Nature* **1992**, *360*, 59-62.
25. Townsend, D. W.; Cammen, L. M.; Holligan, P. M.; Campbell, D. E.; Pettigrew, N. R. *Deep-Sea Res. Part I*, **1994**, *41*, 747-765.
26. Morrisey, D. J.; Underwood, A. J.; Howitt, L.; Stark, J. S. *J. Exp. Mar. Biol. Ecol.* **1992**, *164*, 233-245.
27. Underwood, A. J. *Ecol. Appl.* **1994**, *4*, 3-15.
28. Novitsky, J. A.; Kepkay, P. E. *Mar. Ecol. Prog. Ser.* **1981**, *4*, 1-7.
29. Martens, C. S.; Klump, J. V. *Geochim. Cosmochim. Acta* **1984**, *48*, 1987-2004.

Chapter 5

The Use of Aromatic Acids and Phospholipid-Ester-Linked Fatty Acids for Delineation of Processes Affecting an Aquifer Contaminated with JP-4 Fuel

Jiasong Fang[1], Michael J. Barcelona[1], and Candida West[2]

[1]National Center for Integrated Bioremediation Research and Development, Department of Civil and Environmental Engineering, University of Michigan, Ann Arbor, MI 48109
[2]National Risk Management Research Laboratory, Subsurface Protection and Remediation Division, U.S. Environmental Protection Agency, Ada, OK 74820

A glacio-fluvial aquifer located at Wurtsmith Air Force Base, Michigan, has been contaminated with JP-4 fuel hydrocarbons released by the crash of a tanker aircraft in October of 1988. A comprehensive analysis of the inorganic and organic geochemical constituents and geomicrobiological markers has documented the occurrence of *in situ* biodegradation of hydrocarbons in the aquifer. Concentration profiles of aromatic hydrocarbons, aromatic acids, and phospholipid ester-linked fatty acids (PLFA) in aquifer solids suggest microbially mediated degradation of hydrocarbons and production of aromatic acid metabolites. Microbial community structure as indicated by the PLFA patterns shows an absence of polyunsaturated fatty acids characteristic of microeukaryotes and high proportions of C_{12}-C_{20} fatty acids typical of bacteria. Contamination increased microbial biomass by one order of magnitude and shifted the community to a more anaerobic bacterial consortium.

Jet fuel-4 (JP-4) spillage has been a source of contamination for soils and ground water. Monoaromatic and aliphatic hydrocarbons are major constituents of this fuel. Because the aromatic fraction of the fuel mixture is relatively water soluble, these compounds tend to migrate from contaminated soils into aquifers (*1*). A dissolved plume typically develops downgradient from the point of free product release. Physical (dispersion, volatilization), chemical (sorption, dissolution), and biological processes (microbial degradation) control the subsequent fate and transport of the dissolved contaminants in the subsurface. Aromatic hydrocarbons are subject to degradation by a variety of microbial transformation processes in the subsurface (*2-3*). Degradation of monoaromatic hydrocarbons such as benzene, toluene, ethylbenzene, and xylenes (BTEX) has been demonstrated in numerous laboratory studies. Biodegradation has been documented under aerobic (*4*), denitrifying (*5-12*), dissimilatory iron-reducing (*13*), sulfate-reducing (*14-17*), and methanogenic conditions (*18-21*).

Direct measurement of *in situ* biodegradation of contaminants is often difficult (*22*). The field evidence for potential microbial remediation of organic contaminants involves demonstration of compound mass removal, production of metabolites, functional microbial activities, and indirect measures of redox changes and isotopic composition of dissolved inorganic carbon (23). Mass balance modeling of the disappearance of dissolved parent compounds and reactants (e.g. dissolved oxygen) (24-25) has been used to demonstrate *in situ* attenuation processes. This approach is limited by the natural heterogeneity and geochemical variability of aquifers, and by competing abiotic processes (sorption, dilution, and volatilization). Alternatively, measurements of oxygen and carbon dioxide (26) and stable carbon isotopic composition of evolved CO_2 from biodegradation (27) in soil gas have been utilized as indirect indicators of hydrocarbon degradation. Recently, stable carbon isotope measurements of dissolved inorganic carbon in a shallow aquifer showed evidence for biodegradation of petroleum hydrocarbons under sulfate reducing and methanogenic conditions (28-29). Recent interest has developed on using intermediate metabolites such as aliphatic and aromatic acids (30-32) as indicators of *in situ* biodegradation of hydrocarbons in contaminated aquifers. Comprehensive geochemical studies and numerical modeling have been conducted of an aquifer contaminated with crude oil at Bemidji, Minnesota (28, 31, 33-36).

Documentation of *in situ* biodegradation processes must show the production of degradation intermediates and the responsible microbial processes, in additional to the mass removal of contaminants (parent compounds). This study was undertaken to use biogeochemical markers to document the natural microbial degradation of alkylbenzenes, identify major controlling processes, and determine changes in microbial communities associated with these processes in an aquifer contaminated with JP-4 fuel.

Materials and Methods

Field Site. The study area is located at the KC-135 Crash Site at the former Wurtsmith Air Force Base (WAFB) Oscoda, Michigan (Figure 1). The aquifer is composed of quaternary glacial-fluvial sand and gravel materials in the upper 18 m of the subsurface. The glacial aquifer deposits are composed of medium-fine quartzose sand and gravel deposits underlain by a thick lacustrine clay. The clay unit is brown to gray, relatively impermeable, and cohesive. The groundwater table is at about 3.4 m below land surface (bls). A groundwater divide cuts diagonally across the Base from northwest to southeast. Southwest of the divide ground water flows toward the Au Sable River. North of the divide where the study site is located ground water flows toward the east discharging to Van Etten Lake. Ground-water flow at the KC-135 Crash Site is to the southeast with a hydraulic gradient of ~0.0024. The KC-135 Crash Site is located near the main runway of WAFB (Figure 1). The aquifer was contaminated by JP-4 fuel resultant from the tragic crash of a KC-135 fuel tanker in October 1988. At least 3,000 gallons of JP-4 were spilled at the site, an unknown quantity of which burned as a result of the accident. An existing area of monitoring wells finished at or near the water table had been established in previous investigations at the site. Depth-discrete changes in soil gas and ground water composition were evaluated using drive-point techniques at upgradient, source, and downgradient locations.

Field Sampling. Wells for the present study were installed in November 1995 at the KC-135 Crash Site using either a Geoprobe piston corer (Geoprobe Systems, Salina, Kansas) or a hollow stem auger without lubricants (37). Geoprobe barrels and augers for the drill rigs, and stainless screens were steam-cleaned prior to use. Multi-level sampling (MLS) wells (2.5 cm internal diameter) were emplaced above and below the water table with screen lengths of 0.3 m. Soil samples were taken using either Geoprobe or "Waterloo" coring device (38) with a hollow stem auger (37). Upon retrieval, the core was placed on a table and subsampled. For volatile organic compound (VOC) analysis, about 5 mL (or 7 g) of solids was taken using a plastic disposable syringe with one end cut. The samples were preserved by adding 5 mL of 40% $NaHSO_4$ aqueous solution. Samples for aromatic acids and phospholipid ester-linked fatty acid (PLFA) analysis were taken using a sampling spoon into 473-mL wide-mouth jars. The jars were placed in a freezer until analysis.

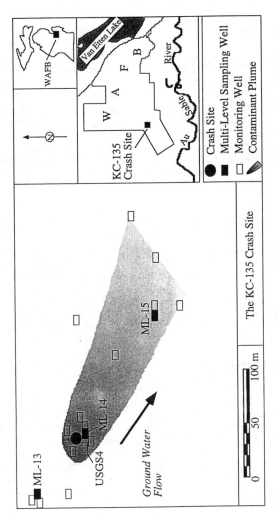

Figure 1 Field site showing the KC-135 Crash Site at Wurtsmith Air Force Base (WAFB) in Michigan.

Once the dissolved oxygen (DO), pH and conductivity values had stabilized (*39*), ground water samples for VOC analysis were taken using a peristaltic pump into 40-mL glass vials fitted with teflon-lined screw caps. Water samples were preserved immediately by adding 1 mL of 40% $NaHSO_4$ solution. Field analyses for pH, DO, redox potential (Eh), sulfide, Fe(II), and alkalinity (as bicarbonate) were made using a CHEMetrics kit (CHEMetrics, Inc., Calverton, VA). Concentrations of nitrate (NO_3^-) and sulfate (SO_4^{2-}) were measured in the laboratory using a Dionex-100 ion chromatograph.

Analytical Procedures.

Volatile Organic Compounds. Solid samples were analyzed for volatile organic compounds on an HP-7694 headspace analyzer interfaced with an HP 5890 Series II gas chromatograph (GC) and HP 5972 mass selective detector (MSD). Samples were spiked with internal standard (*o*-xylene-d_{10} and naphthalene-d_8) prior to analysis. Separation of individual components (43 volatile compounds) was accomplished using an HP-624 GC column: 60 m x 0.25 mm i.d. (film thickness d_f=1.8 μm) (Hewlett Packard, Wilmington, DE). The GC-MSD was calibrated by injection of calibration standards at five different concentration levels (10, 50, 100, 200, and 400 μg/L). Response factors were obtained for all compounds including internal standards. Compounds were identified based on relative retention time (*40*) and mass spectra. Concentrations of each compound were calculated using the internal standard method, and are reported as μg/kg of dry weight soil.

Geochemical Species. Laboratory analysis of nitrate and sulfate in ground water was carried out on a Dionex DX-100 ion chromatograph (IC, Dionex Corp., Sunnyvale, CA) equipped with an AS40 autosampler. A Dionex AS-14S column (250 mm x 4 mm) with a guard column (50 mm x 4 mm) was used for the analysis. Each sample of 2.5 mL was injected into the IC. Results were obtained as an average of duplicate analyses.

H_2 Measurements. Dissolved H_2 was measured by the "bubble method" (*41*). The amount of H_2 in ground water was expressed as nanomoles/liter for the purpose of differentiating microbial processes, as defined by previous researchers (*41*).

Aromatic Acid Metabolites. Soil samples collected for determination of aromatic acids were subjected to procedures as described in Barcelona *et al.* (*32*). Briefly, an aliquot of soil (~ 20 g, wet weight), after spiking with an internal standard (4-fluorobenzoic acid-d_4), was slurried with 20 mL distilled water, adjusted to pH 8, and then agitated for 24 h on a shaker. The slurries were then centrifuged at 2,000 g for 1 h. The supernatant was decanted and then freeze-dried in a Labconco lyophilization system (Labconco Corp., Kansas City, Missouri). The dried residue was transferred to a test tube with three 2-mL methanol rinses. The volume was reduced to ~500 μL using a nitrogen blow-down system, and 50 μL of BF_3/methanol derivatizing reagent (Supelco Inc., Bellefonte, PA) was added to each test tube. The test tubes were sealed with Teflon-lined screw caps and heated at 90°C for 90 min in a Thermolyne Dri-Bath. The test tubes were then removed from the bath, and the reaction was quenched by adding two drops of deionized water to the vials. The methyl esters of aromatic acids were taken up by extraction of the reaction mixture with 1 mL of dichloromethane three times. The volumes were reduced and adjusted to 500 μL for analysis by gas chromatography/mass spectrometry (GC/MS). Blanks were treated in the same manner as samples. Chromatographic separation was achieved using a 30 m x 0.25 mm (d_f=0.25 μm) DB5-MS fused-silica capillary column (Hewlett Packard, Wilmington, DE). Individual components were identified from their mass spectra and by comparison with authentic standards.

Analysis of Phospholipid Ester-Linked Fatty Acids (PLFA). Soil samples were extracted by a modified one-phase chloroform/methanol Bligh and Dyer extraction (*42*). Approximately 40 g of a freeze-dried soil sample was added to a hypovial with 100 mL extraction solution of methanol, chloroform and phosphate buffer (pH 7.4) 2:1:0.8. The extraction mixture was shaken on a wrist action shaker for two hours and allowed to stand overnight in darkness at 4°C. The extraction mixture was then decanted through a cellulose #4 filter into a 250 mL separatory funnel. The soil was washed with 2x10 mL extraction solution. The lipids were partitioned by adding chloroform and water such that the final ratio of chloroform-methanol-water was 1:1:0.9. The aqueous phase was then washed with 2x10 mL chloroform. The organic phase was pooled, and the volume was reduced. Lipid classes were separated by column chromatography into neutral lipids, glycolipids, and phospholipids by sequentially eluting with 4 mL of chloroform, acetone, and methanol, respectively. Ester-linked phospholipid fatty acids were subjected to a mild alkaline trans-methylation procedure (*43*) to produce fatty acid methyl esters (FAMEs). FAMEs were analyzed on a Hewlett-Packard 5890 GC with a flame ionization detector (FID) using a J&W DB-5 column (60 m x 0.25 mm i.d., d_f=0.25 μm). The gas flows were: hydrogen 20 mL/min, air 372 mL/min, and nitrogen (makeup gas) 30 mL/min. The column temperature was programmed from 100°C to 150°C at 10°C/min, then from 3°C to 280°C at 3°C/min. Response factors were obtained for each compound, and the absolute quantitation of individual compound was obtained using a C19:0 fatty acid as internal standard. Fatty acids was identified using an Hewlett-Packard 5890 Series II GC interfaced with an HP 5972 Mass Selective Detector. Double bond position and geometry of monounsaturated fatty acids were determined using procedures of Dunkelblum *et al.* (*44*).

Fatty acids are designated as total number of carbon:number of double bonds. The position of the double bond is indicated with an omega (ω) number representing the carbon closest to the terminus of the alkyl chain of the fatty acid molecule with the geometry of either c (*cis*) or t (*trans*). Terminal methyl-branching is indicated with i (*iso*) and a (*anteiso*). The cyclopropyl ring in the fatty acids is indicated by "cyc".

Results and Discussion

Ground Water Geochemistry. Ground water exhibited distinct geochemical patterns. Dissolved oxygen concentrations ranged from 3.8 mg/L to 8.7 mg/L at the upgradient (ML-13) and downgradient (ML-15) locations, and from 0.56 mg/L to undetectable at ML-14, near the source (Table I). The concentrations of nitrate (NO_3^-) and sulfate (SO_4^{2-}) in ground water ranged from 0.13 mg/L to 4.15 mg/L at the upgradient well ML-13, and from 6.92 mg/L to 7.91 mg/L at the downgradient well ML-15. Ground water at ML-14 contained essentially non-detectable (<0.5 mg/L) nitrate and low concentrations of sulfate. Dissolved iron concentrations were highest (9.5 mg/L) near the water table at ML-14H and were below detection (<0.5 mg/L) in ground water at ML-13 and ML-15 (Table I). Dissolved hydrogen in the source area ranged from 1.08 to 1.43 nmoles/liter in the source area (Table I), and from 0.14 to 0.86 nmoles/liter in the upgradient and downgradient areas. Geochemical evidence, thus, suggests that nitrate-, iron-, and sulfate-reduction were important microbial processes in the aquifer. Methanogenesis was not a dominant anaerobic process in the aquifer, although methane was observed in the soil gas (data not shown).

Microbial Degradation of Hydrocarbons and Production of Intermediate Metabolites. Alkylbenzenes with varying degrees of alkyl substitution (C_1-C_5) were identified in aquifer solids, including eight C_3 isomers, seventeen C_4 isomers, and two C_5 isomers. Other compounds identified include naphthalene, methylnaphthalenes, 1,2,3,4-tetrahydronaphthalene, 1,2,3,4-tetrahydro-2-, 1,2,3,4-tetrahydro-1-, and 1,2,3,4-tetrahydro-5-methylnaphthalenes. Highest concentrations were detected for 1,2,4-trimethylbenzene, followed by *m*- and *p*-xylenes. Solid samples near the water table in the source zone (ML-14, 3.7-4.1 m) contained elevated concentrations of alkylbenzenes, with a secondary maximum at depths of 6.1-6.7 m (Figure 2).

Aromatic acid metabolites identified in contaminated aquifer solids include: benzoic

Table I Ground water geochemical measurements at multi-level sampling wells at the KC-135 Crash Site

Well	Depth[a] (m)	O_2 (mg/L)	Fe^{2+} (mg/L)	NO_3^- (mg/L)	SO_4^{2-} (mg/L)	H_2 (nanomoles/L)	Eh (mv)
ML-13 (C)	11.6	8.5	<0.01	0.13	7.39	0.84	+270
ML-13 (F)	13.5	3.8	<0.01	4.15	7.30	0.16	+268
ML-14 (H)	10.5	0.56	9.5	BLD[b]	0.90	1.43	-58
ML-14 (C)	12.5	BLD	0.07	BLD	BLD	1.08	-96
ML-15 (H)	10.7	8.7	<0.01	3.56	7.91	0.32	+229
ML-15 (C)	11.9	6.5	<0.01	2.4	6.92	0.86	+196

[a] Depth below land surface.
[b] Below limit of detection.

Table II Aromatic acid metabolites (μg/kg) identified in aquifer solids at the KC-135 Crash Site

Compound	Concentration (μg/kg)		
	ML-14 (3.0-3.8 m)	ML-14 (6.1-6.7 m)	ML-15 (3.7-4.6 m)
Benzoic acid	1285.1	325.6	13.3
2-Methylbenzoic acid	87.6	-	
Salicylic acid	21.1	-	
3-Methylbenzoic acid	434.8	-	27.36
4-Methylbenzoic acid	83.3	-	154.5
2,6-Dimethylbenzoic acid	39.1	-	
2,5-Dimethylbenzoic acid	102.0	-	
2,3-/3,5-Dimethylbenzoic acid[a]	170.0	-	
2,4,6-Trimethylbenzoic acid	1348.7		
3,4-Dimethylbenzoic acid	344.5	-	

[a] Not resolved under the analytical conditions used.

acid, *o*-, *m*-, *p*-toluic acids (2-, 3-, and 4-methylbenzoic acids), 2,5-, 2,3-/3,5-, 2,6- and 3,4-dimethylbenzoic acid, and 2,4,6-trimethylbenzoic acid. The most abundant aromatic acids were 2,4,6-trimethylbenzoic acid and benzoic acid (Table II). Highest concentrations of aromatic acids were found in solid samples near the water table in the source area (ML-14, 3.7-4.1 m) (Figure 3). No aromatic acids were detected in soil in ML-13, and only small amounts of the benzoic acids, *m*- and *p*-toluic acids were found in solids from ML-15 (1.5-3.0 m) (Table II).

The alkylbenzene profiles paralleled the distributions of aromatic acids in the solids (Figure 2). Since these aromatic acids are intermediates of anaerobic degradation (*16-17, 45*), are not original constituents of the JP-4 fuel, and were not detected in soil samples from the upgradient area, it may be concluded that the production of aromatic acids was associated with the anaerobic degradation of aromatic hydrocarbons. Concentrations of total BTEX (benzene, toluene, ethylbenzene, and *m*-, *p*-, *o*-xylenes) in ground water have decreased considerably from 1993 to 1996 (Figure 3). For instance, the concentration of BTEX declined from 15,000 µg/L to 3951 µg/L at USGS4 between Febuary 1993 and June of 1996, and from 3750 µg/L to 2257 µg/L at ML-14H between January and June 1996. The oxidation of hydrocarbons may have been coupled to anaerobic processes. Iron-reduction and sulfate-reduction may be the major terminal electron-accepting processes (TEAP) in the aquifer, although methanogenesis cannot be ruled out (Table I).

Response of Microbial Populations to Contamination. The PLFA profiles at ML-14 are presented in Figure 4. Twenty three different fatty acids ranging from C_{12} to C_{20} were identified, including saturated, monounsaturated, branched, cyclopropyl and polyunsaturated fatty acids. The only polyunsaturated fatty acid detected was 18:2ω6c. Branched fatty acids detected in the aquifer solids include terminal branched (*iso* and *anteiso*) and mid-branched fatty acids (10MeC18:0).

The predominance of C_{12}-C_{20} PLFA suggests a dominance of bacterial communities in the aquifer (*46-47*). The only polyunsaturated fatty acids detected at the KC-135 Crash Site were C18:2ω6, suggesting the sparse distribution of microeukaroytes in the subsurface. Terminal-branched fatty acids (*iso* and *anteiso* C15 and C17) are biomarkers for anaerobic bacteria (*47-50*). Parkes and Taylor (*47*) reported elevated concentrations of terminal-branched fatty acids isolated from a mixed culture of sulfate-reducing bacteria. The terminal-branched fatty acid iC16:0 is an indicator of sulfate-reducing bacteria (White, personal communication, 1996.) Cyclopropyl fatty acids have been found in a variety of Gram-positive and Gram-negative bacteria (*51*). These fatty acids are typical biomarkers of anaerobic bacteria (*48*). The concentrations of terminal-branched and cyclopropyl fatty acids increased significantly from uncontaminated depths to the contaminated depths (Figure 4), suggesting stimulated growth of anaerobic microbial populations. The community structure appeared to shift from aerobic microbiota rich in monounsaturated fatty acids (*52*) to a more anaerobic bacterial consortium. This finding is in agreement with ground-water chemical measurements. A mirror-image relationship was observed between the concentration of terminal electron acceptors such as dissolved oxygen (DO) and sulfate in ground water and the concentrations of branched and cyclopropyl fatty acids in the soil (Fig. 5). Organic contamination in the subsurface often results in the complete consumption of available oxygen by indigenous microorganisms and the development of anaerobic conditions (*30*). Groundwater geochemistry changes from aerobic to anaerobic conditions (Fig. 5) and changes from terminal electron acceptors of high energy potential (oxygen or nitrate) to low energy potential (sulfate or carbon dioxide) could have resulted in the shifts in microbial communities (*53*), as evidenced by the changes in the concentrations of branched and cyclopropyl fatty acids.

Bacterial cell numbers in the soil were calculated based on PLFA, assuming that the average bacterium, the size of *E. coli* contains 100 µmol PLFA/g dry weight and 1 g of bacteria is equivalent to 5.9×10^{12} cells (*42*). The total microbial biomass in aquifer soil showed varied profiles with depth indicating considerable microbial heterogeneity in the subsurface. The lowest bacterial populations were found at depth while the highest were

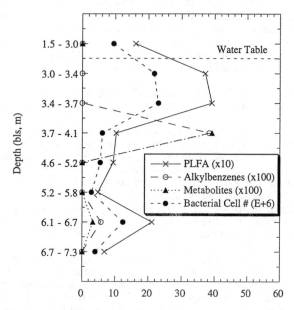

Figure 2   Vertical distribution of total alkylbenzenes (µg/kg), aromatic acid metabolites (µg/kg), and phospholipid ester-linked fatty acids (PLFA) (picomoles/g) in aquifer solids at ML-14, KC-135 Crash Site. Estimated bacterial cell numbers are in 10^6/g dry weight solids; bls=below land surface.

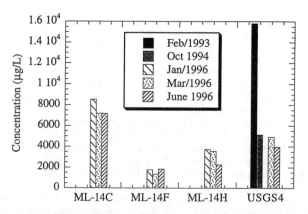

Figure 3   Changes in concentrations of total BTEX in ground water at wells USGS4 and ML-14, between Febuary 1993 to June 1996. The ML-14C, ML-14F, and ML-14H indicates the depths of the screened intervals at the MLS well (C=3.8 m, F=4.2 m, and H=3.2 m below land surface). Data prior to 1996 were analyzed by WW Engineering.

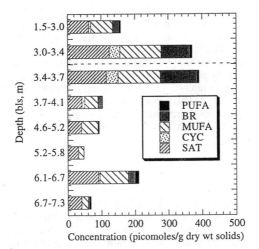

Fig. 4 Vertical profile of different classes of phospholipid ester-linked fatty acids (PLFA) in sand aquifer at ML-14 at the KC-135 Crash Site. The abbreviations represent fatty acids: SAT, saturated, MUFA, monounsaturated; PUFA, polyunsaturated; BR, branched; and CYC, cyclopropyl fatty acids. The horizontal dashed-line indicates water table.

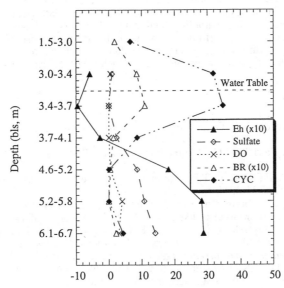

Fig. 5 Distributions of branched (BR) and Cyclopropyl (CYC) fatty acids (picomoles/g dry wt soil) in aquifer soil and redox potential (Eh, mv), sulfate (mg/L) and dissolved oxygen (DO) (mg/L) in ground water at ML-14 at the KC-135 Crash Site.

measured in the shallow contaminated zone (Fig. 3). Estimated bacterial cell numbers increased from 9.5×10^6/g dry wt soil at or near the surface to 2.3×10^7/g dry wt soil near the ground water table at the depth of 3.4-3.7 m and decreased to 4.0×10^6/g dry wt soil deeper in the subsurface (6.7-7.3 m). The increase of an order of magnitude in bacterial biomass in soil near the water table can be attributed to contamination-stimulated bacterial growth, a phenomenon observed previously in sediment from contaminated surface waterways (54) and in soils from creosote-contaminated (55-56) and chlorinated aliphatic compound-contaminated aquifers (57). The correlation of microbial populations with contaminant levels suggests the utilization of contaminants by microorganisms. This contention is further supported by the parallel distributions of microbial populations and aromatic acid metabolites in the aquifer solids.

Conclusions

Distributions of phospholipid ester-linked fatty acid (PLFA) biomarkers, aromatic hydrocarbons, and acid metabolites were determined during a geochemical investigation of a developed jet-fuel contamination plume. The aromatic fuel contaminants and major aromatic acid metabolites showed high concentrations in the predominantly anaerobic source zone in aquifer solids. Coincident with dilution of the contaminant plume, acid metabolite concentrations remained high in downgradient zones. PLFA distributions provided evidence for predominantly anaerobic bacterial degradation processes in the oxygen-depleted source zone. Elevated concentrations of terminal-branched and cyclopropyl fatty acids in the PLFA fractions were coincident with sulfate-reducing conditions.

Acknowledgments. The authors thank Mark Blankenship of USEPA-RSKERL for valuable assistance in the design and execution of the project, Mark Henry and Charles L. Major for assistance in field sampling, and Amy M. Peters and Michael Beebee for laboratory assistance. We gratefully acknowledge R. P. Eganhouse (U.S. Geological Survey) and two anonymous reviewers for their valuable comments and suggestions. This work was supported by cooperative agreement #CR824184 of the U.S. Environmental Protection Agency.

Literature Cited

(1) Arvin, E.; Jensen, B.J.; Gundersen, A.T. *Appl. Environ. Microbiol.* **1989**, *55*, pp. 3221-3225.

(2) Barker, J.F.; Patrick, G.C.; Major, P.D. *Ground Water Monitor. Rem.* **1987**, *7*, pp. 64-71.

(3) Eganhouse, R.P.; Dorsey, T.F.; Phinney, C.S.; Westcott, A.M. *Env. Sci. Tech.* **1996**, *30*, pp. 3304-3318.

(4) Gibson, D.T.; Subramanian, V., *Microbial Degradation of Organic Compounds*, Gibson, D.T., Ed., Marcel Dekker, New York, New York, **1984**, pp. 181-252.

(5) Major, D.W.; Mayfield, C.I.; Barker, J.F. *Ground Water*, **1988**, *26*, pp. 8-14.

(6) Jensen, B.K., Arvin, E., and Gundersen, T. *Organic Contaminants in Wastewater, Sludge and Sediment: Occurrence, Fate and Disposal.* Quagherbeur D.; Temmerman, I.; Angeletti, G., Eds., Elsevier Applied Science, New York, New York, **1990**.

(7) Hutchins, S.R.; Sewell, G.W.; Kovacs, D.A.; Smith, G.A. *Environ. Sci. Technol.* **1991**, *25*, pp. 68-76.

(8) Hutchins, S.R.; Moolenaar, S.W.; Rhodes, D.E. *J. Haz. Mat.* **1992**, *32*, pp. 195-214.

(9) Kuhn, E.P.; Colberg, P.J.; Schnoor, J.L.; Wanner, O.; Zehnder, A.J.B.; Schwarzenbach, R.P. *Environ. Sci. Technol.*, **1985**, *19*, pp. 961-968.

(10) Kuhn, E.P.; Zeyer, J., Eicher, P.; Schwarzenbach, R.P. *Appl. Environ. Microbiol.*, **1988**, *54*, pp. 490-496.
(11) Zeyer, J.; Eicher, P.; Dolfing, J.; Schwarzenbach, R.P. *Biotechnology and Biodegradation.*, Kamely, D.; Chakrabarty, A.; Omenn, G.S., Eds., Portfolio Publishing Company, The Woodlands, Texas. **1990**, pp. 33-40.
(12) Evans, P.J.; Mang, D.T.; Kim, K.S.; Young, L.Y. *Appl. Environ. Microbiol.* **1991**, *57*, pp. 1139-1145.
(13) Lovley, D.R.; Baedecker, M.J.; Lonergan, D.J.; Cozzarelli, I.M.; Philips, E.J.P.; Siegel, D.I. *Nature* (London) **1989**, *339*, pp. 297-300.
(14) Edwards, E.; Wills, L.; Grbiĉ-Galiĉ, D.; Reinhard, M. *In Situ Bioreclamation: Processes for Xenobiotic and Hhydrocarbon Treatment.*, Hinchee, R.E.; Olfenbuttel, R.F., Eds.; Buttworth-Heinemann, Stoneham, MS, **1991**; pp. 463-470.
(15) Lovley, D.R.; Coates, J.D.; Woodward, J.C.; Phillips, E.J. *Appl. Environ. Microbiol.* **1995**, *61*, pp. 953-958.
(16) Beller, H.R.; Grbiĉ-Galiĉ, D.; Reinhard, M. *Appl. Environ. Microbiol.* **1992**, *58*, pp. 786-793.
(17) Beller, H.R.; Spormann, A.M.; Sharma, P.K.; Cole, J.R.; Reinhard, M. *Appl. Environ. Microbiol.* **1996**, *62*, pp. 1188-1196.
(18) Vogel, T.M.; Grbiĉ-Galiĉ, D. *Appl. Environ. Microbiol.* **1986**, *52*, pp. 200-202.
(19) Wilson, B.H.; Smith, G.B.; Rees, J.F. *Environ. Sci. Technol.* **1986**, *20*, pp. 997-1002.
(20) Grbiĉ-Galiĉ, D.; Vogel, T.M. *Appl. Environ. Microbiol.* **1987**, *53*, 254-260.
(21) Edwards, E.A.; Grbiĉ-Galiĉ, D. *Appl. Environ. Microbiol.* **1994**, *60*, 313-322.
(22) Braddock, J.F.; McCarthy, K.A. Env. Sci. Tech. **1996**, *30*, 2626-2633.
(23) National Research Council. *In Situ Bioremediation: When Does It Work?* National Academic Press, Washington, D. C., **1993**.
(24) Borden, R.C.; Bedient, P.B. *Wat. Resources Res.* **1987**, *23*, pp. 629-636.
(25) Chiang, C.Y.; Salanitro, J.P.; Chai, E.Y.; Colthart, J.D.; Klein, C.L. *Ground Water*, **1989**, *27*, pp. 823-834.
(26) Kerfoot, H.B.; Mayer, C.L.; Durgin, P.B.; D'Lugosz, J.J. *Ground Water Monit. Rev.*, **1988**, pp. 67-81.
(27) Aggarwal, P.K.; Hinchee, R.E. *Environ. Sci. Technol.* **1991**, *25*, 1178-1180.
(28) Baedecker, M.J.; Cozzarelli, I.M.; Eganhouse, R.P. *Appl. Geochem.* **1993**, *8*, pp. 569-586.
(29) Landmeyer, J.E.; Vroblesky, D.A.; Chapelle, F.H. *Environ. Sci. Technol.* **1996**, *30*, pp. 1120-1128.
(30) Cozzarelli, I.M.; Eganhouse, R.P.; Baedecker, M.J. *Environ. Geol. Water Sci.*, **1990**, *16*, pp. 135-141.
(31) Cozzarelli, I.M.; Baedecker, M.J.; Eganhouse, R.P.; Goerlitz, D.F. *Geochim. Cosmochim. Acta* **1994**, *58*, pp. 863-877.
(32) Barcelona, M.J.; Lu, J.; Tomczak, D.M. *Ground Water Monitor. Rem.* Spring **1995**, *15*, pp. 114-124.
(33) Eganhouse, R.P.; Dorsey, T.F.; Phinney, C.S. *US Geological Survey Program on Toxic Waste-Ground-Water Contamination -- Proceedings of theTechnical Meeting*, Pensaçola, Florida, March 23-27, **1987** Frank B.J., Ed. USGS Open File Report 87-109, pp. C29-C30.
(34) Eganhouse, R.P.; Baedecker M.J.; Cozzarelli, I.M.; Aiken, G.R.; Thorn T.F. *Appl. Geochem.* **1993**, *8*, 551-567.
(35) Badecker, M.J.; Cozzarelli, I.M. *US Geological SurveyToxic Substances Program -- Proceedings of theTechnical Meeting, Pensacola, Monterey, California, March 11-15*, **1991**, U.S. Geological Survey Water-Resources Investigations Report 91-4034, pp. 627-632.
(36) Essaid, H.I.; Bekins, B.A.; Godsy, E.M.;Warren, E., Baedecker, M.J.; Cozzarelli, I.M. Water Resources Res., **1995**, *31*, pp. 3309-3327.

(37) Barcelona, M.J.; Morrison, R. *Methods for Ground Water Quality Studies*, Proceedings of a National Workshop USDA, CRS in cooperation with USGS and USEPA, Nov. 1-3, **1988**, Arlington, VA. USDA-ARS Univ. of Nebraska, Lincoln, NE. pp. 49-62.

(38) Zapico, M.M.; Vales, S.; Cherry, J. *Ground Water Monit. Rem.* Summer **1987**, pp. 74-82.

(39) Barcelona, M.J.; Wehrmann, H.A.; Varljen, M. *Ground Water.* **1994**, *32*, pp. 12-22.

(40) Eganhouse, R.P.; Dorsey, T.F.; Phinney, C.S.; Westcott, A.M. *J. Chromatogr.* **1993**, *628*, pp. 81-92.

(41) Chapelle, F.H.; McMahon, P.B. *J. Hydrol.* **1991**, *127*, 85-108.

(42) White, D.C.; Bobbie, R.J.; Herron, J.S.; King, J.D.; Morrison, S.J. *Native Aquatic Bacteria: Enumeration, Activity and Ecology*, Costeron, J.W.; Colwell, R.R., Eds., ASTM STP 695. Am. Soc. for Testing and Materials, Philadelphia, Pennsylvania, **1979**, pp. 69-81.

(43) Findlay, R.H.; Dobbs, F.C., *Aquatic Microbial Ecology*; Kemp, P.F.; Sherr, B.F.; Sherr, E.B.; Cole, J.J., Eds., Lewis Publishers, Boca Raton, Florida. **1993**, pp. 271-284.

(44) Dunkelblum, E.; Tan, S.H.; Silk, P.J. *J. Chem. Ecol.* **1985**, *11*, pp. 265-277.

(45) Evans, P.J.; Ling, W.; Goldschmidt, B.; Ritter, E.R.; Young, L.Y. *Appl. Environ. Microbiol.*, **1992**, *58*, pp. 496-501.

(46) Lechevalier, M.P. *Crit. Rev. Microbiol.* **1977**, *5*, pp. 109-210.

(47) Parkes, R.J.; Taylor, J. *Estuarine Coastal Shelf Sci.* **1983**, *16*, pp.173-198.

(48) Guckert, J. B.; Antworth, C.P.; Nichols, P.D.; White, D.C. *FEMS Microbiol. Ecol.* **1985**, *31*, pp. 147-158.

(49) Edlund, A.; Nichols, P.D.; Roffey, R.; White, D.C. *J. Lipid Res.* **1985**, *26*, 982-988.

(50) Dowling, N.J.E.; Widdel, F.; White, D.C. *J. Gen. Microbiol.* **1986**, *132*, pp. 1815-1825.

(51) Goldfine, H. *Advan. Microbiol. Physiol.* **1972**, *8*, 1-58.

(52) Findlay, R.H.; Trexler, M.B.; Guckert, J.B.; and White, D.C. *Mar. Ecol. Prog. Ser* **1990**, *62*, pp. 121-133.

(53) White, D. C. *Arch. Hydrobiol. Beith.* **1988**, *31*, 1-18.

(54) Fang, J.; Findlay, R.H., *J. Microbiol. Meth.* **1996**, *27*, pp. 63-71.

(55) White, D.C., Smith, G.A.; Gehron, M.J.; Parker, J.H.; Findlay, R.H., Martz, R.F.; Fredrickson, H.L. *Dev. Industr. Microbiol.* **1983**, *24*, 201-211.

(56) Smith, G. A.; Nichols, J.S.; Kerger, B.D.; Davis, J.D.; Collins, S.P.; Wilson, J.T.; McNabb, J.F.; White, D.C. *Can. J. Microbiol.* **1986**, *32*, 104-111.

(57) McCarthy, C. M.; Murray, L. *Microb. Ecol.* **1996**, *32*, 305-321.

Chapter 6

Application of Chlorophyll and Carotenoid Pigments for the Chemotaxonomic Assessment of Seston, Periphyton, and Cyanobacterial Mats of Lake Okeechobee, Florida

Nancy M. Winfree[1], J. William Louda[1], Earl W. Baker[1], Alan D. Steinman[2], and Karl E. Havens[2]

[1]Organic Geochemistry Group, Department of Chemistry and Biochemistry, Florida Atlantic University, 777 Glades Road, Boca Raton, FL 33431–0991
[2]Ecosystems Restoration Division, South Florida Water Management District, 3301 Gun Club Road, West Palm Beach, FL 33406

The presence of chlorophyll-$\underline{a}$ and its derivatives in an environmental sample indicates the presence of oxygenic photoautotrophs. The accessory (-$\underline{b}$, -$\underline{c}$) and alternate (bacteriochlorophylls-$\underline{a}$, -$\underline{b}$, -$\underline{c}$, -$\underline{d}$, -$\underline{e}$) chlorophylls and the carotenoids provide additional biomarker specificity. That is, the presence of these pigments gives information about taxonomic structure. In the present study we utilized high performance liquid chromatography (HPLC) / photodiode array (PDA) methodology to investigate the photoautotrophic communities in Lake Okeechobee, Florida. Pigment distributions were measured in unispecific cultures and compared to literature reports on natural populations in order to generate equations for the estimation of the relative abundances of photoautotrophic taxa in seston, epiphytes, and benthic samples from 14 locations in the lake. The taxa and their specific biomarkers considered herein include: anoxygenic Eubacteria (purple and green sulfur bacteria: bacteriochlorophylls-$\underline{a}$ / -$\underline{c}$), Cyanobacteria (myxoxanthophyll), Chlorophyta (chlorophyll-$\underline{b}$, lutein), Chrysophyta (fucoxanthin), and Pyrrhophyta (peridinin). The estimate of each taxon was based upon the relative abundance of the specific biomarker pigment for that group and its quantitative relationship to chlorophyll-$\underline{a}$. As examples of these methods, the technique of pigment-based chemotaxonomy is applied spatially and temporally to the photoautotrophic communities in Lake Okeechobee.

The estimation of phytoplankton biomass using pigment data, with certain inference as to types (taxa), was first performed by Kreps and Verbinskaya during a 1930 study of the seasonal changes in diatom populations in the Barents sea (1). The utilization of pigment 'colorimetry' for phytoplankton studies was popularized in 1934 by Harvey and included the first report of 'patchiness' in natural aquatic ecosystems (2).

Spectrophotometric analyses of photopigment quantity and quality gained increasing utility with the advent of sets of equations with which to estimate the chlorophylls (-a, -b, -c) and total carotenoids in lipid extracts (3-6). Errors in these spectrophotometric methods have been reported as being due to the overlapping spectral bands of coincident pigments (7-11). The sensitivity and, to a certain degree, the selectivity of spectral estimates were enhanced by the utilization of fluorometric determinations (6,12-15). Reports of relatively large errors in this technique, due to overlapping fluorescence bands, have also appeared (16).

The potential of reverse phase (RP)-HPLC was shown in 1979 by Brauman and Grimme with the separation of 15 pigments, including the epimeric forms of chlorophyll-a (17). The application of RP-HPLC to the analysis of phytoplankton communities, stressing the use of alloxanthin as a chemotaxonomic indicator for the cryptophytes, was reported by Gieskes and Kraay in 1983 (18). In that same year, Mantoura and Llewellyn reported on the inclusion of ion-pairing reagents (e.g. tetrabutylammonium acetate, ammonium acetate) for the enhanced separation of highly polar pigments (8). SCOR-UNESCO established a study ("Working Group '78") to compare methods and develop a standard method for the analysis of marine phytoplankton (19).

Applications of HPLC-PDA analyses of pigment distributions for the study of aquatic systems are becoming more common. We adopted these methods to the study of the algal pigments in Lake Okeechobee, the largest lake in the S.E. United States. Algal blooms in Lake Okeechobee have been defined as occurring when chlorophyll-a concentrations meet or exceed 40 µg/L (20-21). These events have been dominated by cyanobacteria, and the largest bloom was recorded in 1987, covering about 700 Km^2 ($\approx$ 42%) of the lake's pelagic surface (22). Generally, since 1992 chlorophyll-a concentrations in the lake have averaged below 25 µg/L (23), and bloom frequencies are declining (20-21). The existence of algal (viz. cyanobacterial) blooms in this, or any other, lake could result in undesirable conditions for both man and the ecosystem.

The present study was undertaken in order to assess the feasibility of applying HPLC/PDA based pigment analyses to the chemotaxonomic survey of the various photoautotrophic communities within Lake Okeechobee. In order to arrive at taxonomic conclusions from pigment data, we first had to address taxon-specific pigment ratios in (a) unispecific algal cultures, and (b) the literature. Using such data, equations could be derived which would allow taxon-specific chlorophyll-a contributions to be estimated. These equations would then provide an estimation of the relative importance of each taxon in any given sample.

The goal of these studies is to provide a rapid, timely and reasonably accurate HPLC-PDA methodology with which to monitor short-term spatial and temporal changes in the photoautotrophic communities of Lake Okeechobee, Florida, with subsequent application to other environments.

Experimental

Site. Lake Okeechobee is a large (1,730 km^2) shallow ($z_{mean} \approx 2.7$m, $z_{range} \approx 0$ - 4.5m) sub-tropical lake located at about 27°00'N x 80°00'W in the southern Florida peninsula. Wind fetch is in excess of 50 km and, coupled with its shallow depth, leads to resuspension of solids (*20-21, 23-25*). The lake has been described as consisting of five distinct ecological zones: (1) northern, (2) central, (3) southern ("edge"), (4) western-transitional, and (5) littoral. Sediments in these areas are predominately mud in the north and central regions, sand in the west/littoral and a mosaic of rock, marl and organic peat in the south (*23-26*). The littoral zone(s), mainly in the west, southwest and northwest fringes, are about 1-2 m in depth and support a wide variety of emergent macrophytes plus epiphyton. The littoral regions of the Lake Okeechobee amount to about 22% of its areal extent(*20*).

Samples. Samples were collected on a quarterly basis from the 14 sites designated in Figure 1. Samples included seston, epiphyton and benthic phototroph mats.

Benthic samples were collected using a coring device constructed of PVC pipe (7.34 cm^2 cross section) and a one-way valve (*27*). These samples were extruded into polyethylene bags immediately upon retrieval and placed on ice until frozen about 2-6 hours after collection. Prior to analysis, benthic mat samples were thawed to ice-bath temperatures (0-2°C) and collected onto GF/F filters in order to remove excess water.

Epiphyte-containing macrophytes were removed from the lake and placed on ice. Epiphytes were removed from the host plant (*Scirpus* sp., *Eleocharis* sp.), collected onto GF/F filters, placed into polyethylene bags, and frozen. The surface area and weight of the host macrophyte were recorded at this time.

Water samples were collected using a depth integrating technique. That is, a small pump, with tubing, attached to a rod was alternately raised and lowered through the water column to within 0.5 m of the surface and the bottom. Water was pumped into dark brown carboys and transported to shore, where the seston was collected by filtration onto GF/F filters and frozen.

Solvents. All solvents were of Fisher Scientific "Optima" grade, or better. Water used was deionized at a resistivity of $\leq$10-12 megohms.

HPLC-PDA. The system utilized in these studies consisted of the following components: a ThermoSeparations Model 4100 quaternary solvent delivery pump; a Rheodyne Model 7120 injector fitted with a 100μL loop; a 3.9 x 150 mm Waters Nova-Pak C18 reverse-phase (RP) column (4μm packing, 7% carbon load, endcapped, 60 A pore size, 120 m^2/m surface area); and a Waters Model 990 dual photodiode array detector.

Spectrophometry. Electronic absorption spectroscopy (*aka* 'UV/Vis') was performed with a Perkin-Elmer Model Lambda-2 spectrophotometer which was monitored daily

Figure 1. Map of Lake Okeechobee, Florida. Sampling stations: (1) South Bay, (2) 3-Pole Bay, (3) Cochran's Pass, (4) Moore Haven, (5) LZ40, (6) LOO5, (7) Fisheating Bay (F.E.B.), (8) Stake, (9) Poles, (10) LOO1, (11) LOO7, (12) LOO8, (13) Coot Bay, (14) Nubbin's Slough. Shaded area denotes littoral zone.

for both wavelength accuracy (holmium oxide, Ho_2O_3) and photometric stability (*vs.* 40 mg/L K_2CrO_4 in 0.05N $KOH_{(aq.)}$:see *28*).

Internal Standard. Copper meso-porphyrin-IX dimethyl ester (Cu Meso-IX-DME: λ_{max}= 394nm) was chosen as the internal standard (IS). Cu Meso-IX-DME did partially coelute with canthaxanthin (λ_{max} = 472 nm). However, the spectral differences between these two pigments allowed for individual quantitation using integration at 394 and 440 nm, respectively.

Analytical standards. Standards were utilized either as pure ('neat') compounds or as part of well known mixtures in unspecific algal and bacterial cultures. Structures given in text can be found in the literature (*8, 29-30*). In all, 53 pigments were utilized as standards. Elution order followed that given by others (*8, 19, 31-33*).

All sample manipulation was in the dark or dim yellow light. Samples were kept cold, dark, and with the exclusion of oxygen whenever feasible.

Extraction. Extraction of lipophilic pigments was performed with 2-5 mL 90% aqueous acetone and grinding for 2-3 minutes, while keeping the tissue grinder/sample at ice-bath temperature. Following extraction, the extract was centrifuged for 3-5 minutes at about 1400g. The supernatant was then filtered through a 0.45μm PFTE filter and the UV/Vis spectrum of the 'crude extract' was recorded.

Chromatography. The RP-HPLC solvent profile was modified from that given by Kraay and others (*33*). The modifications included lengthening the gradient time between the ion pairing section and the elution of chlorophyll-b, plus the addition of a 1 minute 100% ethyl acetate 'flush' at the end of the run. These modifications enhanced the separation of lutein from zeaxanthin and ensured that all lipophilic compounds were removed from the column, respectively.

Samples were prepared for injection according to Mantoura and Llewellyn (*8*). The internal standard solution (IS) was made such that 25-50 ng Cu-Meso-IX-DME was present with the sample on the column. Recoveries of the internal standard (IS_{eluate} / $IS_{injected}$) provided run-to-run correction factors, which averaged near unity (1.11x) and had a standard deviation (s) of 0.07. All pigments during each specific run were adjusted by their individual recovery efficiencies, or by that of the IS.

Lower limits of detection (LLD) were determined to be 2.10 and 0.95 ng for the tetrapyrrole and tetraterpenoid pigments, respectively, using chlorophyll-a and lutein as models. Individual peak areas containing amounts between the LLD and 100+ ng were found to be within the linear response of the system. All analyses were adjusted to yield a maximum of about 100 ng in any single peak.

Quantitation of pigments was by direct application of the Beer-Lambert relationship to integrated (AU*min with 1mL/min equals AU*vol.) peak areas and using literature values (*8, 34*) for extinction coefficients. Integration was performed at 440, 410, and 394 nm for the determination of (1) chlorophylls (-a, -b, -c), bacteriochlorophylls-c plus carotenoids, (2) pheophytins / pheophorbides-a plus bacteriopheophytins-c, and (3) the internal standard plus bacteriochlorophyll / bacteriopheophytin-a, respectively. Extinction coefficients were adjusted for the analytical wavelength relative to maximal absorption, if not originally reported as such (*cf.* " E_{440}" in *ref. 8*).

Sampling variability was examined by analyzing seven replicate water (seston) samples from a single site in Lake Okeechobee once each for chlorophyll-a and fucoxanthin. This produced relative standard deviations (RSD) of 12.8 and 19.1%, respectively.

Instrumental variability was examined by analyzing a single seston sample seven times. Again, choosing chlorophyll-a and fucoxanthin as typical pigments, this gave RSD values of 5.4 and 5.8%, respectively.

Results and Discussion.

Known Pigment Relationships and Formulation of an Equation for Chemotaxonomic Application to Unknown Populations. In order to obtain taxon-specific internal pigment ratios, namely chlorophyll-a to accessory pigments, we analyzed a wide variety of fresh unisspecific algal cultures. In addition, we scanned the literature for published values on the same. Table I is a collection of these values for five major divisions of oxygenic photoautotrophs.

Table I. Ratios of Chlorophyll-a to Accessory Pigments as Obtained through
Analyses of Unialgal Cultures and from the Literature

DIVISION	RATIO OBSERVED				
PIGMENTS	This Study	Ref.# 37	Ref.#32	Ref.#41	Ref.# 40
CYANOPHYTA					
Chl-a/Myxo	12.3	2.73	--	5.0	--
Chl-a/Zea	*ca.*45	--	--	--	8.6
Chl-a/Echin	16.0	--	--	--	--
CHLOROPHYTA					
Chl-a/Chl-b	3.6	--	1.13	--	--
Chl-a/Lut	6.4	--	--	--	6.7
CRYPTOPHYTA					
Chl-a/Allo	--	--	--	2.5	2.5
CHRYSOPHYTA					
Chl-a/Fuco	1.2	--	1.244	--	1.2
Chl-a/Diadin	2.7	--	--	--	--
PYRRHOPHYTA					
Chl-a/Peri	3.8	--	2.02	--	1.1
Chl-a/Diadin	6.5	--	--	--	--

Abbreviations: Chl-a = total chlorophyll-a, Myxo = myxoxanthophyll, Zea = zeaxanthin, Echin = echinenone, Chl-b = chlorophyll-b Lut = lutein, Allo = alloxanthin, Fuco = fucoxanthin, Diadin = diadinoxanthin, Peri = peridinin.

Quantitative pigment-based chemotaxonomy is still being refined (*cf. 42*), and here we are only attempting to gain reasonable estimates on crude taxonomic groupings. Invariant ratios of chlorophyll-a to any of the accessory pigments may in fact not exist, as exemplified by the tremendous range in relative and absolute concentrations of photosynthetic primary and secondary pigments due to physiological state, cell size, photobleaching (*35*) or adaptive bulk synthesis, as is the case for zeaxanthin (see *36-39*) . Previous studies have generated equations for the accessory pigment based estimation of chlorophyll-a contributions by various taxa (*cf. 31-32, 40-42*). In the present study, using averaged values from Table I, we arrived at the following equation:

$$[Chl\text{-}a] = (7.1[Myxo])+(3.6[Chl\text{-}b])+(1.2[Fuco])+(2.3[Peri]) \qquad (1)$$

This equation was utilized to estimate the chlorophyll-a derived from the Cyanobacteria ('Myxo'), Chlorophyta ('Chl-b'), Chrysophyta (esp. Bacillariophyceae: 'Fuco'), and Pyrrophyta (Dinophyceae: 'Peri'). An example of these calculations is given later. In samples, such as sediments, in which chlorophyll-b has undergone considerable degradation, the use of lutein ('Lut') in place of chlorophyll-b is recommended. Here (Table I) we suggest the relationship (6.5[Lut]) for the Chlorophyta estimation in such cases (*cf. 40*).

Figure 2 contains chromatograms generated at 440 nm from the RP-HPLC/PDA analyses of representative species of the Cyanophyta, Chlorophyta, Chrysophyta (Bacillariophyceae), and Pyrrophyta (Dinophyceae). In addition to the prime biomarker pigments given above, certain other accessory pigments confirm the presence of these taxa. Examples for the Cyanobacteria (zeaxanthin, echinenone), Chlorophyta (antheraxanthin, neoxanthin, violaxanthin), Chrysophyta (diadino-xanthin, diatoxanthin: alternate fucoxanthins [19'-hexanoyloxyfucoxanthin *cf. 33*]), and Pyrrophyta (diadinoxanthin) are given in parentheses (*cf. 19, 32-33, 40-44*).

Chemotaxonomic Assessment of Photoautotrophic Populations in Pelagic, Periphyton, and Benthic Samples from Lake Okeechobee, Florida. A selection of representative chromatograms reveals the pigment distributions of the photoautotrophic communities of Lake Okeechobee. These distributions can then be utilized to provide chemotaxonomic estimations of community structure.

Table II is a listing of the divisional estimates made for 27 sampling events of water (pelagic) and epiphytes at 8 of the 14 Lake Okeechobee sites studied.

One of the simplest cases encountered was the late Spring 1995 pelagic sample at LZ40 (Station 5, Figure 1), the open water site in the middle of the lake. Figure 3a reveals that the pigment mixture we found was dominated by contributions from the Chrysophyta. These pigments include chlorophyll-a, chlorophylls-c_1 /-c_2, fucoxanthin, and diadinoxanthin, identical to the diatom example given earlier (Figure 2). Thus, the photoautotrophic community sampled was concluded as being almost 100% Chrysophyte, with a very minor ($\leq$ 1%) contribution from cyanobacteria. The latter was surmised from the presence of zeaxanthin and an unidentified carotenol believed to be aphanizophyll (4-hydroxy-myxoxanthophyll), a potential marker for filamentous cyanobacteria (*45*).

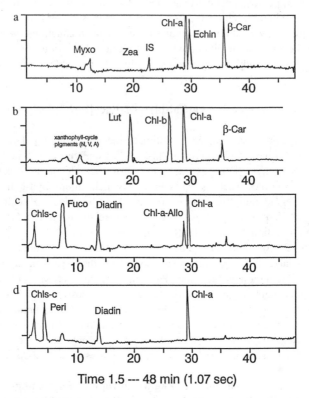

Time 1.5 --- 48 min (1.07 sec)

Figure 2. HPLC chromatograms recorded at 440 nm for four oxygenic photoautotrophic divisions. (a) Cyanophyta, *Microcystis aeruginosa* , (b) Chlorophyta, *Closterium* sp., (c) Chrysophyta, *Synedra* sp., (d) Pyrrhophyta, *Amphidinium carteri*. Pigment code: N,V,A = neoxanthin, violaxanthin, antheraxanthin; Myxo = myxoxanthophyll; Zea = zeaxanthin(position indicated); Chl-$\underline{a}$ = chlorophyll-$\underline{a}$; β-Car = β-Carotene; Lut = lutein; Chl-$\underline{b}$ = chlorophyll-$\underline{b}$; Chls-$\underline{c}$ = chlorophylls-$\underline{c}_1$/-$\underline{c}_2$; Fuco = fucoxanthin; Diadin = diadinoxanthin; Peri = peridinin; IS = internal standard (Cu Meso-IX-DME); Echin = echinenone.

Table II: Representative Taxonomic Structures Calculated for Lake Okeechobee Samples Collected Between December 1994 and September 1995

SITE[a]	SMPL[b]	DATE	PERCENT COMPOSITION[c]				%CHL-a_{calc}[d]
			Cyano	Chlor	Chrys	Pyrr	
(1) S.Bay	S	Dec '94	74	15	11	0	82
	S	Sep '95	82	12	6	0	91
	B	Dec '94	55	25	20	0	115
	B	Mar '95	36	40	24	0	132
	B	Jun '95	55	26	19	0	803
	B	Sep '95	65	33	2	0	393
(4) M.Haven	E	Dec '94	95	0	5	0	394
	E	Mar '95	25	9	66	0	124
	E	Jun '95	30	9	61	0	132
	E	Sep '95	27	6	67	0	195
(6) LOO5	S	Dec '94	81	13	6	0	58
	S	Mar '95	0	0	100	0	28
	S	Jun '95	34	0	56	0	142
	S	Sep '95	86	10	4	0	115
(7) F.E.B.	E	Dec '94	22	23	55	0	74
	E	Mar '95	0	88	12	0	87
	E	Sep '95	0	86	14	0	26
(8) Stake	E	Dec '94	31	39	30	0	84
	E	Mar '95	0	61	39	0	121
	E	Jun '95	0	27	73	0	67
	E	Sep '95	0	43	57	0	72
(9) Poles	E	Dec '94	43	33	24	0	93
	E	Mar '95	0	49	51	0	86
	E	Jun '95	0	30	70	0	76
	E	Sep '95	14	24	62	0	58
(10) LOO1	S	Sep '95	21	57	22	0	53
(11) LOO7	B	Jun '95	0	10	84	6	5,273

[a] Sites are given in Figure 1. [b] Sample code; S = seston, E = epiphyte, B = benthic.
[c] Percent composition of the taxonomic structure calculated according to equation 1. Cyano = Cyanophyta, Chlor = Chlkorophyta, Chrys = Chrysophyta, Pyrr = Pyrrhophyta. [d] "%CHL-a_{calc} " refers to the amount of chlorophyll-a determined from accessory pigment relationships (equation 1) as compared to the chlorophyll-a directly determined by HPLC-PDA ([Chl-a_{calc} / Chl-$a_{"actual"}$] 100).

Additional complexity is added to natural pigment arrays in ecosystems which contain a diverse flora. Figure 3b is the chromatogram for an epiphyte sample collected from bulrush (*Scirpus* sp.) at the "Poles" site (Station 9, Figure 1) during the end of Summer 1995. Here, we found markers for Cyanobacteria (myxoxanthophyll, zeaxanthin, echinenone), Chlorophyta (chlorophyll-b, lutein), and Chrysophyta (chlorophylls-c_1/-c_2, fucoxanthin, diadinoxanthin). Using the equation given above, we calculated the relative percent abundances of these three Divisions to be 14 / 24 / 62, respectively.

The above calculation was made by inserting the concentrations (mg/m^3, ng/g, *etc.*: not shown) determined for each of the chemotaxonomic marker pigments (*viz.* 'Myxo', 'Chl-$\underline{b}$', 'Fuco', 'Peri') into equation #1. In the present case (Poles epiphyte, station 9): Total sample chlorophyll-$\underline{a}$ = (7.1 [12.63 ng/g]) + (3.6 [41.30 ng/g]) + (1.2 [330.66 ng/g]) + (2.3 [0 ng/g]) = (89.67 ng/g) + (148.68 ng/g) + (396.79 ng/g) + (0 ng/g) = 635.14 ng/g. This value should correspond to the amount of chlorophyll-$\underline{a}$ present in the overall community. In order to calculate the taxonomic composition of this community, from individual chlorophyll-$\underline{a}$ contributions, one divides the individual contribution by the total and converts to percentage. The individual calculations are shown in parentheses for the Cyanobacteria (89.67 / 635.14 = 0.14 [100] = 14%), Chlorophyta (148.68 / 635.14 = 0.23 [100] = 23%), and Chrysophytes (396.79 / 635.14 = 0.62 [100] = 62%).

Comparing the amount of chlorophyll-$\underline{a}$ (a) 'calculated' on the basis of observed ratios to accessory pigments (Table I) with (b) that determined *via* HPLC-PDA analyses, a large range of values is found. That is, for the set of data given in Table II, the amount of chlorophyll-$\underline{a}$ calculated to be present on the basis of accessory pigments (equation 1) is often far from the chlorophyll-$\underline{a}$ found in the same HPLC analysis. Considering the benthic samples, these large disparities probably derive from the fact that the chlorophylls (-$\underline{a}$, -$\underline{b}$) *per se* initially degrade faster than the carotenoids (see *46-47*). The percent chlorophyll-$\underline{a}$ values calculated, relative to that present, for the seston and epiphyte samples cover a range of about 25 to 400%, with many samples giving values between 80 - 120%. In theory, the chlorophyll-$\underline{a}$ calculated from the accessory pigment concentrations should equal that found *via* HPLC-PDA analyses (*i.e.* 100%). As discussed earlier, variations in the quantitative relationships of the accessory pigments to chlorophyll-$\underline{a}$ (*cf.31-42*) appear to offer the most plausible reason for these differences within environmental samples.

Figure 3c is the chromatogram of the lipophilic pigments in the benthic sample from the September 1995 South Bay site (Station 1, Figure 1) sampling. The taxonomic structure was calculated to be Cyanobacteria / Chlorophyta / Chrysophyta = 65 / 33 / 2. A small amount of bacteriopheophytin-$\underline{a}$ was found to be present, and the past presence of the purple sulfur bacteria was therefore indicated.

Figure 4a is a histogram of the percent composition of epiphytes growing on spike-rush (_ *Eleocharis* sp.) in the western littoral zone at the Moore Haven site (Station 4, Figure 1). The extremely consistent taxonomic structure of this community from Winter through Summer 1995 is obvious. That is, the ratio of diatoms to cyanobacteria to green algae is relatively constant at about 65 / 27 / 8. At the end of Fall 1994, cyanobacteria were the dominant phototrophs present. The absolute amounts of chlorophyll-a per gram of host macrophyte for the 4-season cycle (12/94-09/95) were 37, 40, 26 and 16 µg/g-dry weight host.

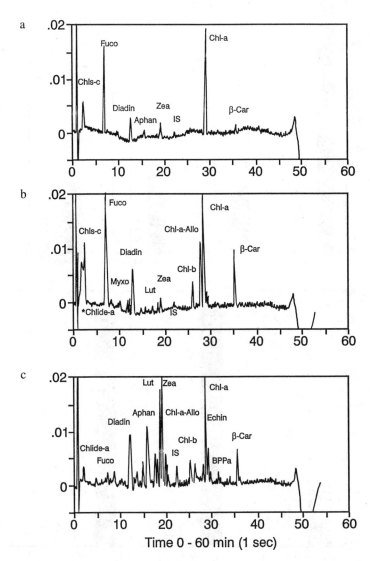

Figure 3. Chromatograms generated at 440 nm for selected samples. (a) Seston from site LZ40 (Station 5) during June 1995 (Aphan = aphanizophyll[tent.]), (b) Epiphyte from "Poles" (Station 9) in September 1995 (asterisk indicates chlorophyllide-a), (c) Benthic sample from South Bay (Station 1) in September 1995 (BPP-a = bacterio-pheophytin-a). Site locations given in Figure 1.

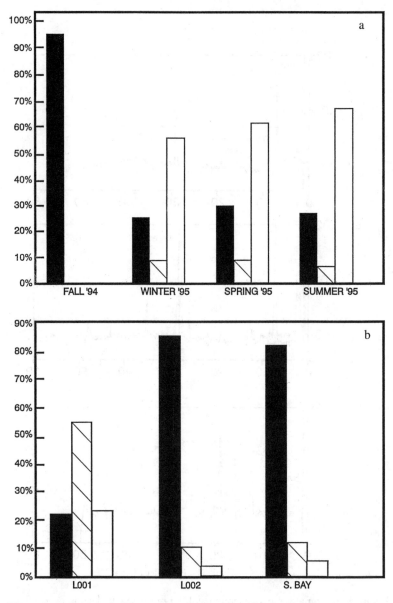

Figure 4. (a) Seasonal pigment-based estimations for the taxonomic structure of epiphytes growing on spike-rush (Eleocharis sp.) at the Moore Haven site, station 4 (see Table II). (b) Pigment-based estimations for the taxonomic structures of the pelagic photoautotrophic communities (seston) collected in the northern (L001, station 10) , western (L005, station 6), and southern (South Bay, station 1) regions of Lake Okeechobee during September 1995. Site locations given in Figure 1. Division codes: Solid = Cyanobacteria, Cross-hatch = Chlorophyta, Open = Chrysophyta.

Figure 4b represents the taxonomic structure calculated for three widely separated pelagic (seston) sites during the end of Summer (September) 1995 sampling event. These sites are L001, L005, and South Bay and are located (Stations 10, 6, and 1, respectively; Figure 1) in the northern, western, and southern regions of the lake, respectively. Here, the southern and western sites were found to be compositionally alike and very different from the northern site. Cyanobacteria were dominant at the southern and western sites while green algae dominated at the northern site.

Conclusions

In this study, we used myxoxanthophyll, chlorophyll-b (lutein), fucoxanthin, and peridinin for the estimation of the amounts of chlorophyll-a contributed to mixed communities by members of the Cyanobacteria, Chlorophyta, Chrysophyta, and Pyrrhophyta, respectively. An equation for these estimations was derived from the study of pure cultures and by comparison to literature values. The basic assumption that the ratios we find in lab-grown or in natural populations (*e.g.* blooms) actually mimic those found in study samples must be made, and, therefore, a good deal of uncertainty is inherent to these efforts. This was shown by comparing the amount of chlorophyll-a calculated from biomarker data (*viz.* accessory pigments) with the amount of chlorophyll-a determined directly by HPLC-PDA analyses. A rather wide range, above and below the theoretical 1:1 relationship, was found. The main problems enter as one considers changes in internal pigment ratios due to the physiological state of individual species and/or entire communities, photooxidative alterations, nutrient limitation, self-shading, and the presence of chlorophyll-a containing species which lack the biomarker accessory pigments being considered.

The use of HPLC-PDA analyses for the generation of pigment-based chemotaxonomic data allows for rapid estimation of photoautotrophic community structure. Currently, this extends only to the Division level, though more specific identifications are possible. Further, this methodology may allow for the more economical and timely processing of greater numbers of samples, relative to visual (microscopic) identification-counting techniques. The technique of pigment-based chemotaxonomy for the investigation of photoautotrophic community structure offers tremendous promise for allowing high resolution spatial and temporal trend studies to be performed. Applications might include: monitoring the onset, spread, and demise of blooms; following algal dynamics in relation to nutrient plumes; assessing the impact of thermal pollution on natural populations; and general photoautotroph surveys.

Acknowledgments

This study was funded under a contract from the South Florida Water Management District, West Palm Beach, Florida. The authors wish to thank the following SFWMD staff for assistance in the field and/or laboratory; Pat Davis, Therese East, Andy Rodusky, Richard Meeker, and Bruce Sharfstein. Mrs. Jackie Morris of Carolina Biological Supply, Inc. (N. Carolina, U.S.A.) is thanked for her assistance in obtaining bulk unispecific cultures of algae and bacteria. The firm of F. Hoffman-LaRoche (Basel, Switzerland) is thanked for generous gifts of authentic

carotenoids. Professor A.S. Holt is thanked for his gift of bacteriochlorophylls-c. FAU
students Mrs. Wenya Fan and Mr. Jie Li are thanked for their chromatographic
purification and/or partial syntheses of known pigments . The diligent efforts of two
anonymous reviewers and the editor are gratefully acknowledged. Major portions of
this study represent the Master's thesis (FAU, April 1996) research of the senior
author (NMW).

Literature Cited

1.   Kreps, E.; Verbinskaya, N. *J. Conseil.* **1930**, *5*, 327-345.
2.   Harvey, H.W. *J. Mar. Biol. Assoc. U.K.* **1934**, *19*, 761-773.
3.   Aronoff, S. *Adv. Food Res.* **1953**, *4*, 133-184.
4.   Richards , F.A. *J. Mar. Res.* **1952**, *11*, 147-155.
5.   Richards, F.A.; Thompson, T.G. *J. Mar. Res.* **1952**, *11*, 156-172.
6.   Strickland, J.D.H.; Parsons, T.R. *Fish. Res. Brd. Can. Bull.* **1968**, *167*, 185-206.
7.   Hall, C.A.S.; Moll, R. In *Primary Productivity of the Biosphere;* Leith, H.; Whittaker, R.H. Eds.;Springer-Verlag; New York, 1975; p 19-53.
8.   Mantoura, R.F.C.; Llewellyn, C.A. *Analyt. Chem. Acta* **1983**, *151*, 297-314.
9.   Parsons, T.R. *J. Mar. Res.* **1963**, *21*, 164-171.
10.   Rai, H. *Veh. Internat. Verein. Limnol.* **1973**, *18*, 1864-1875.
11.   Rieman, B. *Limnol. Oceanogr.* **1978**, *23*, 1059-1066.
12.   Lorenzen, C.J. *Limnol. Oceanogr.* **1967**, *12*, 343-346.
13.   Yentsch, C.S.; Menzel, D.W. *Deep-Sea Res.* **1963**, *10*, 221-231.
14.   Yentsch, C.S.; Yentsch, C.M. *J. Mar. Res.* **1979**, *37*, 471-483.
15.   Yentsch, C.S. *Deep-Sea Res.* **1965**, *12*, 653-666.
16.   Trees, C.C.;Kennicutt-II, M.C.;Brooks, J.M. *Mar. Chem.* **1985**, *17*, 1-12.
17.   Brauman,T.;Grimme, L.H. *J. Chromatogr.* **1979**, *170*, 264-268.
18.   Poole, C.F.;Poole, S.K. *Chromatography Today;* Elsevier, Amsterdam, 1991;1026 pp.
19.   Wright, S.W.;Jeffrey, S.W.;Mantoura, R.F.C.;Llewellyn, C.A.;Bjornland, T.; Repeta, D.;Welschmeyer, N. *Mar. Ecol. Prog. Ser.* **1991**, *77*, 183-196.
20.   Havens, K.E.;Hanlon, C.;James, R.T. *Arch. Hydrobiol./Suppl.* **1995**, *107*, 89-100.
21.   Havens, K.E.;Bierman Jr., V.J.;Flaig, E.G.;Hanlon, C.;James, R.T. Jones, B.L.;Smith, V.H. *Arch. Hydrobiol./Suppl.,* **1995**, *107*, 101-111.
22.   Swift, D.R.; Anclade C.; Kantrowitz, I.H. *Natl. Water Survey,* **1987**, *1987*, 57-64.
23.   Havens, K.E. *Lake Res. Managem.* **1995**, *1995*, 133-138.
24.   Havens, K.E.*Arch. Hydrobiol.,* **1994**, *131*, 39-53.
25.   Smith, V.H.; Bierman Jr., V.J.; Jones, B.L.; Havens, K.E. *Arch.Hydrobiol. / Suppl.,***1995**, *107*, 71-88.
26.   Olila, O.G.; Reddy, K.R. *Arch. Hydrobiol.* **1993**, *129*, 45-65.
27.   Davis,P.; Steinman, A., SFWMD, West Palm Beach, Fla., unpublished data.
28.   Rao, C.N.R.*Ultraviolet and Visible Spectroscopy,* Plenum, New York, 1967; 2nd.Ed.; p.9.

29.   Straub, O. In *Carotenoids;* Isler, O.,Ed., Birkhauser-Verlag: Basel, 1971; pp 771-850.
30.   *Chlorophylls;* Scheer, H., Ed.; CRC Press, Boca Raton, Florida, 1991; 1257pp
31.   Gieskes, W.W.C.; Kraay, G.W. *Mar. Biol.* **1983,**75, 179-185.
32.   Barlow, R.G. Mantoura, R.F.C.; Gough, M.A.; Fileman, T.W. *Deep Sea Res.-II,* **1993,** *40,* 459-477.
33.   Kraay, G.W.; Zapata,M.; Veldhuis, M.J.W. *J. Phycol.,* **1992,** *28,* 708-712.
34.   Fuhrhop, J.H., Smith, K.M. In *Porphyrins and Metalloporphyrins;* Smith, K.M.,Ed.;Elsevier, Amsterdam, 1975; pp. 757 - 789.
35.   Wilhelm,C.; Manns,L. *J.Appl.Phycol.* **1991,** *3,* 305-310.
36.   Demmig-Adams, B. *Biochim. Biophys. Acta* **1990,** *1020,* 1-24.
37.   Millie, D.F., Ingram, D.A., Dionigi, C.P. *J. Phycol.* **1990,** *26,* 660-666.
38.   Leavitt, P.R., Carpenter, S.K. *Can J. Fish. Aquat. Sci.* **1990,** *47,* 1166-1176.
39.   Nelson, J.R. *J. Mar. Res.* **1993,** *51,* 155-179.
40.   Wilhelm, C., Rudolph, I., Renner, W. *Arch. Hydrobiol.,* **1991,** *123,*21-35.
41.   Bianchi, T.S., Findlay, S., Dawson, R. *Estuarine Coastal Shelf Sci.* **1993,** *36,* 359-376.
42.   Millie, D.F., Paerl, H.W., Hurley, J.P. *Can. J. Fish. Aquat. Sci.* **1993,** *50,* 2513-2527.
43.   Bidigare, R.R., Kennicutt-II, M.C., Brooks, J.M. *Limnol. Oceanogr.* **1985,** *30,* 432-435.
44.   Guillard, R.R.L., Keller, M.D., O'Kelley, C.J., Floyd, G.L. *J. Phycol.* **1991,** *27,* 39-47.
45.   Zullig, H. *J. Paleolimnol.* **1989,** *2,* 23-40.
46.   Baker, E.W.; Louda, J.W. In *Biological Markers in the Sedimentary Record*; Johns, R.B.,Ed.; Methods in Geochemistry and Geophysics, 24; Elsevier, Amsterdam, 1986; pp.125-225.
47.   Louda, J.W.; Baker, E.W. In *Organic Marine Geochemistry*; Sohn, M. Ed.; ACS Symposium Series No.305, American Chemical Society, Washington, D.C., 1986; pp. 107-126.

Chapter 7

Higher Molecular Weight Terpenoids as Indicators of Organic Emissions from Terrestrial Vegetation

Monica A. Mazurek[1] and Bernd R. T. Simoneit[2,3]

[1]Institute of Marine and Coastal Sciences, Rutgers, The State University of New Jersey, P.O. Box 231, New Brunswick, NJ 08903–0231
[2]College of Oceanic and Atmospheric Sciences, Oregon State University, 104 Ocean Admin Building, Corvallis, OR 97331–5503

Higher molecular weight terpenoids (C_{15} to C_{40} carbon units) are naturally occurring compounds that are introduced into the atmosphere from terrestrial vegetation. These compounds are released by a variety of mechanisms such as: (1) direct volatilization due to plant metabolic activities; (2) steam distillation or steam stripping that occurs during large-scale (biomass burning, natural wildfires) and small scale (camp fires, fireplaces) combustion of biofuels; and (3) mechanical processes which include abrasion and disintegration of plant leaf surface waxes and structural materials. Mechanism (2) is illustrated with an example. The C_{15} to C_{40} terpenoids encompass a broad range of chemical compositions which makes this class of plant organic matter a diverse category of source material that may be exploited in environmental applications (*e.g.*, the atmosphere). Reviews of the biochemical and phytochemical literature indeed show that many terpenoids in this molecular weight range are specific for major vegetation taxa and in some cases, are exclusive to certain subspecies of vegetation. Consequently, C_{15} to C_{40} terpenoid compounds provide a variety of important molecular probes that can be used in biogeochemical studies of atmosphere-biosphere interactions.

Emissions from terrestrial biota to the atmosphere and the effects of these emissions on atmospheric clarity have been known for over 500 years. Leonardo da Vinci described the phenomenon of decreased visibility in his scientific journals of the early sixteenth century (*1-2*). As an artist who painted the lush Tuscan countryside, da Vinci observed the presence of bluish hazes in the atmosphere. He attributed the hazes to exhalations by the surrounding vegetation and suggested these emissions consisted of moisture given off by plants. Since about the 17th century, the paintings of other landscape artists such as Albert Cuyp, Turner, and Monet, also have documented the presence of such atmospheric bluish hazes (*2*).

[3]Corresponding author

In the mid-1940's a scientific basis for the role of biomass emissions and reactive hydrocarbons in atmospheric chemical processes began to emerge. The work of Haagen-Smit provided evidence for the conversion of reactive hydrocarbons to secondary photochemical species in the presence of oxides of nitrogen (NO_x) together with sunlight (*3-5*). This research laid the foundation for what was to become the field of photochemical smog chemistry, now a very active research area within atmospheric chemistry. In 1960, Went, the director of the Missouri Botanical Garden in St. Louis, MO, commented on the significant gap in understanding the atmospheric fate of terpenoids, a major group of plant natural products (*2*). Went offered examples and evidence of the photooxidation of higher molecular weight terpenoids under natural conditions (*1-2*), summarizing observations of bluish hazes which developed over large biomes away from industrialized areas, particularly in summer on windless days and at times when no fires occurred (*e.g.*, Blue Ridge Mountains in Tennessee, and Blue Mountains of Australia, eucalyptus vegetation). Although Went was not a chemist by training, he conducted a series of experiments exposing a variety of freshly generated volatile terpenoids to dilute ozone in sealed jars placed against black backgrounds (*2*). He noted the formation of hazes with time as a strong light was placed alongside the jars. Freshly crushed pine and fir needles showed rapid formation of bluish hazes, whereas crushed oak or maple leaves did not.

Since the early 1970's atmospheric chemistry has developed as a distinct field of applied chemistry. Many advances in our understanding of the role of reactive hydrocarbons in atmospheric processes are based on laboratory experiments, field measurements, and atmospheric modeling studies. Increasingly, it has become evident that natural volatile organic emissions from terrestrial biota are key factors that are involved in many pressing atmospheric, geochemical, and ecological issues, including air quality (local and regional scale); biogeochemical cycles of organic matter (meso- to global scale); changing oxidative capacity of the atmosphere (climate change implications); and ecosystem response to the changing chemical composition of the Earth's troposphere. Consequently, the organic geochemistry of higher molecular weight terpenoids, coupled with their sources and fates in the atmosphere, may yield valuable insight for many of these areas of uncertainty.

Here we discuss categories of C_{15} to C_{40} terpenoids which have been used as key indicators for studies of: (1) The origin of organic aerosol particles in the troposphere, (2) the influence of biomass burning practices on atmospheric radiative, chemical, and cloud nucleation processes, and (3) the role of aerosol organics on visibility, particularly for nonurban airsheds. This paper presents an overview of the C_{15} to C_{40} terpenoids which can be used as atmospheric molecular tracers and summarizes their utility as indicators for dominant vegetation taxa. We illustrate the precursor-tracer concept with an example and confine our focus principally to nonurban environments.

Terpene Chemistry

Higher molecular weight terpenoids (C_{15} to C_{40}) and the smaller, volatile monoterpenes (C_{10} compounds) consist of one or more 5-carbon units (*i.e.*, isoprene structure) that are arranged usually in a head-to-tail manner:

Isoprene

Myrcene

Other combinations of the basic 5-carbon skeletal unit are possible, but they occur less frequently in nature (6-10). The biosynthesis of C_{10} and higher terpenoids is described by Croteau (11). Biosynthesis of terpenoids follows the mevalonate pathway to yield isopentenyl pyrophosphate, a 5-carbon unit reaction intermediate which ultimately forms the carbon skeletal structure of terpenoids. The exception to this well-established pathway for isoprenoid biosynthesis is isoprene itself, for which the precise synthesis mechanism is an active area of study (12).

Terpenoid Classification. In general, terpenoid compounds are classified in terms of the number of constituent 5-carbon or "isoprene" units, and follow the "isoprenoid rule". Thus, the terpenoids are polyisoprenoid structures that encompass the types of compounds (with the respective carbon skeletons) shown in Table I.

The basic structural arrangements of the terpenoids fall into acyclic, monocyclic, and polycyclic ring classifications, according to a subsidiary classification scheme (9). This secondary classification scheme is necessary because of the preponderance of structural isomers possible for a single higher molecular weight terpenoid. As an example, sesquiterpenoids comprise the largest of the terpenoid classes, consisting of 30 main structural types and almost 70 less common skeletons. Diterpenoids can be placed into 20 main classes of skeletal structures, and naturally occurring monoterpenoids consist of 15 structural classes (10). For a given

Table I. The Terpenoid Classes

Group	Number of isoprene units	Number of carbon atoms	Empirical formula of hydrocarbon analog	Number of main skeletal classes (10)
Hemiterpenes	1	5	C_5H_8	1
Monoterpenes	2	10	$C_{10}H_{16}$	15
Sesquiterpenes	3	15	$C_{15}H_{24}$	30
Diterpenes	4	20	$C_{20}H_{32}$	20
Sesterperpenes	5	25	$C_{25}H_{40}$	3
Triterpenes	6	30	$C_{30}H_{48}$ (also steroids, C_{21} to C_{30})	22
Tetraterpenes	8	40	$C_{40}H_{64}$	3
Polyterpenes	>9	>45	$(C_5H_8)_n, n>9$	

skeletal structure, it is possible to find a variety of functional groups (*e.g.,* C=C, -OH, -CHO, -CO$_2$H), thus forming analogs such as alcohols, aldehydes, ketones, acids, ethers, and lactones. The reader is referred to reviews of the natural product literature for more complete discussions of higher molecular weight terpenoids found in plants (*6-10, 13-14*).

Physical Properties. The wide diversity of molecular weights and functional groups comprising terpenoids cause this class of compounds to be present in the atmosphere as gases, liquids, or solids. Two other factors influencing the physical state of terpenoids in the atmosphere are: (1) meteorological conditions, such as temperature and relative humidity; and (2) the presence of aerosol particulate matter, where terpenoids may exist as surface coatings (especially liquid coatings) or may be internally mixed with other aerosol chemical constituents. Currently, little is known about aerosol mixing state (*i.e.,* internal or external mixtures of individual chemical substances as suspended particles) (*15-17*). The ability of higher terpenoids to exist as several states of matter under natural conditions confounds attempts by atmospheric modelers to accurately describe the higher molecular weight terpenoids within atmospheric chemical models (*18-19*).

Sources of information concerning the physical properties of terpenoids are available from standard handbooks (*20-21*). Other sources also exist that specialize in terpenoid chemistry (*6-8)* and in natural products that are important in the food, flavor, and fragrance industry (*22-24)*. Graedel (*25*) has summarized the physical properties of acyclic, monocyclic, and bicyclic terpenoids (C$_5$ to C$_{10}$ range). Molecular weight and the presence of functional groups influence the volatility (vapor pressure) of terpenoid compounds. As an example, isoprene (C$_5$ compound) has a boiling point of 34°C and a vapor pressure of 560 torr at standard conditions. Monoterpene hydrocarbons (C$_{10}$ compounds) have boiling points generally ranging from 150°C to 185°C and vapor pressures of 3 to 4 torr at standard conditions. Monoterpenoid alcohols, aldehydes, and ketones (C$_{10}$ compounds) have boiling points in the range of 190°C to 220°C, with vapor pressures of 0.1 to 0.4 torr (*25*).

At present, atmospheric measurements of terpenoids have focused primarily on C$_5$ to C$_{15}$ terpenoid alkanes and alkenes, with fewer reports of functionalized or higher molecular weight compounds greater than the sesquiterpenoids class (see following sections). In large part, the lack of field observations for >C$_{15}$ terpenoids can be related to terpenoid volatility and to functional group composition. Most field measurement studies have involved sampling of atmospheric terpenoids using canister methods or Teflon bag enclosures (*26-31*). Inherent sampling bias exists with these collection methods because only gas phase compounds can be efficiently transferred to gas chromatography systems. Detection of alcohols, ketones, aldehydes, and acids is further limited due to the relatively high polarity of these compounds with gas chromatographic sorbent phases. Consequently, a derivatization step is usually required before these classes of functionalized terpenoids are converted to analogs of higher volatility than the original, polar terpenoid (*32,33*). In the derivatized state, these compounds can pass through moderately polar GC column phases (*e.g.*, DB-5) (*32-33*).

Terpenoid Photochemistry. Another factor which has contributed to the remarkably limited number of functionalized higher molecular weight terpenoids

(C_{15} and > compounds) that have been reported in the atmospheric chemical literature is the rapid oxidation of terpenoids due to presence of C-C double bonds. These unsaturated carbon bonds are highly unstable under ambient conditions in the troposphere. The dominant fate of atmospheric terpenoid compounds (especially monoterpenoids) is reaction with ozone or with hydroxyl or nitrate radicals (*18-19,25,34-40*) (cf. Fig. 1). Atmospheric lifetimes of terpenoids are only a few hours (*25,35,38*). Recently, it has been shown that biogenic hydrocarbons, including terpenoids, are oxidized to alkyl nitrate species in regions that are influenced by biomass emissions. Thus, biogenic hydrocarbons are thought to be a significant reservoir for reactive nitrogen compounds in the atmosphere (*41-43*)

Emission Processes. Higher molecular weight terpenoids are released into the atmosphere by a variety of mechanisms including: (1) direct volatilization due to plant metabolic activities; (2) steam distillation or steam stripping that occurs during large-scale (biomass burning, natural wildfires) and small scale (camp fires, fireplaces) combustion of biofuels; and (3) mechanical processes which include abrasion and disintegration of plant leaf surface waxes and structural materials. The relative importance of each of these mechanisms are poorly understood, especially when extended to global emission inventories of higher molecular weight terpenoids from vegetation. This lack of information is largely due to the limited number of field experiments which have focused on measurements of C_{15} to C_{40} terpenoids present in the atmosphere.

Two recent articles have reviewed the literature relating to hemi-, mono-, and sesquiterpenes emitted from various vegetation sources (*38,44*). Total global production of natural volatile organic compounds from vegetation were estimated to range from 400 to 1150 Tg C yr^{-1}, where isoprene was estimated to contribute 175 to 500 Tg C yr^{-1} and monoterpenes roughly 300 to 500 Tg C yr^{-1} (*44*). Functionalized monoterpenes, or oxy-hydrocarbons (oxyHCs) were estimated to have global emissions of 260 Tg C yr^{-1} for reactive oxyHCs (atmospheric lifetimes < 1 day) and 260 Tg C yr^{-1} for less reactive oxyHCs (*44*). Estimates of the emission of C_{15} to C_{40} terpenoid compounds to the global troposphere from natural sources, either as hydrocarbons or oxygenated analogs, have not been reported. Consequently, the influence of this significant class of natural compounds from vegetation on the chemistry and physical properties of the atmosphere is uncertain.

Direct volatilization due to plant metabolic activities. Terpenoids are lipophilic substances secreted by specialized plant tissues and are present in the essential oils, resins and waxes of plants. The essential oils contain volatile terpenoids such as the mono- and sesquiterpenoids, whereas the resins represent mixtures of volatile and non-volatile terpenoids that include sesquiterpenoids, diterpenoids (major constituents), triterpenoids, and steroids (C_{21} to C_{30}), and the waxes contain mainly triterpenoids and steroids with the aliphatic components (*10,22,45*). The physiology of essential oil and resin secretion from plants is described in the botanical literature (*45-47*). Generally, the roles of essential oils and resins in plants are as aids in plant reproductive processes (*e.g.,* pollination, chemical attractants) and as a healing response to plant injury (*45-46*).

Higher molecular weight terpenoids and monoterpenoids are secreted from the upper (adaxial) surfaces of leaves through resin ducts in needle-bearing plants, or from oil glands, glandular hairs, or glandular epidermis of leaf-bearing plants. In gas exchange experiments involving eucalyptus vegetation isoprene emission was found to occur from the underside (abaxial) of leaves, mainly through gas phase exchange involving the leaf stomata (*26*). Other researchers also found isoprene emission to occur chiefly through leaf stomata (*48*). Similar work involving emission rate studies for the C_{10} to C_{15} terpenoids is reported by Winer *et al.* (*30*), although the distinction between abaxial and adaxial mechanisms was not explored. In general, emission rates for C_{10} and higher terpenoids are thought to be influenced by environmental factors such as light, temperature, humidity, and season (*38, 48-49*).

Mechanical processes. Plant senescence and surface abrasion are processes that contribute to the comminution of plant cuticular layers, yielding small fragments that become airborne as aerosol particulate matter. Higher molecular weight terpenoids have been measured in surface abrasion products from composited samples of vegetation (*50*) and as ambient particles collected from rural and urban atmospheres (*33,51-59*).

Steam distillation or steam-stripping during combustion. Biomass combustion is an important source of particles to the global atmosphere (*60-62*). The injection of higher molecular weight terpenoids into smoke occurs primarily by direct volatilization/steam stripping and by thermal alteration based on combustion temperature. These compounds are present in the smoke as the intact terpenoid species, or may have been thermally altered to recognizable functionalized, or partially-aromatized molecular analogs. The degree of alteration increases as the burn temperature rises and the moisture content of the fuel decreases. Diterpenoids are especially good indicators of smoke from burning of gymnosperm wood (*63-66*). These compounds and their derivatives at various stages of thermal alteration have been found in other ambient aerosols (*51,67*) and in smoke from slash and wood burning (*64,65,68*). A typical example of the diterpenoids in ambient atmospheric particulate matter is shown in Figure 2 for a sample from the rural northwestern United States. The major resin acids detectable are pimaric and abietic acids, with the predominent alteration products of dehydroabietic acid (major), bisnordehydroabietic acid and retene. Diterpenoids from gymnosperm wood combustion have been found in urban atmospheres (*33*), and have been used successfully to estimate the contribution of wood smoke emissions to metropolitan Los Angeles (*69,70*). This steam-stripping during biomass combustion with concomittant precursor alteration will be illustrated further after the following section.

Terpenoid Molecular Markers

Despite the hundreds of higher molecular weight terpenoids known to be present in vegetation, remarkably few have been detected in atmospheric samples. Most studies have focused on monoterpenes or isoprene. Here, we give a brief overview of the terpenoids as general background for the reader.

Monoterpenoids (C₁₀ Compounds). Monoterpenoids are alkenes or less commonly alkanes (*i.e.,* contain zero, one, or more C-C double bonds) or have heteroatoms with oxygen as the primary element found in functional groups that are attached to the basic C_{10} isoprenoid molecular structure. Monoterpenoids are emitted in large amounts by vegetation and generally are comprised mainly of pinenes, carenes,

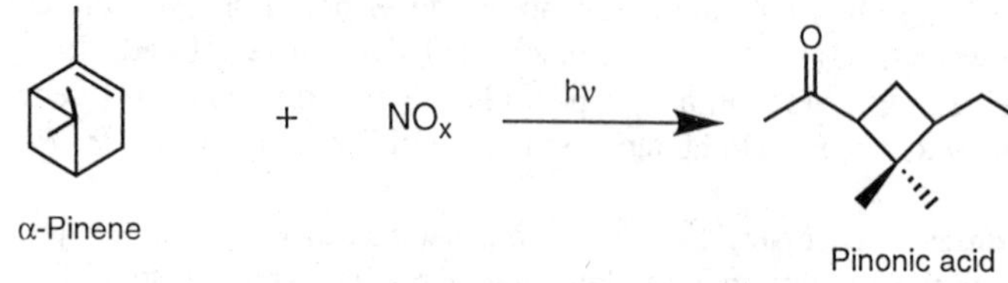

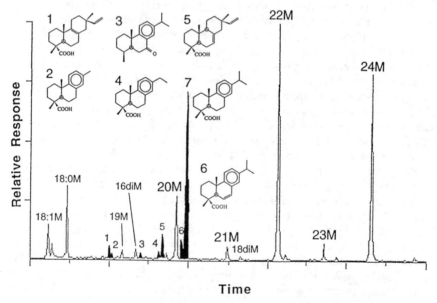

Figure 1.   Oxidation of α-pinene by NO_x under UV to pinonic acid (*39,40*).

Figure 2.   Partial GC-MS total ion current trace showing the major diterpenoids found in ambient aerosol particles of the rural Northwestern United States (1 = 8,15-Pimaradien-18-oic acid, 2 = 16,17-Bisnordehydroabietic acid, 3 = 7-Oxodehydroabietin, 4 = 16-Nordehydroabietic acid, 5 = Isopimaric acid, 6 = Abieta-6,8,11,13-tetraenoic acid, and 7 = Dehydroabietic acid, the numbers 16-24 refer to the n-alkanoic or dioic acids all analyzed as methyl esters).

camphene, sabinene, β-phellandrene and terpinolene. Field measurements of monoterpenoids present in the atmosphere of rural and agricultural areas have been reported (*25,27,31,44,71-75*). Emissions inventories for biogenic monoterpenoids for the continental U.S. are available (*9-13*). The chief interest in monoterpenes stems from the interaction of these compounds with atmospheric oxidant species such as NO_x, O_3 and OH radical. The presence of C-C double bonds within the molecules accounts for the short atmospheric lifetimes of monoterpenoids, which usually are on the order of several hours or less. Hence, monoterpenoids are key molecular tracers in studies of atmosphere-biosphere processes. The primary focus of these studies relate to natural processes that influence ambient levels of tropospheric ozone at rural and semirural continental sites.

Monoterpenoids are an important class of natural products that has been observed during biomass and fossil fuel (coal) combustion. Steam stripping is the injection mechanism which releases the monoterpenoids directly into the atmosphere with little or no thermal alteration of the original molecule. Both unaltered and thermally altered terpenoids have been studied in smoke plume aerosols in order to document discrete molecular tracers that can be linked in a direct and unique manner to the biofuel burned. The presence and abundance of a terpenoid compound and its alteration products in smoke aerosols have been used to study environmental processes during combustion of biomass fuels (*64,68*).

Sesquiterpenoids (C_{15} Compounds). Sesquiterpenoid compounds have been studied less in atmosphere-biosphere applications compared to monoterpenoids, but they are an important group of semivolatile components which occur in plants. The C_{15} terpenoids span a large range of compositions containing the basic isoprenoid molecular structure and many functional group possibilities as well as stereochemical configurations that produce numerous isomeric compounds having the same molecular formulas. Cadinene derivatives have been identified in rural airsheds and consisted of cadalene, calamenene and tetrahydrocadalene (*49*) (Fig. 3). Studies of the atmospheric concentrations of sesquiterpenoids occurring within agricultural regions of the California Central Valley and in metropolitan Los Angeles have also been performed (*30,76-78*). These investigators identified mainly β-caryophyllene, α-humulene and various unknowns. They also proposed photochemical oxidation pathways of dominant sesquiterpenes and examined their potential impact on ozone formation in these regions of California. Furthermore, sesquiterpenoids have been found in rural background sites and linked to organic emissions into the atmosphere from vegetation with high contents of essential oils (*51,67,79*). Sesquiterpenoids have not been reported as dominant markers in smoke aerosol particulate matter that was collected during biomass or fossil fuel combustion. This is probably due to the fact that sesquiterpenoids, like monoterpenoids, are more concentrated in the gas phase.

Diterpenoids (C_{20} Compounds). Diterpenoid compounds are abundant natural products that are associated with coniferous vegetation in the northern hemisphere, where they are found in the resins, gums, and exudates. Diterpenoids have been measured in forest regions as well as in semi-urban and urban airsheds located near

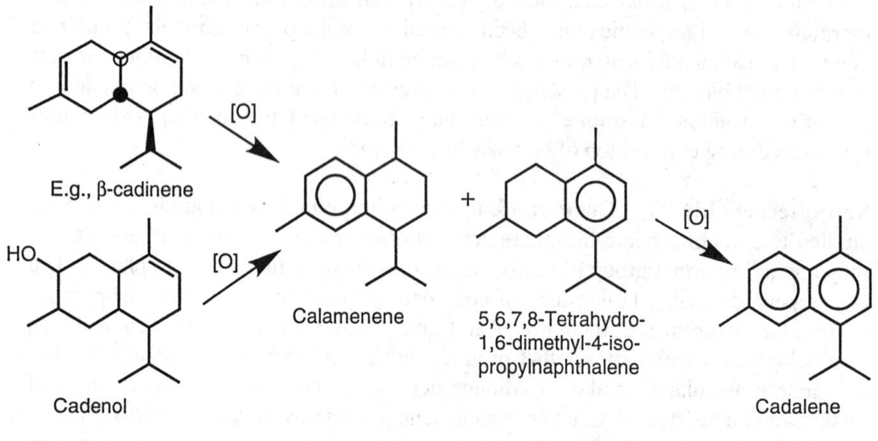

Figure 3. Precursor-product relationship of the cadinene sesquiterpenoids found in the atmosphere (*51*).

zones of coniferous vegetation or where biomass combustion occurs (*33,51,64,67*). Diterpenoid acids are the primary molecular components found in smoke particles sampled from the combustion of conifer wood and of large forests containing gymnosperms (*64, cf. Fig. 2*). They are steam stripped directly with some oxidative alteration into smoke (*68*). Temperate vegetation of the southern hemisphere (*e.g., Podocarpaceae, Araucaraceae*) produces significant amounts of tetracyclic diterpenoids such as the classes of compounds having either the kaurane or podocarpane basic molecular skeletons.   The introduction of these tetracyclic diterpenoids into the atmosphere is currently under study (*80*).

Triterpenoids (C_{30} Compounds).   This group of high molecular weight terpenoids is more commonly linked to higher plant taxa such as angiosperms. Triterpenoids are found in the surficial wax layers of leaves and in bark. They are observed as aerosol particulate matter due to their low volatility at ambient temperatures. The major mode of injection is by mechanical abrasion of waxy layers from leaf surfaces and by vaporization during burning (*64,68*). The amyrins (Figure 4-I) are the major precursors found in smoke aerosol particles (*81*).

A special subgroup of triterpenoids, the phytosterols (Fig. 4-VIII) have been measured in samples of fine (nominal particle diameter, d_p, < 2.1 mm) and total (coarse plus fine) atmospheric particles in rural, semirural and urban locations (*54*). Phytosterols vary in molecular weight and carbon number distribution depending on the specific vegetation taxa and on admixture of cholesterol which is generally enriched in urban airsheds.   Aerosols in rural and remote regions contain phytosterols skewed to higher molecular mass ($C_{29} > C_{27}$) relative to urban areas (*54*). Therefore, phytosterols can be used as molecular indicators of emissions of plant material from vegetation sources. The related sterol, cholesterol, is used as a tracer for algal and faunal (animal) contributions in rural and remote regions and for cooking and rendering in urban areas (*33*).

Tetraterpenoids and Polyterpenoids (C_{40} and Higher Compounds).   The tetraterpenoids comprise mainly the carotenoid pigments. These compounds have not been found in atmospheric samples, although their geological equivalents (*e.g., carotane*, Fig. 4-IX) have been identified as petroleum markers in aerosols of China (*57*). Polyterpenoids are found as biopolymers and their degradation products may become airborne during pyrolysis (*e.g., biomass burning*). For example, solanesol ($C_{45}H_{74}O$), a polyterpenoid in tobacco, has been reported in cigarette smoke and proposed as a specific tobacco tracer (*82*).

Terpenoid Precursor - Tracer Concept

The precursor to product approach of organic geochemistry is illustrated here to delineate example tracers (biomarkers) for the environmental process of biomass combustion. These analyses are carried out by GC-MS analysis and molecular markers are identified by comparison with authentic standards and characterized mixtures. Unknown compounds are characterized by interpretation of their mass spectrometric fragmentation pattern.

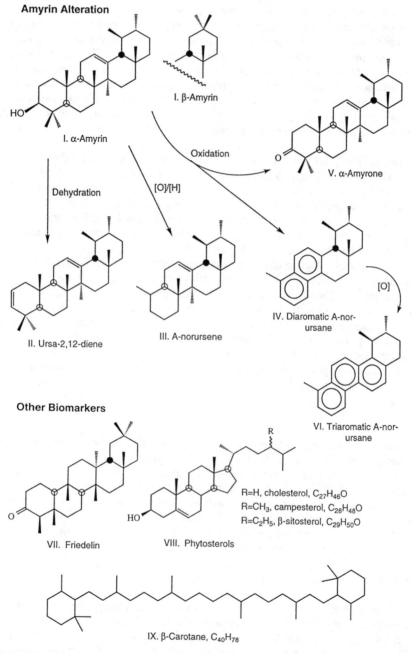

Figure 4.   Thermal alteration products from amyrin precursors in the Amazon smoke sample (*81*) and other biomarkers discussed in the text.

The example discussed here is a smoke sample from burning of composited vegetation in Amazonia (*81*). The various lipid fractions extracted from the particulate matter are comprised mainly of *n*-alkanoic acids, *n*-alkanes, polynuclear aromatic hydrocarbons (PAH), and triterpenoids. The analysis of the hydrocarbon fraction is shown in Fig. 5. The *n*-alkanes range from C_{17} to C_{37} with a $C_{max} = 29/31$ and high odd carbon number preference (Fig. 5c). This is similar as reported for aerosols from the Amazon region, indicating that alkanes from burning are indistinguishable from plant wax alkanes in the ambient aerosol (*58*).

More specific homologous aliphatic tracers for combustion are the *n*-alk-1-enes, *n*-alkan-2-ones and α,ω-alkanedioic acids (*81*). For example, the major series of *n*-alk-1-enes present ranges from C_{17} to C_{35}, with $C_{max} = 22$ (minor at 29/31) and a slight even carbon number predominance (CPI = 0.8, Fig. 5b). Alkenes are not dominant components in aerosols or plant waxes. The origin of these compounds is inferred to be from the biomass fuel. Based on their carbon number distribution the alkenes are derived primarily from *n*-alkanols by dehydration (*n*-alkan-1-ols are easily dehydrated by high temperatures to *n*-alk-1-enes) and overlapping to a minor degree with alkenes derived from the *n*-alkanes by oxidation during incomplete combustion (compare Fig. 5b with c and a, respectively). The even carbon number dominance and $C_{max} = 22$ are inherited from the *n*-alkanol distribution, and the minor dominance of *n*-C_{29} and *n*-C_{31} reflects the dehydrogenation of the *n*-alkanes.

The aliphatic hydrocarbon fraction contains a group of derivatives from amyrins, the natural product precursor (peaks 1-6, Fig. 5a, structures are shown in Fig. 4). These are various triterpadienes (peaks 4-6) (*e.g.*, II. ursa-2,12-diene), noroleanene, norursene (peak 1, III), diaromatic A-noroleananes (peak 2) and A-norursanes (peak 3, IV), and minor triaromatic A-noroleananes and A-norursanes (VI) (*81*). These compounds are not known as natural products and are therefore indicators for combustion of biomass containing amyrin precursors. The α- and β-amyrins (I) are the predominant biomarkers in the total smoke lipids and α- and β-amyrones (V) (mild oxidation products of amyrins) and friedelin (VII) are also significant (81). Phytosterols from plant waxes are trace components in this smoke sample and consist mainly of β-sitosterol (VIII, R=C_2H_5) with lesser amounts of other C_{29} and C_{28} isomers. The phytosterol distribution is characteristic as was observed for other smoke emissions from biomass combustion (*54,68*).

Conclusions

Higher molecular weight terpenoids are an important class of natural plant products. However, the applications of these compounds in atmospheric chemistry and more broadly, in biogeochemical process studies, is just beginning to emerge. In particular, C_{15} to C_{40} terpenoids, are important to understanding dynamic biosphere/atmosphere interactions. Such interactions include the role of C_{15} to C_{40} terpenoids in atmospheric photochemical oxidation pathways, the biogeochemical cycles of light elements, and ecosystem responses to the changing chemical composition of the Earth's troposphere (global climate change, acid deposition, injury to vegetation caused by air pollution). The C_{15} to C_{40} terpenoids have utility as important molecular tracers in air quality management and visibility studies. They

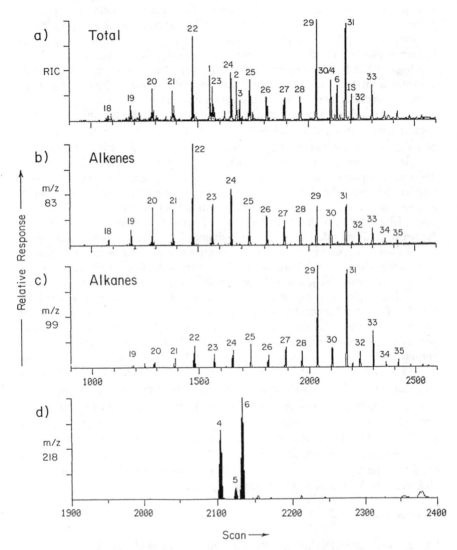

Figure 5.   Salient features of the GC-MS data for the aliphatic hydrocarbons
              (F1) from a smoke sample of Amazon biomass combustion: (a) total
              ion current trace, (b) alkenes, key ion m/z 83, (c) alkanes, key ion
              m/z 99, (d) triterpenes, key ion m/z 218 (numbers 18-35 refer to the
              carbon chain length of homologous compound series and 1-6 are
              biomarkers discussed in the text and shown in Fig. 4) (*81*).

provide molecular evidence for contributions of organic matter from vegetation to urban atmospheres and allow for estimations of mass emissions from biomass sources. Finally, C_{15} to C_{40} terpenoids have potential utility in atmospheric transport and air mass characterization studies, because vegetation species have unique terpenoid assemblages which may be employed as molecular markers for ecological zones.

Acknowledgments

This research was supported in part by the U.S. Environmental Protection Agency under agreement R-823990-01. This manuscript has not been subject to the EPA's peer and policy review, and hence does not necessarily reflect the views of the EPA.

Literature Cited

1.   Went, F. W., *Nature*, **1960**, *187*, 641-643.
2.   Went, F. W., *Proceedings of the National Academy of Sciences*, **1960**, *46*, 212-221.
3.   Haagen-Smit, A. J., *Industrial and Engineering Chemistry*, **1952**, 44, 1342-1346.
4.   Haagen-Smit, A. J., Darley, E. F., Zaitlin, M., Hull, H., and Noble, W., *Plant Physiology*, **1952**, *27*, 18-34.
5.   Haagen-Smit, A. J. and Fox, M. M., *Industrial and Engineering Chemistry*, **1956**, *48*, 1484-1487.
6.   Simonsen, J. L., *The Terpenes. Volume I: The Simpler Acyclic and Monocyclic Terpenes and their Derivatives,* Cambridge University Press, Cambridge, U.K., 1947.
7.   Simonsen, J. L., *The Terpenes. Volume II: The Dicyclic Terpenes and their Derivatives*, Cambridge University Press, Cambridge, U.K., 1949.
8.   Simonsen, J. L., *The Terpenes. Volume III: The Sesquiterpenes, Diterpenes and their Derivatives*, Cambridge University Press, Cambridge, U.K., 1952.
9.   Pinder, A. R., *The Chemistry of the Terpenes*, Chapman and Hall, Ltd., London, England, 1960.
10. Devon, T. K. and Scott, A. I., *Handbook of Naturally Occurring Compounds Vol. 2*, *Terpenes*, Academic Press: New York, NY, 1972.
11. Croteau, R., *Chemical Reviews*, **1987**, *87*, 929-953.
12. Sharkey, T. D., Loreto, F., and Delwiche, C. F., The biochemistry of isoprene emission from leaves during photosynthesis; in *Trace Gas Emissions by Plants*, Sharkey, T. D., *et al.,* eds., Academic Press: San Diego, CA, pp.153-184, 1991.
13. Simoneit, B. R. T., Cyclic terpenoids of the geosphere. Chapter 2; in *Biological Markers in the Sedimentary Record, Methods in Geochemistry and Geophysics, 24,* Johns, R. B., ed., Elsevier Science Publishers: Amsterdam, NL, pp. 43-99, 1986.
14. Goodwin, T. W. and Mercer, E. I., *Introduction to Plant Biochemistry*, Pergamon Press: Oxford, UK, 1972.
15. White, W. H., Chapter 4. NAPAP Report 24-- Visibility: Existing and

Historical Conditions--Causes and Effects; National Acid Precipitation
Assessment Program: Washington, DC, Volume III, 1990, 129 pgs.

16.   Sloan, C. S., *Atmos. Environ.*, **1984**, *18*, 871-878.

17.   Pilinis, C. and Pandis, S. N., Physical, Chemical and Optical Properties of
       Aerosols; Springer Verlag: Heidelberg, Physical, Chemical and Optical
       Properties of Aerosols, 1995.

18.   Pandis, S., Paulson, S., Seinfeld, J. H., and Flagan, R. C., *Atmos. Environ.*,
       1991, *25A*, 997-1008.

19.   Paulson, S., Pandis, S. N., Baltensperger, U., Seinfeld, J. H., Flagan, R. C.,
       Palen, E. J., Allen, D. T., Schaffner, C., Giger, W., and Portmann, A.,
       Journal of Aerosol Science, 1990, *21*, S245-S248.

20.   Merck & Company, *The Merck Index*, Merck & Company, Inc.: Whitehouse
       Station, NJ, 1996.

21.   Chemical Rubber Company, *Handbook of Chemistry and Physics*, CRC
       Press: Boca Raton, FL, 1996.

22.   Poucher, W. A., *Poucher's Perfumes, Cosmetics and Soaps. Volume 1: The
       Raw Materials of Perfumery*, Chapman and Hall: London, 1991.

23.   Opdyke, D. L. J., *Food & Cosmetics Toxicology*, **1974**, *12*, 807-1016.

24.   Opdyke, D. L. J., *Food & Cosmetics Toxicology*, **1975**, *13*, 681-923.

25.   Graedel, T. E., *Reviews of Geophysics and Space Physics*, **1979**, *17*, 937-947.

26.   Guenther, A. B., Monson, R. K., and Fall, R., *Journal of Geophysical
       Research*, **1991**, *96*, 10799-10808.

27.   Lamb, B., Guenther, A. B., Gay, D., and Westberg, H., *Atmos. Environ.*,
       1987, *21*, 1695-1705.

28.   Peters, R. J. B., Duivenbode, J. A. D. V. R., Duyzer, J. H., and Verhagen, H.
       L. M., *Atmos. Environ.,* **1994**, *28*, 2413-2419.

29.   Simon, V., Clement, B., Riba, M.-L., and Torres, L., *Journal of Geophysical
       Research*, 1994, *99*, 16,501-16,510.

30.   Winer, A. M., Arey, J., Atkinson, R., Aschmann, S. M., Long, W. D.,
       Morrison, C. L., and Olszyk, D. M., *Atmos. Environ.*, **1992**, *26A*, 2647-2659.

31.   Lerdau, M., Dilts, S. B., Westberg, H., Lamb, B., and Allwine, E. J., *Journal
       of Geophysical Research*, **1994**, *99*, 16,609-16,615.

32.   Mazurek, M. A., Mason-Jones, M., Mason-Jones, H., Salmon, L. G., Cass, G.
       R., Hallock, K. A., and Leach, M., J., *J. Geophys. Res.*, in press, **1997**.

33.   Rogge, W. F., Mazurek, M. A., Hildemann, L. M., Cass, G. R., and Simoneit,
       B. R. T., *Atmos. Environ.*, **1993**, *27A*, 1309-1330.

34.   Finlayson-Pitts, B. and Pitts, J. N., Jr., *Atmospheric Chemistry:
       Fundamentals and Experimental Techniques*, John Wiley and Sons: New
       York, NY, pp. 977-993, 1986.

35.   Atkinson, R., *Atmos. Environ.*, **1990**, **24A**, 1-41.

36.   Horie, O. and Moortgat, G. K., *Atmos. Environ.,* **1991**, *25A*, 1881-1896.

37.   Pandis, S. N., Harley, R. A., Cass, G. R., and Seinfeld, J. H., *Atmos. Environ.*,
       1992, *26A*, 2269-2282.

38.   Fehsenfeld, F., Calvert, J., Fall, R., Goldan, P., Guenther, A. B., Hewitt, C.
       N., Lamb, B., Liu, S., Trainer, M., Westberg, H., and Zimmerman, P., *Global
       Biogeochemical Cycles*, **1992**, *6*, 389-430.

39.   Atkinson R., Aschmann S.M., Winer A.M. and Pitts J.N. Jr., *Environ. Sci.
       Techn.*, **1985**, *19*, 159-163.

40.   Atkinson R., Aschmann S.M., Winer A.M. and Pitts J.N. Jr., *Environ. Sci. Techn.*, **1984**, *18*, 370-375.
41.   Singh, H., O'Hara, D., Herlth, D., Sachse, W., Blake, D., Bradshaw, J., Kanakidou, M., and Crutzen, P., *J. Geophys. Res.*, **1994**, *99*, 1805-1819.
42.   Beine, H. J., Jaffe, D. A., Blake, D. R., Atlas, E., and Harris, J., *J. Geophys. Res.*, **1996**, *101*, 12613-12619.
43.   Winer, A. M., Atkinson, R., and Pitts, J., James N., *Science*, **1984**, *224*, 156-159.
44.   Guenther, A., *J. Geophys. Res.*, **1995**, *100*, 8873-8892.
45.   Fahn, A., *Secretory Tissues in Plants*, Academic Press: New York, NY, 1979, 302 pgs.
46.   Martin, J. T. and Juniper, B. E., *The Cuticles of Plants*, Edward Arnold Ltd.: Edinburgh, U.K., 1970, 347 pgs.
47.   Schnepf, E., *Dynamic Aspects of Plant Ultrastructure*, McGraw-Hill: New York, NY, 331-357, 1974.
48.   Fall, R. and Monson, R. K., *Plant Physiology*, **1992**, *100*, 987-992.
49.   Monson, R. K. and Fall, R., *Plant Physiology*, **1989**, *90*, 267-274.
50.   Rogge, W. F., Hildemann, L. M., Mazurek, M. A., Cass, G. R., and Simoneit, B. R. T., *Environ. Sci. Technol.*, **1993**, *27*, 2700-2711.
51.   Simoneit, B. R. T. and Mazurek, M. A., *Atmos. Environ.*, **1982**, *16*, 2139-2159.
52.   Simoneit, B. R. T., Cox, R. E., and Standley, L. J., *Atmos. Environ.*, **1988**, *22*, 983-1004.
53.   Cox, R. E., Mazurek, M. A., and Simoneit, B. R. T., *Nature*, **1982**, *296*, 848-849.
54.   Simoneit, B. R. T., Mazurek, M. A., and Reed, W. E., Characterization of organic aerosols over rural sites: phytosterols; in *Advances in Organic Geochemistry 1981*, Bjorøy, M., et al., eds., J. Wiley & Sons Limited: Chichester, U.K., pp. 355-361, 1981.
55.   Simoneit, B. R. T., *Atmos. Environ.*, **1984**, *18*, 51-67.
56.   Simoneit, B. R. T., Crisp, P. T., Mazurek, M. A., and Standley, L. J., *Environ. Internat.*, **1991**, *17*, 405-419.
57.   Simoneit, B. R. T., Sheng, G.-Y., Chen, X.-J., Fu, J.-M., Zhang, J.-A., and Xu, Y.-P., *Atmos. Environ.*, **1991**, *25A*, 2111-2129.
58.   Simoneit, B. R. T., Cardoso, J. N., and Robinson, N., *Chemosphere*, **1990**, *21*, 1285-1301.
59.   Simoneit, B. R. T., Cardoso, J. N., and Robinson, N., *Chemosphere*, **1991**, *23*, 447-465.
60.   Andreae, M. O., Biomass burning: its history, use and distribution and its impact on environmental quality and global climate; in *Global Biomass Burning: Atmospheric, Climatic; and Biospheric Implications*, Levine, J.S., editor, The MIT Press, Cambridge, MA, 1990, pp. 3-21.
61.   Levine, J. S., Global biomass burning: Atmospheric, climatic, and biospheric implications; in *Global Biomass Burning: Atmospheric, Climatic; and Biospheric Implications*, Levine, J.S., editor, The MIT Press, Cambridge, MA, 1990, pp. 25-30.

62.   Levine, J. S., Cofer, I. I. I., Wesley R., Cahoon, J., Donald R., and Winstead, E. L., *Environ. Sci. Technol.*, **1995**, *29*, 120A-125A.
63.   Ramdahl, T., *Nature*, **1983**, *306*, 580-582.
64.   Standley, L. J. and Simoneit, B. R. T., *Environ. Sci. Technol.*, **1987**, *21*, 163-169.
65.   Standley, L. J. and Simoneit, B. R. T., *J. Atmos. Chem.*, **1994**, *18*, 1-15.
66.   Simoneit, B. R. T., Radzi bin Abas, M., Cass, G. R., Rogge, W. F., Mazurek, M. A., and Hildemann, L. M. (1995): Natural organic compounds as tracers for biomass combustion in aerosols. Technical paper presented at the *Chapman Conference on Biomass Burning and Global Change*, sponsored by the American Geophysical Union, March 13-17, 1995.
67.   Simoneit, B. R. T., *J. Atmos. Chem.*, **1989**, *8*, 251-275.
68.   Simoneit, B. R. T., Rogge, W. F., Mazurek, M. A., Standley, L. J., Hildemann, L. M., and Cass, G. R., *Environ. Sci. Technol.*, **1993**, *27*, 2533-2541.
69.   Rogge, W. F., Hildemann, L. M., Mazurek, M. A., Cass, G. R., and Simoneit, B. R. T., *J. Geophys. Res.*, **1996**, *101*, 19,379-19,394.
70.   Schauer, J. J., Rogge, W. F., Hildemann, L. M., Mazurek, M. A., Cass, G. R., and Simoneit, B. R. T., *Atmos. Environ.*, **1996**, *30*, 3837-3855.
71.   Whitby, R. A. and Coffey, P. E., *J. Geophys. Res.*, **1977**, *82*, 5928-5934.
72.   Juuti, S., Arey, J., and Atkinson, R., *J. Geophys. Res.*, **1990**, *95*, 7515-7519.
73.   Lamb, B., Westberg, H., and Allwine, G., *J. Geophys. Res.*, **1985**, *90*, 2380-2390.
74.   Trainer, M., Williams, E. J., Parrish, D. D., Buhr, M. P., Allwine, E. J., Westberg, H. H., Fehsenfeld, F. C., and Lui, S. C., *Nature*, **1987**, *329*, 705-707.
75.   Altshuller, A. P., *Atmos. Environ.*, **1983**, *17*, 2131-2165.
76.   Corchnoy, S. B., Arey, J., and Atkinson, R., *Atmos. Environ.*, **1992**, *26B*, 339-348.
77.   Arey, J., Winer, A. M., Atkinson, R., Aschmann, S. M., Long, W. D., Morrison, C. L., and Olszyk, D. M., *J. Geophys. Res.*, **1991**, *96*, 9329-9336.
78.   Arey, J., Winer, A. M., Atkinson, R., Aschmann, S. M., Long, W. D., and Morrison, L. C., *Atmos. Environ.*, **1991**, *25A*, 1063-1075.
79.   Mazurek, M. A., Newman, L., Daum, P. H., Cass, G. R., Salmon, L. G., Winner, D. A., Mason-Jones, M. C., and Mason-Jones, H. D., and Hallock, K., Visibility-reducing organic aerosols in the vicinity of Grand Canyon National Park: 2. Molecular composition. Technical paper presented at the 1995 Annual Meeting of the American Association for Aerosol Research, October 9-13, Pittsburgh, PA, 1995.
80.   Simoneit, B. R. T., unpublished results.
81.   Radzi bin Abas, M., Simoneit, B. R. T., Elias, V., Cabral, J. A., and Cardoso, J. N., *Chemosphere*, **1995**, *30*, 995-1015.
82.   Ogden, M. W. and Maiolo, K. C., *Environ. Sci. Technol.*, **1989**, *23*, 1148-1154.

Fossil Biomarkers

Chapter 8

Applications of Biomarkers for Identifying Sources of Natural and Pollutant Hydrocarbons in Aquatic Environments

John K. Volkman[1], Andrew T. Revill[1], and Andrew P. Murray[2]

[1]Division of Marine Research, CSIRO, GPO Box 1538, Hobart,
Tasmania 7001, Australia
[2]Australian Geological Survey Organisation, GPO Box 378, Canberra,
Australian Capital Territory 2601, Australia

Over the past decade, many thousands of compounds have been
identified in crude oils using high resolution capillary gas
chromatography-mass spectrometry (GC-MS) and other techniques.
Some of these compounds have distinctive chemical structures which
are closely related to the organic compounds produced by plants,
bacteria and algae. These 'biomarkers' are used in petroleum exploration
to identify the likely source rocks from which different oils were
derived. The same compounds can also be valuable for identifying
sources of petroleum contamination in the marine environment, for
recognising natural constituents and as tracers for physical and
biological processes that affect hydrocarbon distributions such as
dissolution, evaporation, biodegradation, bioturbation, and uptake by
biota. Some applications of petroleum biomarkers to environmental
studies of hydrocarbon pollution are reviewed in this paper.

Hydrocarbons are ubiquitous in modern marine environments. Those derived from
large oil spills due to shipping accidents have tended to capture the attention of the
media, public and regulators. However, there are many other sources, and in order to
differentiate between them it is necessary to examine the types and distribution of
hydrocarbons present. Petroleum geochemists have developed a wide variety of
biomarkers for use in petroleum exploration studies (1,2) which can also be useful for
identifying contamination of modern environments by crude oil and petroleum-derived
products (e.g. 3,4,5). Biomarkers are defined here as those compounds, or groups of
compounds, which have distinctive structures which can be related, through reasonable
transformation pathways, to compounds produced by living organisms. The term
"biomarker" is also used in a different context in the field of ecotoxicology, and so
alternatives such as "signature lipid" (especially in studies of Recent environments),
"chemical fossil" or "molecular fossil" (in ancient sediments), "biogeochemical
marker", molecular marker and "chemical fingerprint" are also used by geochemists.

This paper provides an introduction to the wide range of biomarkers now being used by petroleum geochemists together with some examples of the use of these compounds for identifying the sources, transport and fates of natural and pollutant hydrocarbons in aquatic environments. For more information on the composition of oils and the geochemical applications of biomarkers the reader is referred to textbooks by Peters and Moldowan (*1*), Tissot and Welte (*6*) and Engel and Macko (*7*).

Analytical Methods
Crude oils are composed primarily of hydrocarbons with small amounts of polar functionalized compounds such as porphyrins, and compounds containing sulfur, oxygen or nitrogen (collectively referred to as the NSO fraction or "resins plus asphaltenes"). Aliphatic hydrocarbons contain either no double bonds (alkanes or saturates) or one or more double bonds (alkenes). Alkanes and alkenes may be straight-chain (*e.g. n*-heptadecane, *n*-C_{17}; structure I, Figure 1), branched (*e.g.* pristane and phytane; structures VI,V, respectively) or cyclic (*e.g.* steranes and hopanes; structures XVI-XXI and XXIV-XXXIV, respectively). Alkenes are minor constituents of most oils. Aromatic hydrocarbons commonly contain from 1 to 5 carbocyclic fused rings; examples include naphthalene, phenanthrene, chrysene and perylene (structures XLIII-XLV, XLVIII, respectively).

Many methods have been proposed to measure the amount of petroleum in the marine environment (*8*). The concentration of total hydrocarbons can be measured by direct weighing of the hydrocarbon fraction of the total solvent extract. However, accurate data can be difficult to obtain when the sample size is small. Also, elemental sulfur must be removed from the extract, usually by treatment with activated copper. Thin layer chromatography-flame ionisation detection (TLC-FID) provides an alternative with a reproducibility of about ± 8% (*9,10*). The TLC-FID technique is particularly well suited to the analysis of samples contaminated with heavier oils, biodegraded oils or those which have already lost light-ends due to evaporation or water washing. However TLC-FID is not suited to the analysis of highly volatile oils and condensates.

Spectroscopic methods can provide a rapid measure of oil in seawater, but they provide little indication of the source of the oil, are difficult to calibrate and, hence, can be inaccurate. Total polycyclic aromatic hydrocarbons can be estimated using ultraviolet fluorescence spectroscopy (UVF) which involves the excitation of the sample with UV light and then measuring the light emitted at longer wavelength by the excited multi-ring aromatic compounds in the sample (*11*). UVF is a very sensitive technique, but it provides little information about oil composition. Some limited fingerprinting information can be obtained using synchronous scanning of the excitation and emission wavelengths.

High resolution capillary gas chromatography with its high sensitivity and resolving power (particularly when 50 or 60 m long narrow-bore fused silica capillary columns are used) coupled with mass spectrometry can provide a great variety of detailed information on the structures of compounds present. The mass spectrometer can be a simple bench-top quadrupole or ion trap instrument or a sophisticated MS-MS or triple quadrupole system. Source-specific biomarkers (steranes, diasteranes, hopanes, methylhopanes etc.; see Figure 1 for structures) can be measured by established GC-MS techniques such as selected ion monitoring (*12-14*), as described below, or by using metastable reaction monitoring with MS-MS systems. These

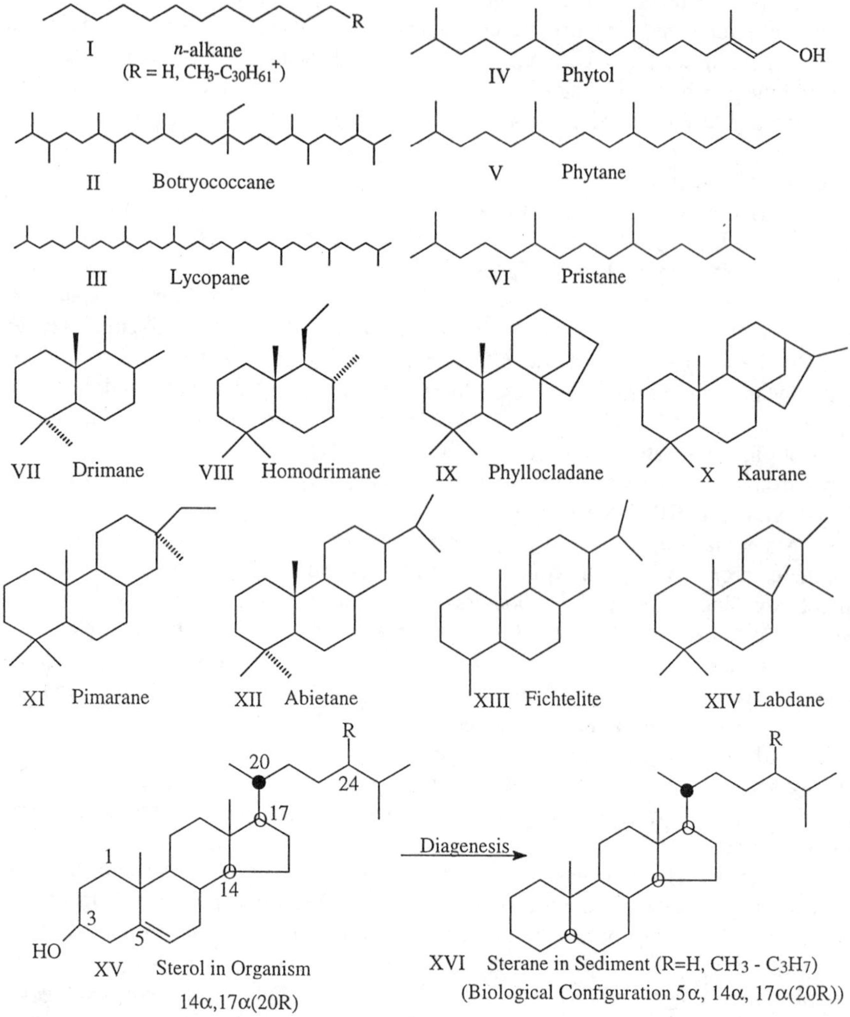

I   n-alkane
          (R = H, CH$_3$-C$_{30}$H$_{61}$$^+$)

IV   Phytol

II   Botryococcane

V   Phytane

III   Lycopane

VI   Pristane

VII   Drimane   VIII   Homodrimane   IX   Phyllocladane   X   Kaurane

XI   Pimarane   XII   Abietane   XIII   Fichtelite   XIV   Labdane

XV   Sterol in Organism
          14α,17α(20R)

Diagenesis →

XVI   Sterane in Sediment (R=H, CH$_3$ - C$_3$H$_7$)
          (Biological Configuration 5α, 14α, 17α(20R))

Figure 1. Chemical structures of some biomarkers commonly encountered in crude oils and contaminated sediments as well as some of their biological precursor molecules such as phytol (IV), sterols (XV) and bacteriohopanetetrol (XXIII). Note that "R" refers either to a hydrogen or alkyl group substituent of general formula C$_n$H$_{2n+1}$ or to the stereochemistry at a chiral centre in the molecule (contrast XVII with XVIII which show S and R stereochemistries at position 20 in the sterane side-chain).

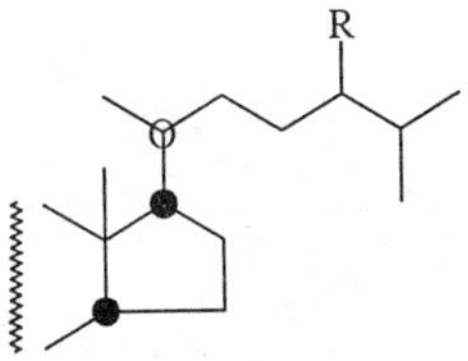

XVII 14β, 17β(20S)

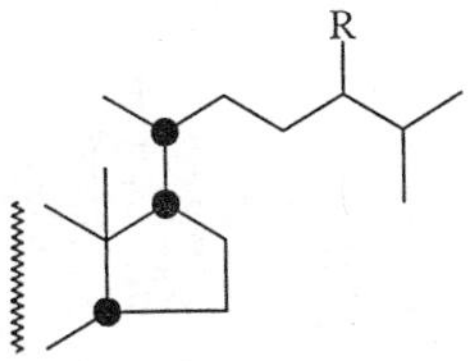

XVIII 14β, 17β(20R)

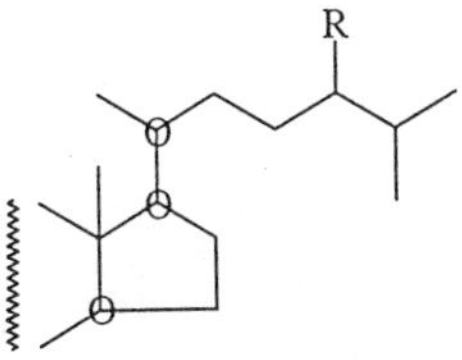

XIX 14α, 17α(20S)

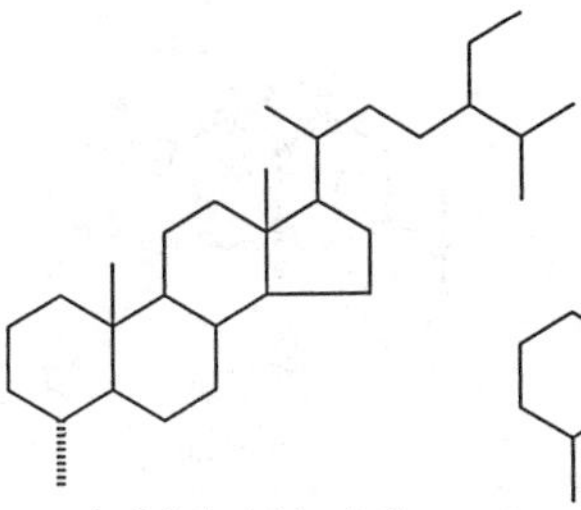

XX 4α-Methyl-24-ethylsterane

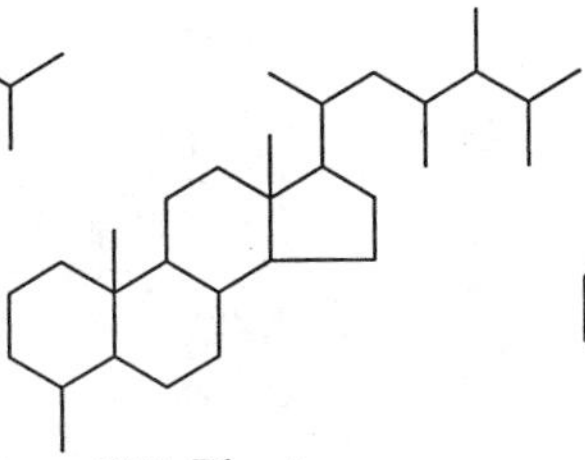

XXI Dinosterane

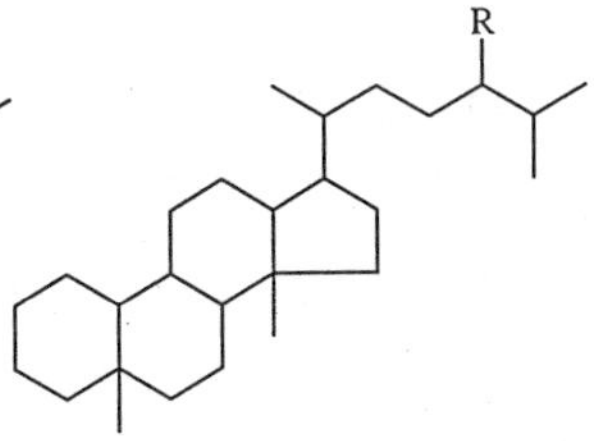

XXII Diasteranes

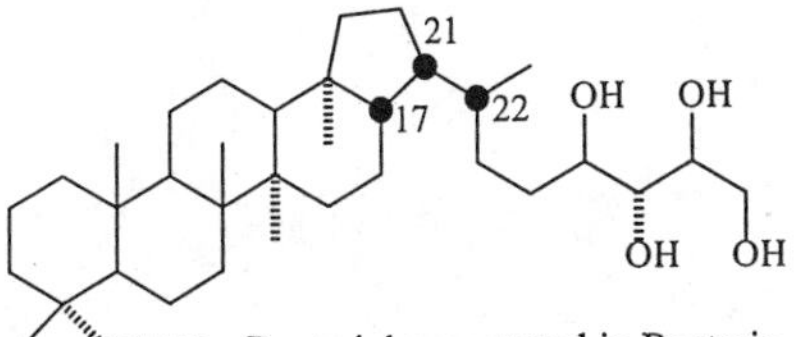

XXIII Bacteriohopanetetrol in Bacteria

Diagenesis →

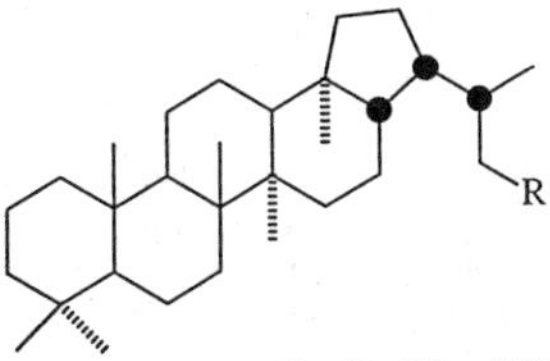

XXIV
Hopanoids in Sediment (R = H, CH_3 - C_6H_{10})

(Biological Configuration 17β ,21β(22R))

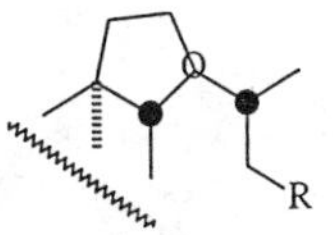

XXV 17β, 21α(22R)
Moretane

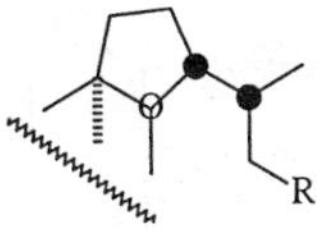

XXVI 17α, 21β(22R)

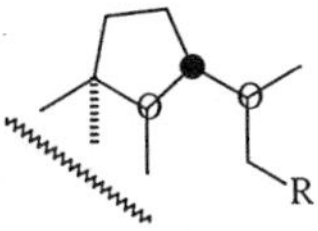

XXVII 17α, 21β(22S)

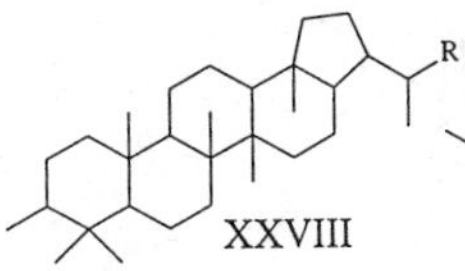

XXVIII
3-methylhopanes

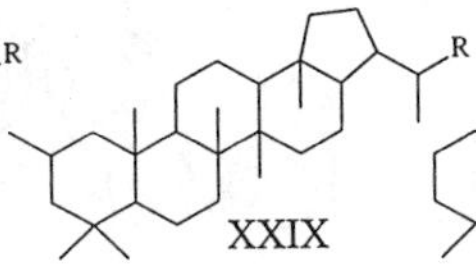

XXIX
2-methylhopanes

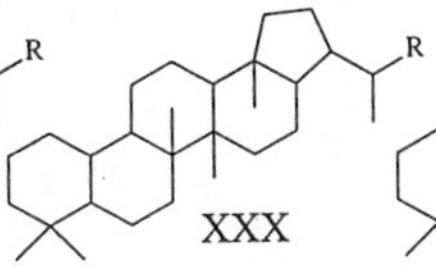

XXX
25-norhopanes

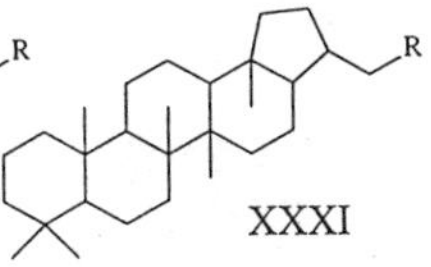

XXXI
30-norhopanes

Figure 1. *Continued.*

Continued on next page

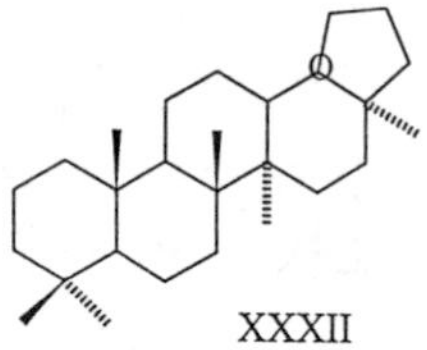

XXXII
18α(H)-22,29,30-trisnorneohopane
(Ts)

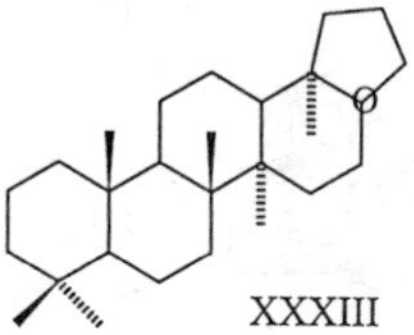

XXXIII
17α(H)-22,29,30-trisnorhopane
(Tm)

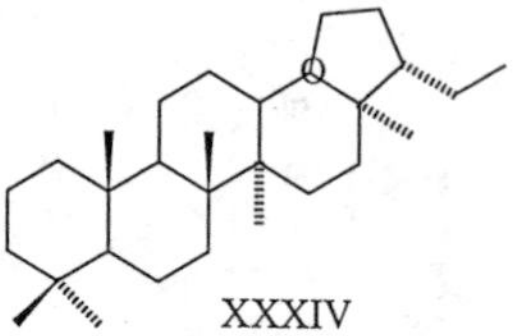

XXXIV
18α(H)-30-norneohopane
(C29 Ts)

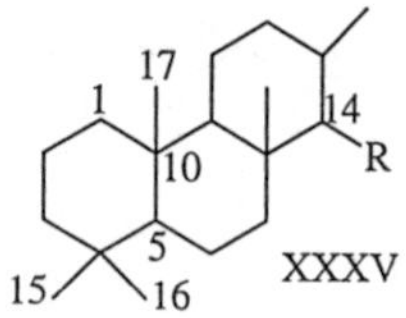

XXXV
Cheilanthanes
(R = H, CH3-C12H25)

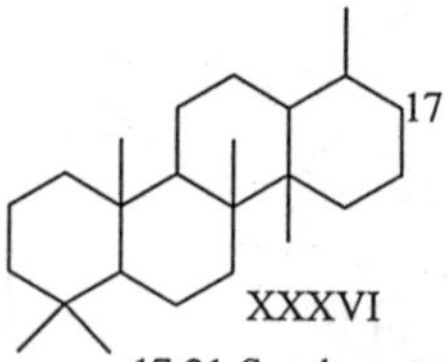

XXXVI
17,21-Secohopane

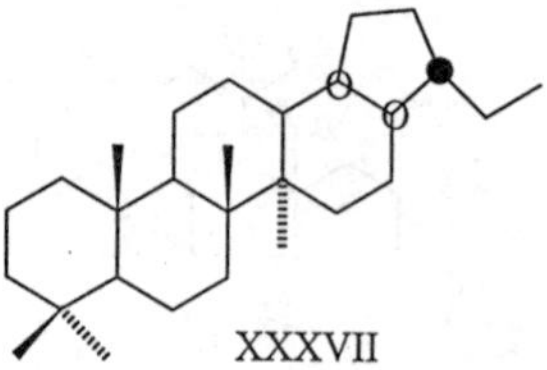

XXXVII
28,30-Bisnorhopane

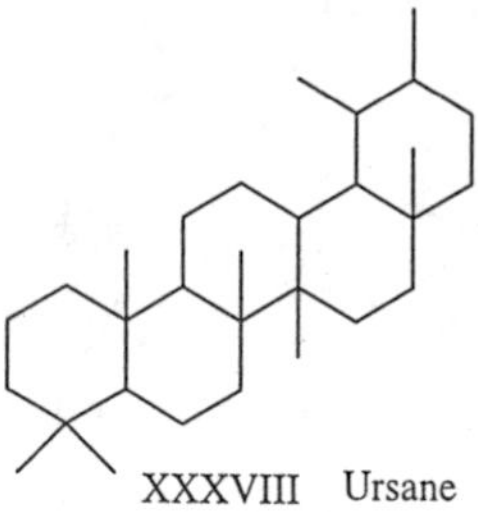

XXXVIII   Ursane

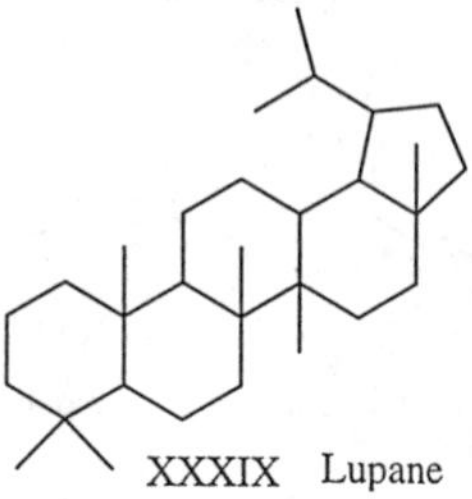

XXXIX   Lupane

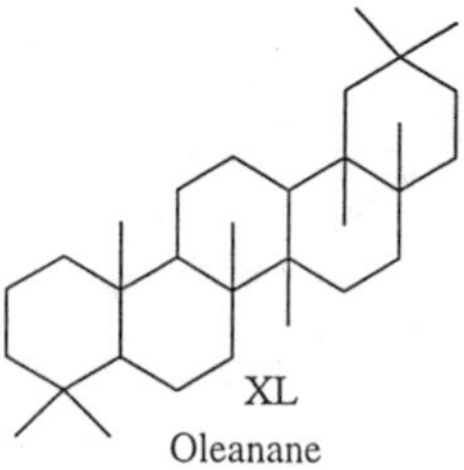

XL
Oleanane

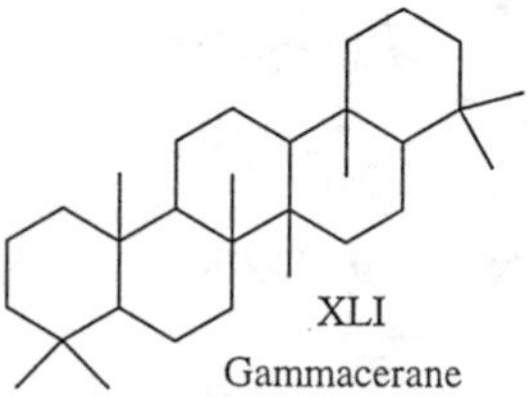

XLI
Gammacerane

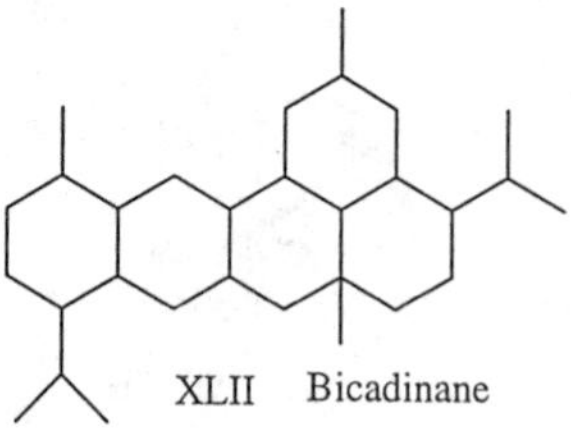

XLII   Bicadinane

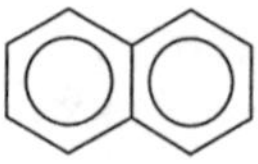

XLIII   Naphthalene

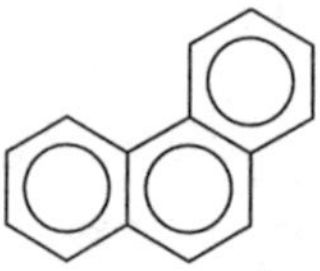

XLIV   Phenanthrene

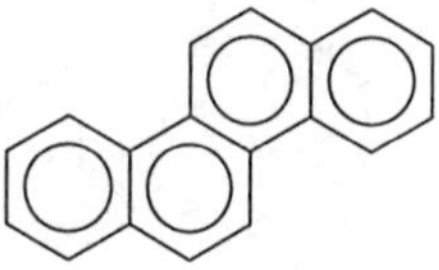

XLV   Chrysene

Figure 1. *Continued.*

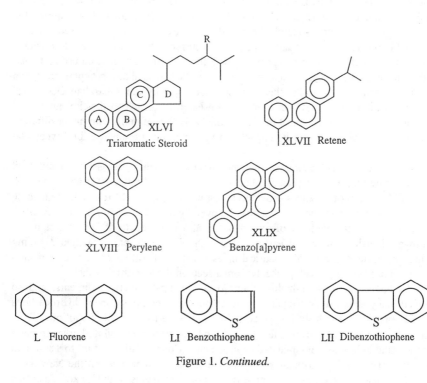

XLVI
Triaromatic Steroid

XLVII Retene

XLVIII Perylene

XLIX
Benzo[a]pyrene

L Fluorene

LI Benzothiophene

LII Dibenzothiophene

Figure 1. *Continued.*

analyses take longer and are more expensive than techniques such as UVF, but they are particularly useful where there is the possibility of multiple sources of hydrocarbons or when there is a legal requirement to prove that oils have the same source.

Sources of Petroleum Hydrocarbons in Marine Waters and Sediments

Hydrocarbons enter the sea from a large number of sources. Figures published by the U.S. National Research Council (*15*) in 1985 showed that, on a global scale, municipal and industrial inputs make up the largest contribution (31.1%), followed by tanker operations (22%), other transportation (12.5%), tanker accidents (12.5%), atmospheric fallout (9.4%), natural sources (7.8%) and refinery wastewater (3.1%). Offshore production contributes about 1.6% of the 3.2 million tons of oil entering the sea in an average year. It should be noted that these estimates are based on data collected in the 1970s, and an assessment of the current situation using modern data is long overdue. A database of compositions and distributions of petroleum products that contribute to coastal pollution needs to be established to facilitate comparisons when new spills occur. In Australia, this is being undertaken by the Australian Geological Survey Organisation (AGSO) in Canberra. Research is needed to develop integrated physical, biological and chemical models of the likely fate of petroleum under different environmental conditions and its impacts on the environment, aquatic food-webs and water quality.

Wilson *et al.* (*16*) estimated that the amount of hydrocarbons from natural oil seeps entering the sea each year was between 0.2×10^6 and 6.0×10^6 tons. The U.S. National Academy of Sciences provided an estimate of 0.6×10^6 tons although some observations imply that the amounts may be even higher. For example, Harvey *et al.* (*17*) observed a massive layer of oil-rich water at a water depth of 200 meters in the southwest North Atlantic and eastern Caribbean. These waters contained 3-12 mg oil/liter and extended over 800 nautical miles which implied that more than 1 million tons of oil had been released, probably from a seep on the Venezuelan shelf.

Complex mixtures of hydrocarbons occur in most aquatic sediments, and so detailed analysis is usually required. The amount and composition of hydrocarbons present provides an historical record of the severity of oil pollution. Total concentrations can range from a few µg/g dry wt. (ppm) in non-polluted areas to parts per thousand in heavily contaminated environments. Concentrations of up to 1800 ppm have been reported in petroleum-contaminated surface sediments from the New York Bight (*18*), and in excess of 5000 ppm in estuarine sediments near Hobart, Tasmania (*14*). Alkenes from microalgae (*19*) are often the most abundant single components in uncontaminated sedimentary hydrocarbon distributions. Some hydrocarbons are also produced from diagenetic transformation of functionalized lipids present in the sediment. For example, phytenes, hopenes, retene (XLVII) and sterenes are produced from bacterial and chemical degradation of naturally occurring phytol (IV), bacteriohopanetetrols (XXIII), resin acids and sterols (XV) respectively.

Examples of Commonly Used Biomarkers for Oil Spill Fingerprinting and Identifying Sources of Hydrocarbons

Although a wide range of hydrocarbons are available for oil spill fingerprinting, most attention has been given to some of the better-studied biomarkers used by petroleum geochemists which are readily analysed using selected ion monitoring GC-MS techniques. These include isoprenoid alkanes (m/z 113 and 183), steranes (m/z 217 and 218), 4-methylsteranes (m/z 231; XX, XXI), hopanes (m/z 191), 25-norhopanes (m/z 177; XXX) and 2-methylhopanes (m/z 205; XXIX). In many cases, the distributions can look quite similar, and the analyst must rely on small differences in the abundances of different components plus the presence or absence of specific compounds such as gammacerane (XLI), dinosterane (XXI), oleanane (XL), bicadinanes (XLII), botryococcane (II) and the like to establish whether the samples have the same origin or not. Considerable detail about these biomarkers is provided in the book by Peters and Moldowan (*1*), so only those features of most relevance to pollution studies are provided here. Note that biomarkers represent a very small proportion by weight of the oil and, thus, a range of compounds should be considered, not just a few biomarkers in isolation.

Steranes and Diasteranes. Steranes are widely used to fingerprint petroleum (*20*). These compounds are produced by a sequence of oxidation-reduction reactions in sediments from the sterols derived from living organisms (*e.g.* XV→XVI). These processes tend to leave the sterol-derived side-chain intact, and so different proportions of C_{27}, C_{28} and C_{29} constituents can be used to characterize the oil (*21,22*). Sterane mass spectra have major fragment ions at m/z 217 and 218 which are commonly used to characterize the sterane distribution. An example is shown in Figure 2. Steranes occur in a number of isomeric forms having different stereochemistries at positions 5, 14, 17, 20 and 24 (*cf.*, XVI-XIX). All but the latter isomers are separable on non-polar capillary columns so that sterane distributions are quite complex. The ratio of 5α(H),14α(H),17α(H) to 5α(H),14β(H),17β(H) steranes (ααα/αββ) is often used as an index of thermal maturity, but most crude oils have similar proportions of these isomers which limits the use of this ratio in environmental studies. A ratio of 20S and 20R isomers of about 1.0 is typical of mature crude oils. If the 20R isomer greatly predominates in a contemporary sediment sample, this could indicate the presence of hydrocarbons from eroded sediments which are thermally immature. A high abundance of C_{21} and C_{22} steranes relative to C_{27}-C_{29} steranes is typical of oils of high thermal maturity.

A predominance of C_{29} steranes, such as in Australia's Bass Strait oils (peaks 9-12; Figure 2), is usually associated with source rocks containing primarily higher plant organic matter although some exceptions are known. The predominance of C_{27} steranes (peaks 3,4,6,7) and only slightly lower abundances of C_{28} steranes and C_{29} steranes is typical of oils derived from marine algal source rocks such as those from the Middle East (*e.g.* Kuwait Crude; Figure 2). The distributions of 4-methylsteranes, as shown by mass chromatograms of the m/z 231 ion, can provide additional evidence for algal-derived organic matter since the presence of the four major isomers of dinosterane (XXI) as well as 4α-methyl-24-ethyl-C_{30} steranes (XX) is usually associated with marine depositional environments (*23*). Lacustrine sediments commonly, but not always, contain a much greater abundance of the latter 4α-methylsterane. Diasteranes (XXII) are found in most oils and are usually fingerprinted using the m/z 259 mass chromatogram although they are also readily discernible in m/z 217 mass chromatograms (peaks 1,2,5,8; Figure 2). A high diasterane abundance is

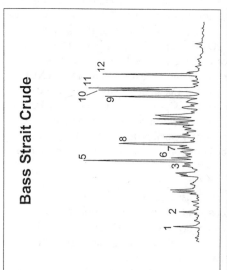

Kuwait Crude

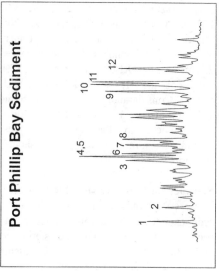

Bass Strait Crude

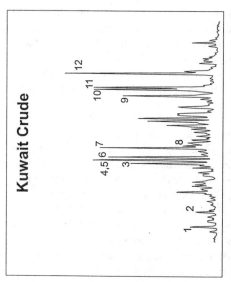

1:1 Mixture

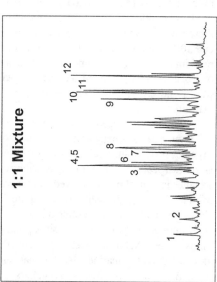

Port Phillip Bay Sediment

Figure 2. Mass chromatograms for m/z 217 showing the steranes and diasteranes in oils from Kuwait and Bass Strait, a 1:1 mixture (wt./wt.) of these two oils and a contaminated sediment from Port Phillip Bay, Victoria, Australia. The Kuwait and Bass Strait crude oils contain similar amounts of the C_{29} $5\alpha,14\alpha,17\alpha$-20S-24-ethylcholestane (peak **9**; 44 and 40 ppm respectively), but the Kuwait crude oil contains much less of the C_{29} 20S-diacholestane (peak **5**; 11 and 40 ppm respectively), so the C_{27}:C_{29} ratio of the steranes and diasteranes in the polluted sediment is very different from either oil, but clearly similar to a 1:1 mixture of the two oils. Peak identifications: **1** - C_{27} 20S-diacholestane (XXII); **2** - C_{27} 20R-diacholestane (XXII); **3** - 20S-5α,14α,17α-cholestane (XIX); **4** - 20R- 5α,14β,17β-cholestane (XVIII); **5** - C_{29} 20S-24-ethyldiacholestane (XXII); **6** -20S-5α,14β,17β-cholestane (XVII); **7** - 20R-5α,14α,17α-cholestane (XVI); **8** - C_{29} 20R-24-ethyldiacholestane (XXII); **9** - 20S-5α,14α,17α-24-ethyl-cholestane (XIX); **10** - 20R-5α,14β,17β-24-ethylcholestane (XVIII); **11** - 20S-5α,14β,17β-24-ethylcholestane (XVII); **12** - 20R-5α,14α,17α-24-ethylcholestane (XVI).

often associated with oils from clastic source rocks containing clays which catalyse the steroid backbone rearrangement. The Bass Strait oils provide a good example of this.

Figure 2 illustrates the use of sterane fingerprinting to ascertain the sources of hydrocarbon pollution in polluted sediments from Port Phillip Bay (Australia). These profiles are consistent with an approximately 1:1 mixture of the locally produced Bass Strait Crude oil from the offshore Gippsland Basin and an oil of Middle Eastern, carbonate origin such as those from Kuwait. The latter are used in Australia as a feedstock for lubricating oils because the Bass Strait Crudes are too waxy for this purpose. Of particular note is the presence in the sediment of a high proportion of C_{27} steranes (peaks 3, 4, 6 and 7) mainly from the Middle East oil in conjunction with high amounts of the C_{29} diasteranes (peaks 5 and 8) mainly from the Bass Strait crude oils. This example highlights the wide variation in absolute concentrations of particular biomarkers in different oil types which must be considered when determining the source of oil polution based on biomarker ratios.

Hopanes and Methylhopanes. Hopanes are found in almost all ancient sediments and crude oils (*20,24*). They are derived from oxygenated analogues such as the bacteriohopanetetrols (*e.g.* XXIII) found in most bacteria and cyanobacteria (*24*). Their distribution is usually recorded using a *m/z* 191 mass chromatogram (Figure 3). In mature samples, the 17α(H),21β(H)-isomers (XXVI, XXVII) greatly predominate over the 17β(H),21α(H) isomers (moretanes; XXV). However, the isomer commonly found in living organisms is 17β(H),21β(H) (*e.g.* XXIV), so that in polluted sediments all three isomers can be found as well as hopenes (containing double bonds at various positions in the ring system) derived from indigenous bacteria.

An unusually high proportion of the C_{29} hopane (XXXI, where R=CH$_3$) is often associated with oils derived from carbonate source rocks oils, which includes most of those from the Middle East (peak 3, Figure 3). These oils also show a slightly enhanced abundance of the C_{35} extended hopanes (peaks 17,18; Figure 3) compared with the C_{34} homohopanes (peaks 15, 16; Figure 3). The ratio of the two C_{27} hopanes, Ts (18α(H)-22,29,30-trisnorneohopane; XXXII) and Tm (17α(H)-22,29,30-trisnorhopane; XXXIII) can be a sensitive indicator of thermal maturity (*e.g. 55*) when comparing oil or sediment samples from the same source. The ratio varies considerably in oils and often provides a good parameter for environmental studies. For example, in Middle East oils the Ts/Tm ratio can vary from 0.4 in Kuwait crude oil to 0.5 in Arabian heavy crude oil to 1.5 in Arabian light crude oil. In contrast, the ratio for a typical Bass Strait oil is about 0.7. Some oils also contain significant amounts of the C_{29} analogue of Ts (*i.e.* 18α(H)-30-norneohopane; XXXIV) which elutes just after the C_{29} 17α(H),21β(H)-hopane (peaks 4 and 3, respectively; Figure 3).

Figure 3 shows a *m/z* 191 mass chromatogram obtained by GC-MS analysis of aliphatic hydrocarbons in the surface sediment from Port Phillip Bay, Victoria referred to above. Hopanes having the 17α(H),21β(H)-stereochemistry (*e.g.* XXVI) predominate indicating a substantial contribution from petroleum. Hopanes and hopenes derived from prokayotic organisms in the sediment having 17β(H),21β(H)-stereochemistry (XXIV) are barely detectable. In this case the absolute abundances of hopanes in the Kuwait crude oil is very much higher than in the clastic sediment-derived Bass Strait oil so that even with a 1:1 mixture, the contribution from the latter to the hopane distribution is barely discernible. The simulated mixing shown here indicates but one of many possible mixtures. The actual pattern obtained will depend on the mixing ratio and on the type of Bass Strait and Middle Eastern crudes contributing. For example, not all Middle Eastern crude oils have the high C_{29} 30-norhopane abundance of the Kuwait Crude. Furthermore, some crude oils from individual fields in the Bass Strait area have C_{29}/C_{30} hopane ratios nearer to unity.

Another hopane which has proven useful in oil spill fingerprinting is the C_{28} hopane $17\alpha(H),18\alpha(H),21\beta(H)$-28,30-bisnorhopane (also referred to as dinorhopane; XXXVII) which is particularly abundant in some Californian oils derived from the Miocene Monterey formation. The presence of this hopane in contaminated sediments from Prince William Sound showed that Monterey Formation oils imported into the region prior to the development of the Alaskan North Slope oils were sources of pollution before the *Exxon Valdez* oil spill (5).

2- and 3-Methylhopanes (XXIX, XXVIII). The 2- and 3-methylhopanes are fingerprinted using their major fragment ion at *m/z* 205. Carbonate-derived oils contain significant amounts of 2-methylhopanes, although they are not restricted to this source facies (20,25). Methylhopanes are not common in Australian crude oils. Australia produces most of its lubricating oils from Middle East crudes, and the distributions of hopanes and methylhopanes seen in many Australian estuarine sediments reflect pollution from these lubricating oils (4,14,26). In most instances, the main source of pollution seems to be lubricating oils from road run-off, spillages and illegal dumping.

25-Norhopanes and 30-Norhopanes (XXX, XXXI). 25-Norhopanes are sometimes termed 'demethylated hopanes' because they lack the methyl group at C-10 of conventional hopanes, but otherwise they have the same structure. They are readily fingerprinted using a *m/z* 177 mass chromatogram. 25-Norhopanes are abundant in some biodegraded oils, and many authors now assert that when present as the complete $C_{26}-C_{34}$ series, they are formed by bacterial transformation of hopanes during severe biodegradation of the crude oil in the reservoir (27-29). There is no evidence, as yet, that they are formed in the aquatic environment from biodegradation of spilled oil. A series of 30-norhopanes spanning the carbon number range $C_{28}-C_{34}$ is commonly found in carbonate-derived oils. These compounds are resistant to biodegradation, and, thus, their abundance relative to hopanes is often enhanced in severely biodegraded crude oils. These compounds are readily discerned in the Middle East oil (*e.g.* peak 7; Figure 3) and in the polluted sediment, but the C_{30} 30-norhopane is almost undetectable in the Bass Strait oil.

Tricyclic Alkanes. Tricyclic alkanes fit into several structural categories. One widely occurring group in crude oils is the extended cheilanthanes which have a long isoprenoid chain attached to the C ring of the tricyclic system (XXXV; Figure 1). This gives rise to a carbon number distribution from C_{19} to C_{30} (or higher) which is readily determined from the *m/z* 191 mass chromatogram used to fingerprint hopanes. Several isomers occur in immature sediments, but in crude oils the $13\beta,14\alpha$-isomers predominate with 22R and 22S doublets occurring for higher homologs. The abundance of extended tricyclic alkanes in Tasmanite oil shale led several authors to

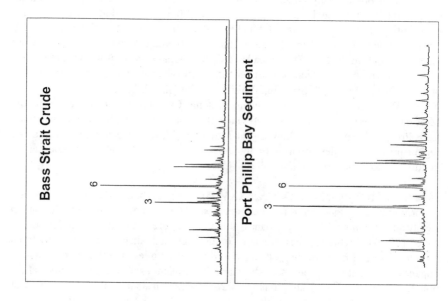

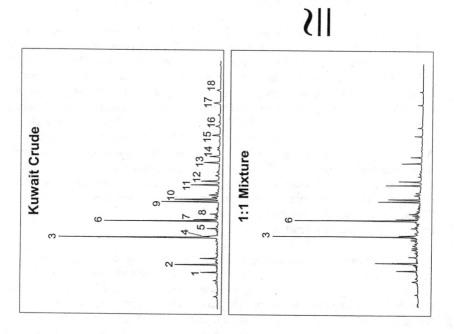

Figure 3. Mass chromatograms for m/z 191 showing the distribution of hopanes in oils from Kuwait and Bass Strait, a 1:1 mixture (wt./wt.) of these two oils and a contaminated sediment from Port Phillip Bay, Victoria, Australia. The absolute abundances of hopanes are much lower in the Bass Strait oil (78 ppm of the C_{29} hopane and 121 ppm of the C_{30} hopane) than in the Kuwait crude (605 ppm of the C_{29} hopane and 391 ppm of the C_{30} hopane), and, thus, its contribution to the sediment hopane distribution is substantially masked by hopanes from the Middle East oil. Peak identifications: **1** - Ts C_{27} 18α(H)-22,29,30-trisnorneohopane (XXXII); **2** - Tm C_{27} 17α(H)-22,29,30-trisnorhopane (XXXIII); **3** - C_{29} 17α(H),21β(H)-30-norhopane (XXXI, stereochemistry as in XXVI); **4** - C_{29}Ts 18α(H)-30-norneohopane (XXXIV); **5** - C_{29} 17β(H),21α(H)-30-norhopane (normoretane; XXV) ; **6** - C_{30} 17α(H),21β(H)-hopane (XXVI); **7** - C_{30} 17α(H)-30-nor-29-homohopane; **8** - C_{30} 17β(H),21α(H)-hopane (moretane; XXV); **9** - **18**; C_{31} - C_{35} 22S (XXVII) and 22R (XXVI) epimers of extended 17α(H),21β(H)-hopanes.

suggest an algal origin (*30*), but recent work (*31*) based on stable carbon isotope signatures suggests a bacterial source. Their abundance relative to hopanes is quite variable, a feature which can provide a convenient fingerprint in environmental studies. For example, differences in the abundance of two C_{26} tricyclic alkane isomers relative to the C_{24} tetracyclic terpane were used to identify oils derived from the *Exxon Valdez* spill (*5*).

Another class of tricyclic alkanes includes the diterpenoids derived from higher plants. Examples include pimarane (XI), isopimarane and abietane (XII) skeletons derived from land plants (*32*). Fichtelite (XIII), retene (XLVII) and iosene are well known constituents of lignites. Such compounds have yet to be used in environmental studies.

Bicadinanes. Bicadinanes (XLII) were first detected in Far East oils (*33*) and are typically labelled W, T and R in m/z 191 (Figure 4) and m/z 369 mass chromatograms. This shorthand nomenclature is still in use with new isomers denoted as T' or T1 etc. These compounds occur infrequently in oils and are thought to originate from cracking of polycadinene polymers derived from Angiosperm plant resins of the Dipterocarpaceae family. They are significant constituents in oils from Malaysia, Bangladesh and Indonesia. Their presence in the tar balls collected on northern, western and southern coastlines of Australia (Figure 4) provided compelling evidence that they must have been derived from oil seeps to the north (*34*) since these plants have not been found in Australia. Note, in this example, the use of m/z 412 (molecular ion for bicadinanes and C_{30} triterpanes) to confirm the identification of these compounds. These tar balls can be further divided into categories according to differences in biomarker (*35-37*) and stable carbon isotope characteristics (*38*). Key parameters were pristane/phytane ratios and the presence of 4-methylsteranes having 24-ethyl side-chain substitution common to lacustrine oils, oleanane (XL) from higher plant lipids, the unusual isoprenoid alkane botryococcane (II) derived from microalgae and bicadinanes (XLII) from plant resins. Very few of these tar balls are derived from oil spills.

Aromatic and Heteroaromatic Hydrocarbons. Aromatic hydrocarbons can be very useful for fingerprinting crude oils, although the high water solubility and volatility of the lower molecular weight benzenes, naphthalenes and phenanthrenes can lead to major changes in the composition of spilled crude oils in a matter of days. The distribution of isomers found in a crude oil is largely determined by thermal maturity, and, thus, differences between distributions can be quite minor.

Linear alkylbenzenes (LABs) are common constituents of sewage and wastewaters (*39*), largely due to their presence in widely-used anionic surfactants based on linear alkylbenzenesulfonates. Studies of urban-derived pollution have tended to concentrate on 4-6 ring polycyclic aromatic hydrocarbons since these petroleum constituents are well documented carcinogens (*40*). The distributions of polycyclic aromatic hydrocarbons (PAH) from pyrolytic sources tend to be dominated by high molecular weight non-alkylated components such as chrysene, benzo[*a*]pyrene (XLIX) etc. Some PAH such as retene (XLVII) from resin acids and perylene (which appears to be formed in sediments; XLVIII) also have natural sources.

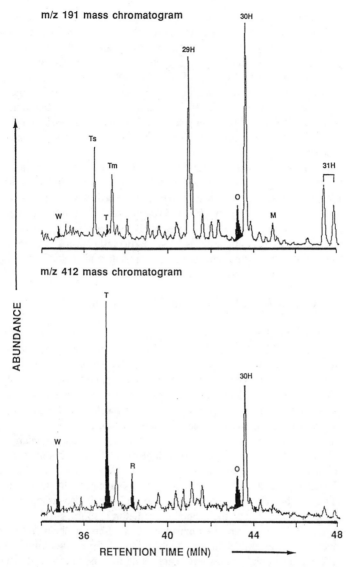

Figure 4. Mass chromatograms for m/z 191 and 412 (molecular ion) showing the distribution of hopanes in a tar ball collected from Lancelin, Western Australia. Bicadinanes (XLII) are labelled W, T and R; oleanane (XL) is labelled O; 29H refers to the C_{29} $17\alpha(H),21\beta(H)$-30-norhopane (XXXI), 30H to the C_{30} $17\alpha(H),21\beta(H)$-hopane (XXVI); and 31H to 22S (XXVII) and 22R (XXVI) epimers of C_{31} $17\alpha(H),21\beta(H)$-homohopanes; M refers to the C_{30} $17\beta(H),21\alpha(H)$-hopane (moretane; XXV). The occurrence of similar distributions in some Southeast Asian oils confirmed that these bitumens originated from oil seeps to the north of Australia. (Reproduced with permission from ref. 34. Copyright 1992 Elsevier Science Ltd.)

At least seven types of aromatic steroid hydrocarbons are commonly found in crude oils (41). Triaromatic steroid hydrocarbons (e.g. XLVI) with one and two methyl groups in the ABC ring system are usually abundant and readily detected by mass chromatograms for m/z 231 and 245 respectively. Mass chromatograms for m/z 231 were used to trace aromatics from the Exxon Valdez oil spill (42), but such chromatograms often show only small variations between different oils, and, thus, we suggest that they be used with caution in environmental studies. Small amounts of aromatic steroids having an m/z 217 base peak can interfere with peak assignments in sterane mass chromatograms if aromatic and aliphatic hydrocarbons are not separated before GC-MS analysis.

Organic sulfur compounds such as benzothiophene (LI), dibenzothiophene (LII) and their alkylated homologs are common constituents of sulfur-rich oils. They have been found in urban runoff and sewage discharges (39), and they were particularly useful in the Exxon Valdez oil spill study (5,42). They are readily monitored in SIM GC-MS using m/z 184+14n ions (where n = number of alkyl substituents). The ratio of alkylphenanthrenes to alkyldibenzothiophenes was particularly useful because it remained relatively constant with time after the spill, implying similar solubilities and degradation rates for the two compound classes.

Other parameters available for identifying hydrocarbon sources

n-**Alkane Distributions.** Straight-chain alkanes are not biomarkers *sensu stricto* since their structures contain no distinguishing features (other than chain-length) and they can be derived from multiple sources. However, the overall distribution can provide information about likely sources. With few exceptions, oils contain distributions of *n*-alkanes showing little odd or even chain length predominance. Lacustrine oils tend to have a higher proportion of waxy long-chain C_{25}-C_{35} *n*-alkanes, although higher proportions of shorter chain alkanes are formed with higher maturity.

n-Alkanes with a chain-length range of C_{15} to C_{35} are present in nearly all contemporary sediments. Plant wax inputs to sediments of hydrocarbons can be considerable, particularly in environments close to land. These have an odd-even predominance of 8-10 so that a range of odd-even predominances can be found in sediments depending on the relative weightings of the different hydrocarbon inputs. There is also some evidence that there might be natural sources of *n*-alkane distributions showing no odd or even predominance (4,43). Shorter-chain *n*-alkanes such as *n*-C_{15} and *n*-C_{17} can be abundant due to inputs from microalgae and cyanobacteria. *n*-Alkenes from biological sources are found in most marine sediments, and the polyunsaturated alkene *n*-$C_{21:6}$ (heneicosahexaene) usually predominates since it is abundant in many microalgae. However, alkenes are rapidly decomposed in the presence of oxygen, so high concentrations of these labile compounds indicate the presence of living or intact algal cells in the sample. The usefulness of *n*-alkanes as source indicators has improved with the introduction of compound-specific isotope ratio monitoring techniques (e.g. 38) which provide stable carbon isotope data for individual hydrocarbons.

Isoprenoid Hydrocarbons. The isoprenoid alkanes, pristane (C_{19}; VI) and phytane (C_{20}; V) are common constituents of most sediments, but their relative abundances vary greatly. They are present in most petroleums, usually as major constituents of a much wider distribution of isoprenoid alkanes (*e.g. 44*). The ratio of pristane to phytane can be used to fingerprint the oil and, in favourable cases, can be used to identify the source of an oil spillage. For example, a pristane/phytane ratio of about 1 is typical of oils derived from marine organic matter deposited under reducing conditions (*45*). Natural sources exist for both pristane and phytane, so the presence of these compounds does not provide unequivocal evidence for oil contamination. Other naturally-occurring isoprenoid hydrocarbons found in sediments include the ubiquitous squalene and 2,6,10,15,19-pentamethyleicosane (*46*), as well as lycopane (III) which occurs mainly in anoxic environments (*47*). Highly branched isoprenoid alkenes from diatoms can be the dominant biogenic hydrocarbons present in many marine sediments (*e.g. 19,48*).

Unresolved Complex Mixture (UCM). Perhaps one of the more convincing indications of petroleum contamination in a water or sediment sample is the presence of an unresolved complex mixture (UCM) of hydrocarbons which shows up as a baseline rise in the gas chromatogram of total hydrocarbons. Using chemical degradation techniques, Gough and Rowland (*49*) showed that the UCM consists primarily of linear chains connected at branch points giving rise to "T-shaped" molecules. The UCM can provide limited fingerprinting information, especially for oil-oil correlation of biodegraded oils (*50*). Some authors have suggested that some UCMs may arise from recent biotic sources, but we have no data to support this view. Volkman *et al.* (*46*) observed a narrow range UCM in hydrocarbon fractions isolated from a pristine Antarctic lake. However, this was shown to derive from organic sulfur compounds present in the hydrocarbon fraction. The presence of a UCM and petroleum biomarkers can also indicate hydrocarbons from the weathering of ancient rocks (*51*).

Hydrocarbons present in the UCM are difficult to degrade (*52*), and, thus, gas chromatograms of biodegraded crude oils often show a pronounced UCM. However, this may be difficult to discern in chromatograms of condensates or highly paraffinic oils unless the vertical scale is expanded so that the *n*-alkane peaks are greatly off-scale. This is an important consideration when analysing environmental samples thought to be contaminated with light oil. Accurate measurement of the amount of hydrocarbons represented by the UCM is a problem, especially when high boiling lubricating oils are involved where the baseline often does not return to zero at the maximum GC oven temperature. In such cases, TLC-FID provides an alternative measure of total hydrocarbons. Non-resolvable hydrocarbons often dominate in municipal wastewaters since lubricating oils are a major contributor (*39*).

Bicyclic Alkanes. Bicyclic alkanes are found in the lower molecular weight fraction of crude oils and are usually characterized by mass chromatograms of the base peak at *m/z* 123 and molecular ions (*e.g. m/z* 208 for drimanes [$C_{15}H_{28}$; VII] and *m/z* 222 for homodrimanes [VIII]). Compounds having the drimane skeleton occur widely in sediments and oils (*53*), but variations in their distribution are too limited to be of much value for environmental studies. However, some other bicyclic alkanes based on the labdane skeleton (XIV) might be useful markers for fingerprinting in certain cases (*e.g. 32,54*).

Tetracyclic Alkanes. Several series are known, but these tend to be relatively minor constituents of most oils. Several diterpenoids based on the kaurane (X) and phyllocladane (IX) skeletons are found in Southern hemisphere oils derived from higher plant organic matter and probably originate from unsaturated analogues found in conifers (*32*). Tetracyclic terpanes of the 17,21-secohopane series (*e.g.* XXXVI) occur in most oils and typically encompass the C_{24} to C_{27} carbon number range with C_{24} usually predominant. The latter appears to be a marker for oils derived from carbonate and evaporitic source rock environments (*1*).

Pentacyclic Alkanes. Oleanane (XL), ursane (XXXVIII) and lupane (XXXIX) are triterpanes commonly associated with oils derived predominantly from higher plant organic matter. They are particularly abundant in some terrestrial oils from Tertiary source rocks such as those from China and Nigeria (*55*). These compounds appear as single peaks in *m/z* 191 mass chromatograms and only rarely are more abundant than the hopanes. Oleanane is readily recognized as a peak eluting before the C_{30} 17α(H),21β(H)-hopane on non-polar capillary columns (peaks 'O' and 30H, respectively; Figure 4). However, unlike the hopanes, oleanane shows no corresponding peak in the m/z 369 mass chromatogram.

Constraints on the Use of Biomarkers for Identifying Hydrocarbon Sources

Effects of Physical Weathering. When petroleum products are released into the environment, significant compositional changes occur due to water washing, evaporation, chemical oxidation and biodegradation. Short-chain alkanes and simple aromatics are rapidly lost, but sterane and hopane distributions are amongst the last compounds to be affected. Consequently, these polycyclic alkanes are particularly useful in correlation studies. It is commonly assumed that all volatile petroleum products are rapidly lost to evaporation. However, uptake by the biota can be important (*e.g 3,56,57*). At high concentrations, these low molecular weight constituents can be toxic to aquatic organisms, and they taint the flesh making it unfit for human consumption.

Effects of Biodegradation. Most of the constituents of crude oils can be biodegraded by bacteria, although the rates vary greatly. These compounds are either converted to new compounds which are usually oxygenated and, hence, more polar, or completely remineralized to CO_2. Bacteria tend to utilize the simpler molecules at faster rates so that one sees a progressive change in composition with increasing extent of biodegradation. The effects of biodegradation on the composition of a typical paraffinic crude oil were summarized by Volkman *et al.* (*27,58*), who studied a suite of oils that had been biodegraded in subsurface reservoirs. Similar results have been obtained for oil spilled in tropical estuarine ecosystems (*59*). The rates of biodegradation generally follow the order *n*-alkanes > alkylcyclohexanes > acyclic isoprenoid alkanes such as pristane and phytane > bicyclic alkanes > steranes > hopanes > diasteranes, although some variations are seen in different environments. Further refinement of this scheme was recently presented by Fisher *et al.* (*60*) who studied the fate of a light oil leaking into a tropical mangrove ecosystem.

The ratio of n-C_{17} alkane to pristane decreases markedly in the early stages of biodegradation, and, hence, it serves as a useful index of the extent of biodegradation. The relative resistance of steranes and hopanes to biodegradation has prompted the recent use of the concentration of the C_{30} 17α(H),21β(H)-hopane relative to other hydrocarbons as an index of biodegradation as demonstrated by the *Exxon Valdez* post-spill study (*61*). In this oil spill, n-alkanes and simple isoprenoids in surface tars were completely degraded within 2-3 years (*5*).

Aromatic hydrocarbons are also biodegraded by bacteria (*58*), either by cleavage of the aromatic ring or by oxidation of the alkyl substituents (*62*). Fission of the aromatic ring system produces smaller compounds such as pyruvate which can be used as an energy source for bacterial growth (*63*). Low molecular weight aromatics such as alkylbenzenes are rapidly degraded followed by simple alkylated naphthalenes and phenanthrenes. The rate of biodegradation under oxic conditions decreases with increasing number of aromatic rings and with increasing degree of alkylation. It is also influenced by the positions of the alkyl substituents (*e.g. 58,64,65*). Such biogeochemical processes can lead to rapid and dramatic changes to the composition of spilled oils. A good illustration of this was the study by Eganhouse *et al.* (*66*) of an aquifer contaminated by oil from a subsurface pipeline break. However, polycyclic aromatic hydrocarbons (PAH) with three or more aromatic rings are more resistant to biodegradation, and the metabolites formed from them can be more toxic than the parent PAH (*63*). In the *Exxon Valdez* spill, naphthalenes were rapidly depleted followed by the fluorenes whereas the relative abundance of chrysene increased relative to other PAH (*5*).

New Biomarkers and Stable Carbon Isotopes. From on-going studies of crude oils, sediments and present-day microalgae and bacteria, many new biomarkers are being identified which continue to improve our understanding of the origins of the organic matter deposited in ancient sediments. Once the origins of these compounds have been validated, they may also prove useful for fingerprinting spilled oil particularly when combined with new GC-isotope ratio MS methods for determining stable carbon isotope signatures of individual compounds separated by capillary gas chromatography (*38,67*). Hydrocarbons from different algal, bacterial or plant sources have characteristic $^{13}C/^{12}C$ ratios. Thus, source assignments can be made with more confidence than would be possible based solely on chemical structure data.

Acknowledgments

We thank Daniel Holdsworth for laboratory assistance and R. Alexander, R. Kagi, D. McKirdy, R. Summons, J. Farrington, S. Rowland and K. Burns for many stimulating discussions over many years. Rick Requejo and Marlon Kennicutt II provided helpful comments on an earlier draft of this manuscript. Special thanks to Bob Eganhouse for organising the symposium and for his unstinting efforts to produce a high quality proceedings volume. A. Murray publishes with the permission of the Director, AGSO.

Literature Cited

1. Peters, K. E. and Moldowan, J. M. *The Biomarker Guide. Interpreting Molecular Fossils in Petroleum and Ancient Sediments*; Prentice Hall: New Jersey, **1993**; 363 pp.
2. Johns, R. B. (Ed.) *Biological Markers in the Sedimentary Record*. Elsevier: Amsterdam, **1986**; 364 pp.
3. Rowland, S. J. and Volkman, J. K. *Mar. Environ. Res.* **1982**, *7*, 117-130.
4. Volkman, J. K., Holdsworth, D. G., Neill, G. P. and Bavor, H. J. Jr. *Sci. Tot. Env.* **1992**, *112*, 203-219.
5. Bence, A. E., Kvenvolden, K. A. and Kennicutt, M. C. II. *Org. Geochem.* **1996**, *24*, 7-42.
6. Tissot, B. P. and Welte, D. H. *Petroleum Formation and Occurrence*. Second Edition; Springer-Verlag: Berlin, **1984**; 699 pp.
7. Engel, M. H. and Macko, S. A. *Organic Geochemistry. Principles and Applications*. Plenum Press: New York, **1993**; 861 pp.
8. Adlard, E. R. *J. Inst. Petrol.* **1982**, *58*, 63-74.
9. Karlsen, D. A., and Larter S. R. *Org. Geochem.* **1991**, *17*, 603-617.
10. Volkman, J. K. and Nichols, P. D. *J. Planar Chromatogr.* **1991**, *4*, 19-26.
11. Mason, R. P. *Mar. Pollut. Bull.* **1987**, *18*, 528-533.
12. Albaigés, J. and Albrecht, P. *Int. J. Environ. Anal. Chem.* **1979**, *6*, 171-190.
13. Jones, D. M., Rowland, S. J. and Douglas, A. G. *Mar. Pollut. Bull.* **1986**, *17*, 24-27.
14. Volkman, J. K., Rogers, G. I., Blackman, A. J. and Neill, G. P. In *AMSA Silver Jubilee Commemorative Volume*; Wavelength Press: Chippendale, NSW, **1988**; pp. 82-86.
15. NRC. *Oil in the Sea, Inputs, Fates and Effects*. National Academy Press: Washington, **1985**.
16. Wilson, R. D., Monaghan, P. H., Osanik, A., Price, L. C. and Rogers, M. A. *Science* **1974**, *184*, 857-865.
17. Harvey, G. R., Requejo, A. G., McGillivary, P. A. and Tokar, J. M. *Science* **1979**, *205*, 999-1001.
18. Farrington, J. W. and Tripp, B. W. *Geochim. Cosmochim. Acta* **1977**, *41*, 1627-1641.
19. Volkman, J. K., Barrett, S. M. and Dunstan, G. A. *Org. Geochem.* **1994**, *21*, 407-413.
20. Seifert, W. K. and Moldowan, J. M. In *Biological Markers in the Sedimentary Record*; Johns R. B., Ed.; Elsevier: Amsterdam, **1986**; pp. 261-290.
21. Huang, W.-Y. and Meinschein, W. G. *Geochim. Cosmochim. Acta* **1976**, *40*, 323-330.
22. Volkman, J. K. *Org. Geochem.* **1986**, *9*, 83-99.
23. Summons, R. E., Volkman, J. K. and Boreham, C. J. *Geochim. Cosmochim. Acta* **1987**, *51*, 3075-3082.

24. Ensminger, A., van Dorsselaer, A., Spykerelle, C., Albrecht, P. and Ourisson, G. In *Advances in Organic Geochemistry 1973*; Tissot, B. and Bienner, F., Eds.; Editions Technip: Paris, **1984**; pp. 245-260.
25. Summons, R. E. and Jahnke, L. L. In *Biological Markers in Sediments and Petroleum*; Moldowan, J. M., Albrecht, P. and Philp, R. P. Eds.; Elsevier: Amsterdam, **1992**; pp. 182-200.
26. Volkman, J. K., Miller, G. J., Revill, A. T. and Connell, D. W. In *Environmental Implications of Offshore Oil and Gas Development in Australia   The Findings of an Independent Scientific Review*; Swan, J. M., Neff, J. and Young, P. C., Eds.; Australian Petroleum Exploration Association: Sydney, **1994**; pp. 509-695.
27. Volkman, J. K., Alexander, R., Kagi, R. I. and Woodhouse, G. W. *Geochim. Cosmochim. Acta* **1983**, *47*, 785-794.
28. Peters, K. E. and Moldowan, J. M. *Org. Geochem.* **1991**, *17*, 47-61.
29. Peters, K. E., Moldowan, J. M., McCaffrey, M. A. and Fago, F. J. *Org. Geochem.* **1996**, *24*, 765-783.
30. Simoneit, B. R. T., Leif, R. N., Aquino Neto, F. R., Azevedo, D. A., Pinto, A. C., and Albrecht, P. *Naturwissenschaften*, **1990**, *77*, 380-383.
31. Revill, A. T., Volkman, J. K., O'Leary, T., Summons, R. E., Boreham, C. J., Banks, M. R. and Denwer, K. *Geochim. Cosmochim. Acta.* **1994**, *58*, 3803-3822.
32. Alexander, R., Noble, R. A. and Kagi, R. I. *APEA J.* **1987**, *27*, 63-72.
33. Grantham, P. J., Posthuma, J., and Baak, A. In *Advances in Organic Geochemistry, 1981*; Bjorøy, M. *et al.*, Eds.; John Wiley: Chichester, **1983**; pp. 675-683.
34. Currie, T. J., Alexander, R. and Kagi, R. I. *Org. Geochem.* **1992**, *18*, 595-601
35. McKirdy, D. M., Cox, R. E., Volkman, J. K. and Howell, V. J. *Nature* **1986**, *320*, 57-59.
36. McKirdy, D. M., Summons, R. E., Padley, D., Serafini, K. M., Boreham, C. J. and Struckmeyer, H. I. M. *Org. Geochem.* **1994**, *21*, 265-286.
37. Volkman, J. K., O'Leary, T., Summons, R. E. and Bendall, M. R. *Org. Geochem.* **1992**, *18*, 669-682.
38. Dowling, L. M., Boreham, C. J., Hope, J. M., Murray, A. P. and Summons, R. E. *Org. Geochem.* **1995**, *23*, 729-737.
39. Eganhouse, R. P. and Kaplan I. R. *Environ. Sci. Technol.* **1982**, *16*, 541-551.
40. Wakeham, S. G., Schaffner, C., and Giger, W. *Geochim. Cosmochim. Acta*, **1980**, *44*, 403-413.
41. Mackenzie, A. S., Hoffmann, C. F. and Maxwell, J. R. *Geochim. Cosmochim. Acta* **1981**, *45*, 1345-1355.
42. Hostettler, F. D. and Kvenvolden, K. A. *Org. Geochem.* **1994**, *21*, 927-936.
43. Eganhouse, R. P., Simoneit, B. R. T. and Kaplan I. R. *Environ. Sci. Technol.* **1981**, *15*, 315-326.
44. Volkman, J. K. and Maxwell, J. R. In *Biological Markers in the Sedimentary Record, Methods in Geochemistry and Geophysics, 24*, Johns, R. B., Ed.; Elsevier: Amsterdam, **1986**; pp. 1-42.
45. Didyk, B. M., Simoneit, B. R. T., Brassell, S. C. and Eglinton, G. *Nature* **1978**, *272*, 216-222.

46. Volkman, J. K., Allen, D. I., Stevenson, P. L. and Burton, H. R. In *Advances in Organic Geochemistry 1985*; Leythaeuser, D. and Rullkötter, J., Eds.; Pergamon Press: Oxford, **1986**; pp. 671-681.
47. Wakeham, S. G., Freeman, K. H., Pease, T. K., and Hayes, J.M. *Geochim. Cosmochim. Acta* **1993**, *57*, 159-165.
48. Volkman, J. K., Farrington, J. W., Gagosian, R. B. *Org. Geochem.* **1987**, *11*, 463-467.
49. Gough, M. A. and Rowland, S. J. *Nature* **1990**, *344*, 648-650.
50. Revill, A. T., Carr, M. R. and Rowland, S. J. *J. Chromatogr.* **1992**, *589*, 281-286.
51. Rowland, S. J. and Maxwell, J. R. *Geochim. Cosmochim. Acta* **1984**, *48*, 617-624.
52. Gough, M. A., Rhead, M. M. and Rowland, S. J. *Org. Geochem.* **1992**, *18*, 17-22.
53. Alexander, R., Kagi, R. I., Noble, R. and Volkman, J. K. *Org. Geochem.* **1984**, *6*, 63-70.
54. Alexander, R., Larcher, A. V., Kagi, R. I. and Price, P. L. *APEA J.* **1988**, *28*, 310-324.
55. Ekweozor, C. M., Okogun, J. I., Ekong, D. E. U., and Maxwell, J. R. *Chem. Geol.* **1979**, *27*, 11-28.
56. Connell, D. W. *Bull. Environ. Contam. Toxicol.* **1978**, *20*, 492-498.
57. Hellou, J. and Warren, W. G. *Mar. Env. Res.* **1997**, *43*, 11-25.
58. Volkman, J. K., Alexander, R., Kagi, R. I., Rowland, S. J. and Sheppard, P. N. In *Advances in Organic Geochemistry 1983*; Schenck, P. A.; de Leeuw, J. W. and Lijmbach, G. W. M., Eds.; Pergamon Press: Oxford, **1984**; pp. 619-632.
59. Oudot, J. and Dutrieux, E. *Mar. Env. Res.* **1989**, *22*, 91-106.
60. Fisher, S. J., Alexander, R. and Kagi, R. I. , In. *ACS Division of Environmental Chemistry, Preprints of papers presented at the 212th ACS National Meeting, Orlando, Florida*; ACS: Washington, **1996**; pp. 211-213.
61. Butler, E. L., Douglas, G. S., Steinhauer, W. G., Prince, R. C., Aczel, T., Hsu, C. S., Bronson, M. T., Clark, J. R. and Lindstrom, J. E. In *On-Site Bioreclamation. Processes for Xenobiotic and Hydrocarbon Treatment*; Hinchee, R. E. and Olfenbuttel, R. F., Eds.; Butterworth-Heinemann: Boston, **1991**; pp. 515-521.
62. Ribbons, D. W. and Eaton, R. W. In *Biodegradation and Detoxification of Environmental Pollutants*; Chakrabarty, A. M., Ed.; CRC Press: Florida, **1982**, pp. 59-84.
63. Cripps, R. E. and Watkinson, R. J. In *Developments in Biodegradation of Hydrocarbons*; Watkinson, R. J., Ed.; Applied Science: London, **1978**; pp. 113-134.
64. Walker, J. D. and Colwell, R. R. *Appl. Microbiol.*, **1974**, *27*, 1053-1060.
65. Watkinson, R. J. (Ed). *Developments in Biodegradation of Hydrocarbons*; Applied Science: London, **1978**.
66. Eganhouse, R. P., Dorsey, T. F., Phinney, C. S. and Westcott A. M. *Environ. Sci. Technol.* **1996**, *30*, 3304-3312.
67. Bjorøy, M., Hall, P. B. and Moe, R. P. (1994). *Org. Geochem.* **1994**, *22*, 355-381.

Chapter 9

Biogeochemical Marker Profiles in Cores of Dated Sediments from Large North American Lakes

R. A. Bourbonniere[1], S. L. Telford[1], L. A. Ziolkowski[1], J. Lee[1],
M. S. Evans[2], and P. A. Meyers[3]

[1]National Water Research Institute, Environment Canada, P.O. Box 5050,
Burlington, Ontario L7R 4A6, Canada
[2]National Hydrology Research Institute, Environment Canada, 11 Innovation
Boulevard, Saskatoon, Saskatchewan S7N 3H5, Canada
[3]University of Michigan, C. C. Little Building, Ann Arbor, MI 48109–1063

Hydrocarbon and fatty acid components of sediments from two large remote lakes in northern Canada, Lake Athabasca and Great Slave Lake, are compared with those from two large lakes in the heavily populated Great Lakes Basin, Lakes Ontario and Erie. Studies concentrated on modern sediments dating back 70-150 years, the period during which anthropogenic effects due to development would be expected for these basins. Normal alkane and fatty acid biogeochemical markers are useful for tracking the impacts of land-use changes to the sediments of these lakes. Carbon preference indices indicate the predominance of natural sources, and point to events and trends related to petroleum contamination in all four lakes, especially the Great Lakes. Normal alkane and fatty acid biogeochemical markers yielded different assessments of the relative amounts of aquatic sources of organic matter, suggesting that diagenetic processes alter the aquatic fatty acid profiles and affects their reliability as source indicators. The ratio of unsaturated to saturated n-C_{16} acids appears to be diagnostic of depositional conditions that affect early diagenesis of fatty acids. Application of the biogeochemical marker approach to these lakes highlighted the differences in the degree of development in their watersheds.

The extractable lipid fraction of lake sediments contains hydrocarbon and fatty acid components that are a combination of biologically synthesized lipids and diagenetically modified materials. Anthropogenically derived materials, primarily hydrocarbons, which result from petroleum transport and use, also contribute to the lipids extracted from modern sediments *(1, 2)*. Changes in the downcore distribution patterns of *n*-alkanes can be used to indicate historical changes in a lake's watershed due to anthropogenic impacts and/or natural causes *(1-4)*. Normal fatty acid distributions have likewise been used to indicate impacts of land use and climate changes on lakes *(5, 6)*.

This paper compares recent investigations of the depositional history of two large lakes in a remote area of northern Canada, Lake Athabasca and Great Slave Lake, with those from two large lakes in the heavily populated Great Lakes Basin, Lakes Ontario and Erie. Our studies have concentrated on modern sediments dating back 70-150 years, the period during which anthropogenic effects due to development would be expected for these basins. Previous work has been reported on these Great Lakes cores for a variety of natural and anthropogenic components. The purpose of this paper is to compare source and diagenetic impacts on the hydrocarbon and fatty acid distribution in the sediments of the northern lakes with the Great Lakes. This is done using biogeochemical markers derived by examining *n*-alkane and fatty acid concentrations from the northern lakes sediments and re-examining these same parameters by means we have not used previously for the Great Lakes cores.

Total organic carbon (TOC) profiles of sediment cores give an overall view of the changes that occur in a lake's watershed over time, representing both anthropogenic and natural perturbations to the mix of organic components entering the lake. The biogeochemical markers used in this paper are derived from the distributions of *n*-alkanes and *n*-fatty acids. Carbon Preference Indices (CPI), which are derived from *n*-alkane distributions *(7, 8)*, can be used to indicate the degree of contamination from petroleum and its products. Terrestrial and aquatic plant contributions to lake sediments can be determined from both *n*-alkane and fatty acid distributions. These indirectly measure processes that can influence the input of terrestrially derived organic matter (*e.g.* deforestation) and those that influence primary productivity (*e.g.* nutrient inputs). The ratio of two commonly occurring fatty acids, *n*-hexadecenoic acid (n-$C_{16:1}$) and *n*-hexadecanoic acid (n-$C_{16:0}$) can be used to indicate the degree to which diagenic processes (*e.g.* microbial degradation) affect organic matter after deposition *(2, 3, 9-10)*.

Sampling and Methodology

Coring. Sediments were collected from Lake Athabasca (1992) and Great Slave Lake (1994) using a 10 cm i.d. gravity corer through the ice in late winter. Sediments from Lake Ontario (1981) and Lake Erie (1982) were collected as 6.5 cm i.d. subcores from box cores taken aboard the *CSS Limnos* during summer. Locations of the four lakes are shown in Figure 1 and site information for each are given in Table I. In all cases several replicate cores were collected. Cores were kept cold (4±2 ° C) until extrusion could be done. Sectioning was done, within a few hours of collection, with cores aligned vertically using an hydraulically controlled extruder. Sections of sediment from 0.5 - 2 cm thick were placed into pre-cleaned glass jars, and the sediment samples were frozen immediately. Section thicknesses were selected according to the expected sedimentation rate for each site and generally were thinner on the top of the core and thicker further down.

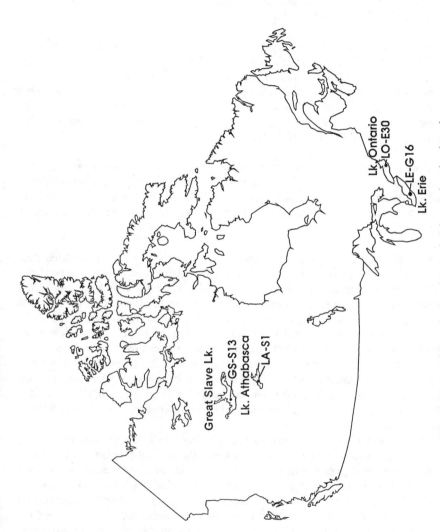

Figure 1. Map of Canada showing study lakes and coring locations.

Cores from each lake were evaluated for overall quality according to slice by slice descriptions recorded during extrusion. Two cores were selected from the replicates collected at each lake for use in this study. One was used for geochronology and bulk physical and chemical determinations; the other was used for organic geochemical analyses. Information used for core selection and the depth of the upper oxidized interval at each coring location are included in Table I.

Geochronology. One of the cores selected from each site was freeze-dried, dated using the ^{210}Pb method *(11-13)* and confirmed by the ^{137}Cs method *(11, 12, 14)*. The average modern mass sedimentation rate determined from the ^{210}Pb profile was applied to the downcore cumulative dry weight distribution, assigning deposition dates to the midpoints of the intervals sampled. All data presented here are plotted according to these ^{210}Pb deposition dates to allow easier comparisons among the cores. Geochronological parameters are listed for each site in Table I.

Total organic carbon. TOC was determined by dry catalytic oxidation of carbonate-free sediment (de-calcified using sulfurous acid) at 950-1000°C on aliquots of the same cores used for dating. Quantitation was done using NDIR detection on either Leco IR12 or Perkin-Elmer instruments. Precision of replicates was within 3% of the mean at the 20 mg/g level.

Extractable lipids. Hydrocarbons, fatty acids and other lipids were isolated from sediments by 2 x 24 h Soxhlet extraction with azeotropic toluene/methanol. The combined Soxhlet extracts constitute the "unbound" or extractable lipid fraction of the organic matter in the sediments. These are concentrated, saponified to break esters, methylated with BF_3/methanol and fractionated to isolate several lipid classes on alumina/silica gel columns *(11,15)*. The first eluent from the columns, hexane/toluene (85:15), contains hydrocarbons and the second eluent, toluene (100%), contains fatty acid methyl esters. Other classes were not considered for this work. After concentration both fractions were analyzed by capillary gas chromatography with flame ionization detection (FID). Samples were injected, on-column, using 30 m x 0.32 mm i.d. fused silica columns (SE30, DB1 or DB5 -- 0.25 μm film thickness). The gas chromatographs used were either a Carlo Erba 4160 or Hewlett Packard 5890.

Quantitation was accomplished by adding internal standards to the sediment samples before the extraction procedure. For *n*-alkanes the internal standard was either *n*-hexatriacontane (*n*-C_{36}) or *n*-octatriacontane (*n*-C_{38}), both of which were shown to occur below detection limits or in negligible quantities during preliminary tests of a few sediment samples. For fatty acids, either *n*-heptadecanoic acid (*n*-$C_{17:0}$) or *n*-nonadecanoic acid (*n*-$C_{19:0}$) was used as the internal standard after testing for negligible occurrence. Inclusion of the internal standards at the beginning of the procedure by injecting them onto the sediment assures that any losses resulting from subsequent operations are accounted in the final results. Quantitation of the integrated gas chromatographic results for all of the *n*-alkanes and fatty acids determined were done by applying FID response factors relative to the internal standards used. These response factors were calculated from daily runs of mixtures of authentic standards covering the range of analytes determined in each class. For the few analytes not represented in the mixtures, the response factors were estimated by averaging the factors determined for the adjacent homologs.

Table I. Location, Identification and Geochronological Parameters from the Four Coring Sites

Lake and Date (Site ID)	Latitude N	Longitude W	Water Depth (m)
L. Athabasca, 1992, (LA-S1)	59° 02.7'	110° 13.4'	12
Great Slave L., 1994, (GS-S13)	61° 24.2'	114° 59.8'	61
L. Ontario, 1981, (LO-E30)	46° 32.2'	76° 54.1'	233
L. Erie, 1982, (LE-G16)	42° 00.0'	81° 36.1'	24

Site ID	[1]GCH Core	[1]BGM Core	[2]Oxid. Layer cm	Mass Sed. Rate g/cm²/yr	[3]Date at [137]Cs Maximum	Geochronology Reference
LA-S1	1C	1D	5-6	0.072	1957	*11*
GS-S13	13C	13E	4-5	0.043	1966	*13, 14*
LO-E30	RNC	LIP	3-4	0.044	1960	*12*
LE-G16	SC5	SC8	7-8	1.360	1963	[4]JAR

[1]GCH - Geochronology Core ID, BGM - Biogeochemical Marker Core ID from field notes
[2]Thickness of brown oxidized layer at top of each core overlying grey sediment
[3]Date from ^{210}Pb method at which the 1963 ^{137}Cs maximum occurs
[4]Robbins, J. A., NOAA / GLERL, Ann Arbor, MI, unpublished data

Quality Assurance. Standard laboratory practices include the determination of total procedural blank for 1 extraction in 12, and running 1 of several laboratory reference standards in every 12 extractions. This latter practice served to test the robustness of the procedures by showing that only small variations occurred even though several analysts carried out the laboratory work. As well, results from replicate extractions of the laboratory standards provide an estimate of the reproducibility of the procedures. Typical results for reproducibility of fatty acid determinations are illustrated by data from 6 determinations of a Lake Ontario laboratory reference standard analyzed by 3 technicians over a 2 year period. The precision for total fatty acids (from total FID area), n-$C_{16:0}$, and n-tetracosanoic acid (n-$C_{24:0}$) were 11%, 9% and 7% respectively, expressed as relative standard error.

Data reported are corrected, when necessary, for small amounts of cross contamination occasionally found in total procedural blanks. Rarely were individual n-alkane or fatty acid components identified in the blank runs in significant quantities. Most often no correction was needed at all. The practical detection limit based on 5 g sample size was 10 ng/g for individual n-alkanes and 5 ng/g for individual fatty acids based on analysis of the signal-to-noise ratio from several representative chromatograms.

Tests conducted during method development showed that 95% of the extractable lipids were isolated during the first 24 hours of Soxhlet extraction, but a second 24 hour extraction was always done to allow for sample-to-sample variations. The ratio of "unbound" fatty acids (extractable as defined here) to "bound" fatty acids (extractable by saponification after "unbound" was removed) from a Lake Ontario surficial sediment was 33:1. This was not tested for other sediments.

The Great Lakes sediments used for this study were freeze-dried before extraction with the chamber temperature maintained at 20° C. Interest in more volatile components of sediments for related work prompted the adoption of wet extraction techniques for the northern lakes sediments and involved only minor modifications to the procedures (11). Volatility tests in our laboratory showed no effect of prolonged evaporation for n-alkanes greater than n-C_{14}, so a significant effect of freeze-drying is unlikely for the compounds used here as biogeochemical markers.

Total Organic Carbon

Concentrations of TOC increase in the upper parts (post 1940) of cores taken from each of the lake basins (Figure 2). The most recent increase is much less for the northern lakes, but can be considered higher than background for all lakes. In addition, one of the Great Lakes cores (LO-E30) shows evidence for a prior increase in TOC beginning about 1860 (Figure 2). The earlier increase has been attributed to deforestation coincident with settlement in the lower Great Lakes basin (3, 16-19). The later increase in TOC for all cores coincides with increases in development and industrial activity, and is especially evident in the Great Lakes cores. TOC concentrations in the contemporary sediments from the northern lakes are the same (10-15 mg/g) as the pre-settlement levels in the lower Great Lakes. The cores from Lake Athabasca and Lake Erie exhibit several excursions in the TOC profiles (Figure 2). The coring sites for these two lakes were shallower than the other two sites (Table I) and may have been influenced by wave action during storms.

n-Alkane Source Indicators

Natural biological precursors of sedimentary organic matter generally contain a wide range of n-alkanes. These components typically exhibit a strong predominance of odd carbon numbered homologs over those of even chain length (2). Thermogenic (petroleum generation) processes that act on sedimentary organic matter after deep burial modify the n-alkane distribution so that no carbon number predominance remains (20). This lack of odd over even predominance can be used in environmental geochemistry to assist in identifying sediments contaminated with petroleum or its by-products.

Two principal sources dominate natural inputs of hydrocarbons to modern lake sediments; photosynthetic algae from aquatic sources, and vascular land plants. The former are represented by n-alkane distributions dominated by C_{17} (2, 11, 21) while the latter contain large proportions of C_{27}, C_{29}, and C_{31} n-alkanes that originate in epicuticular wax coatings (2, 5, 22-25). In modern sediments, n-alkanes from natural sources are usually augmented by those from contamination sources.

Carbon Preference Indices for *n*-Alkanes. The CPI, as originally proposed by Bray and Evans *(7)* and Cooper and Bray *(26)*, has been used widely in petroleum geochemistry as a maturity indicator. Using a mathematical expression that reduces the ratio of odd carbon to even carbon chain length *n*-alkanes to a single value, oil-containing mature sediments tend to show no predominance and yield a CPI value of unity. Immature sediments contain largely unmodified *n*-alkanes and show higher odd predominance (CPI >1).   In a recent paper *(8)*, a mathematical shortcoming of the original Bray and Evans expression was described, and an improved expression was proposed. We have adopted the newer expression for use in this paper.

In practice CPIs should be referenced to the same range of *n*-alkanes to compare values among samples. Our analytical procedures allowed reliable quantitation of *n*-alkanes from C_{15} through C_{35} in all samples from the lake sediments studied. Within this wide range of *n*-alkanes are included all of the aquatic, terrestrial and petroleum alkanes described above which commonly are found in the lipid extracts of modern sediments *(2, 20)*. We define here three CPIs using the formulas in equation 1. The overall CPI (equation 1a) incorporates all natural *n*-alkanes contributed by aquatic algae and vascular land plants as well as contaminants from transport and use of petroleum and its products in the watershed.

$$CPI = (\sum Odds\ C_A \dots C_B\ +\ \sum Odds\ C_C \dots C_D)\ /\ 2\ (\sum Evens\ C_E \dots C_F) \qquad (1)$$

Overall CPI_{15-35} (CPI):   A=15, B=33, C=17, D=35, E=16, F=34   (a)
Low CPI_{15-25} (LCPI):   A=15, B=23, C=17, D=25, E=16, F=24   (b)
High CPI_{25-35} (HCPI):   A=25, B=33, C=27, D=35, E=26, F=34   (c)

If CPI is calculated for only the lower half of the range determined (equation 1b), the resulting low carbon preference index (LCPI) is more influenced by algal and bacterial sourced biogenic *n*-alkanes and lighter petroleum products such as fuel oils *(20)*. CPI calculated from only the higher end of the *n*-alkane distribution (HCPI, equation 1c) should be influenced by inputs from natural higher land plant sources as well as heavier petroleum products such as crude and lubricating oils *(20)* and possibly combustion products. Examples of end-member values from the literature and this work are given in Table II .

High Carbon Preference Index. The CPI value of around 2 is characteristic of the organic matter deposited over the past 100-150 years at three of the four locations studied. Only the LO-E30 core showed consistently lower values, centering around a CPI of 1.4. The downcore patterns of HCPI at all four sites (Figure 3) parallel closely the downcore CPI patterns except that values are shifted higher about 1 unit. For this reason only the former will be discussed.

The fact that HCPI values >1.5 are common at all sites suggests a mixed natural and contaminant source of alkanes to all four lakes. For all lakes, HCPI values >3 occur only episodically, which implies that natural higher plant sources of alkanes, though always a major contributor to these sediments, are sometimes deposited in greater proportions. A possible explanation for such episodic increases in HCPI is a greater contribution from topsoil containing degraded leaf litter eroded from the watershed, caused by forest clearance, urbanization or periods of unusual rainfall. This is a plausible explanation for the increasing trend shown for LO-E30 between 1940-1960. The urban population of Canada nearly doubled from 8 to 14 million between 1941 and 1961 *(27)*, much of this growth

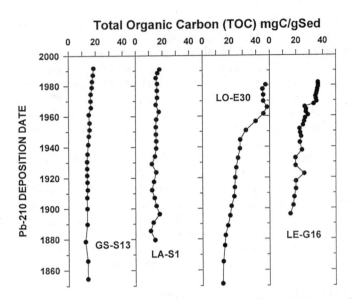

Figure 2. Downcore concentrations of TOC in the four dated cores of sediments from large North American lakes identified in Figure 1 and Table I.

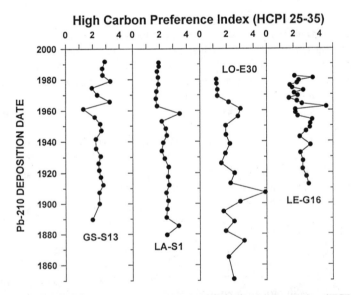

Figure 3. Downcore distributions of High Carbon Preference Index (HCPI) for the n-alkane range: C25 - C35, from sediments of four large North American lakes identified in Figure 1 and Table I.

occurred in southern Ontario. Likewise, the increase between 1940 and 1950 in the LE-G16 core could reflect similar urbanization in the Lake Erie watershed (Figure 3).

The overall HCPI trend for GS-S13 over the past 100 years is a slightly increasing one except for some lower and higher excursions between 1960 and 1980 (Figure 3). The suggestion from this pattern is that higher land plant hydrocarbons have been gradually increasing in relative proportion, but that occasional periods of lower input of these *n*-alkanes or incidents of contamination by heavier petroleum products influenced the HCPI distribution.

Lower values of HCPI, approaching that of unity, occur occasionally in the data from GS-S13 and LE-G16, and over the last decade for LO-E30. Such low HCPI values suggest sporadic incidents of petroleum contamination for the former two lakes and a very recent increase in the supply of petroleum related contaminants to Lake Ontario sediments. A three-fold increase between 1940 and 1970 in the content of hydrocarbons constituting an unresolved complex mixture (UCM) in the LO-E30 sediments was reported previously *(4)*. The LA-S1 core (Figure 3) shows a distinct shift to consistently lower (<2) HCPI from a prior near constant value of >2. This pattern suggests influence from oil sands surface mining and upgrading activities in the region which experienced major development beginning in the 1960s *(30)*, see also Table II.

Low Carbon Preference Index. In Figure 4 the LCPI data from the northern lakes show very consistent values over the past 100 years. LCPI values for the LA-S1 core are a nearly unchanging 1.5 except for a brief excursion to 1.9 in the 1940s. Likewise the GS-S13 core yielded values from 1.3-1.5 for all sections except for a period when the LCPI of the sedimentary organic matter doubled in the 1920s suggesting a shift to increased autochthonous biogenic sources of *n*-alkanes (Table II). The Great Lakes cores exhibit more variable and generally lower values of LCPI. The LO-E30 core shows all values <1.5 over the past 150 years trending downward to unity since the 1920s with the exception of an excursion to even preference (<1) in the 1940s and one to higher odd preference in the early 1960s (Figure 4). The LE-G16 core shows a LCPI trend opposite to the LO-E30 core, increasing more sharply from a value of about 1 in 1940 to 1.8 in 1980, suggesting increasing autochthonous biogenic sources.

Terrestrial and Aquatic Indicators. Terrestrial plants and aquatic algae have very different *n*-alkane distributions, longer chain ($>C_{20}$) alkanes being more typical of the leaf waxes of land plants and short chain *n*-alkanes are more common in aquatic algae *(21-23, 25)*. The sum of *n*-alkane concentrations: ($C_{27} + C_{29} + C_{31}$) can be used as a measure of the inputs of *n*-alkanes from terrestrial plants, and the sum: ($C_{15} + C_{17} + C_{19}$) indicates *n*-alkanes from aquatic sources. Previous work *(3)* used the "terrestrial/aquatic ratio" for hydrocarbons (TAR$_{HC}$), a simple ratio of these two sums (Table II), as a measure of the relative amounts that these sources contribute to sediments. The advantage of using a ratio is that large episodic changes in total organic contribution to sediments are normalized and data from several cores can often be plotted on the same scale. Trends in TAR$_{HC}$ can be used to indicate changes in source to the sediments, but they work better when the trends are mostly due to changes in terrestrial sources. When aquatic sources predominate, the TAR$_{HC}$ decreases to values of <1 (Table II) and changes are less evident. Other shortcomings of ratios are that opposing trends are masked and small absolute trends are dampened. For this work we use the simple sums as biogeochemical markers.

Table II. Examples of End-Member Values for *n*-Alkane
Biogeochemical Marker Ratios

	[1]CPI	[1]LCPI	[1]HCPI	[1]TAR$_{HC}$	Reference
L. Ontario Phytoplankton	8.9	10	1.5	0.02	This work
Algal Ooze	5.2	-	-	-	*28*
[2]White Spruce Needles	1.8	1.2	2.2	10	*29*
[3]Spruce Needles	1.5	1.4	1.5	29	This work
Continental Plants	4 - 7	-	-	-	*20*
[4]Oil Sands Mine Drainage	1.3	1.5	1.2	2.8	This work
[5]Oil Sands Refinery Effluent	1.2	1.3	1.1	0.65	This work
Oil Shales	1.1	-	-	-	*20*

[1]See text for definitions.
[2]*Picea glauca* from Matthei Botanical Gardens, Ann Arbor, MI.
[3]*Picea* sp. from Burlington, ON, near heavy industrial area.
[4]Suspended particulates in strip mine drainage, Ft. McMurray, AB.
[5]Suspended particulates from process water discharge, Ft. McMurray, AB.

The profiles of terrestrial and aquatic indicators for Lake Ontario (Figure 5) show that this lake previously experienced lesser inputs of both land based and algal *n*-alkanes. In the early part of the 20th century some small episodic increases occurred in both aquatic and terrestrial *n*-alkanes. Around 1940 the inputs of both types of *n*-alkanes began to increase and by 1960 the increased inputs became significant. More recently, inputs of *n*-alkanes from vascular plant sources have exceeded by 2-3 times the input of aquatic *n*-alkanes (Figure 5). These trends follow the growth in urban population in the Great Lakes basin as mentioned previously *(19)*. Increases in biogenic silica and total phosphorus have been shown to occur at about the same time in another of the replicate cores from this Lake Ontario site *(18)* and decrease of total phosphorus loadings to the lake in the 1970s *(31)* could explain the leveling off that is evident in the aquatic *n*-alkane profile (Figure 5). The longer chain *n*-alkanes likely originate from multiple sources related to anthropogenic activities, such as soil erosion, sewage inputs, and petroleum inputs from urban runoff, industrial emissions, shipping, etc. The suggestion of petroleum inputs are consistent with recent UCM increases in this core *(4)* and HCPI data (Figure 3), and points to a limitation in the use of these longer chain *n*-alkanes as terrestrial markers without also obtaining evidence of petroleum influence.

Lake Erie has experienced increased anthropogenic phosphorus loadings throughout the 20th century, but especially since the 1940s *(32-33)* and greater aquatic plant contributions to the sedimentary *n*-alkanes is evidenced by the decreasing trend of TAR$_{HC}$ after 1970 *(3)*. Profiles of the terrestrial and aquatic *n*-alkane markers are very complex with considerable cyclic variability evident for both sources throughout the 20th century (Figure 5). Note that the cycles for both sources are in phase, suggesting that overall

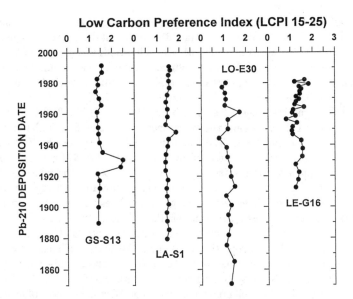

Figure 4. Downcore distributions of Low Carbon Preference Index (LCPI) for the n-alkane range: C15 - C25, from sediments of four large North American lakes identified in Figure 1 and Table I.

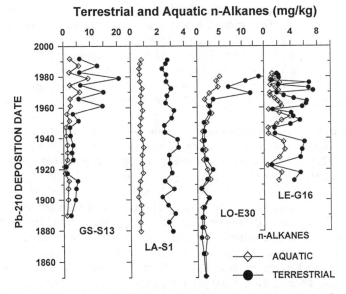

Figure 5. Downcore distributions of n-alkane terrestrial and aquatic biogeochemical markers from sediments of four large North American lakes identified in Figure 1 and Table I. Aquatic = (C_{15} + C_{17} + C_{19}), Terrestrial = (C_{27} + C_{29} + C_{31}).

watershed effects may be largely responsible for these fluctuations. A comparison of the n-alkane data with historical lake level data *(33)* reveals that the periods of high n-alkane inputs from either source are associated somewhat with periods of higher water levels. This suggests that erosion of bluffs *(34)* and agricultural land can provide inputs of terrestrial n-alkanes and nutrients which can increase phytoplankton productivity. Although cultural loadings of phosphorus to Lake Erie have increased dramatically since the 1940s *(33)* the aquatic n-alkane signal is equally high at times before 1940 as it is after that date (Figure 5).

Great Slave Lake shows in-phase cyclic inputs of n-alkanes from both terrestrial and aquatic sources since the 1960s. The fluctuating behavior began earlier (1950) for the terrestrial n-alkanes (Figure 5). A few of the values for the GS-S13 core are the highest values for both aquatic and terrestrial n-alkanes of all the cores. This lake receives input from the Peace River system which flows through major forested regions of British Columbia and Alberta which have seen recent growth in forest products industries. These may be the sources of the terrestrial n-alkanes in the sediment. In contrast to the GS-S13, core the LA-S1 core shows the lowest and least variable contents of both terrestrial and aquatic n-alkanes of all the four cores (Figure 5). Lake Athabasca sediments have received a constant input of aquatic n-alkanes for the last 100 years and a nearly constant input of terrestrial n-alkanes, with some cyclic behavior over the same period. The Athabasca river drains forestry, agricultural and oil producing regions of Alberta, but most of its flow exits into the Peace River system after entering only the far southwestern part of Lake Athabasca *(35)*.

Fatty Acid Source and Diagenesis Indicators

Downcore profiles for certain normal fatty acids can also be used to infer changes in source input to lake sediments. Fatty acids are about ten times more degradable than hydrocarbons *(36-37)* and much more abundant in these sediments *(3, 11)*. Thus, there is a greater potential for fatty acid profiles to be modified by post-depositional processes such as microbial reworking *(38-39)*. This property can be used to advantage in assessing the extent of microbial reworking, but also indicates that caution must be used when ascribing source changes on the basis of labile biogeochemical markers *(29)*.

Terrestrial and Aquatic Indicators. Like the n-alkanes, long chain fatty acids are dominant components of the waxy coatings of leaves, flowers and pollen originating from land plants *(25)*, and shorter chain acids are produced by all plants but are the dominant lipid components of algae *(40)*. Aquatic source normal fatty acids can be represented by the sum: ($C_{12:0} + C_{14:0} + C_{16:0}$), and terrestrial sources by the sum of three longer chain acids: ($C_{24:0} + C_{26:0} + C_{28:0}$). A fatty acid ratio (TAR_{FA}) was defined and used in a similar way to the TAR_{HC} ratio *(3)*. Higher values of TAR $_{FA}$ may indicate increased input of terrigenous sources of lipid matter to the sediments, but they may also indicate preferential degradation of shorter chain fatty acids relative to longer ones *(29, 36, 41-43)*.

Terrestrial fatty acid inputs to the GS-S13 and LA-S1sites have been constant at the same 10 mg/kg level for 80 and 100 years respectively (Figure 6). As well, the LO-E30 core shows constant input of these fatty acids from 1850 through 1930, with only a slight increase for the next two decades. Likewise The LE-G16 core received constant input of terrestrial fatty acids from 1910-1960. The Great Lakes cores contain from 3-10 times as much terrestrial fatty acids as the northern lakes. Both the LO-E30 and LE-G16 cores exhibit significant increases in terrestrial sourced fatty acids beginning in 1950 and 1960 respectively (Figure 6). The most recent trends for the terrestrial fatty acids for both of these Great Lakes are significant decreases, leaving behind a subsurface maxima. As terrestrial fatty acids are less susceptible to degradation *(29, 36, 41-43)*, we can treat the profiles as indicating source changes with more certainty. If we do so, these profiles fit very well with the history of urbanization in these watersheds as described previously.

The lower portions of all cores are characterized by low levels of aquatic fatty acids, the lowest levels in each core (Figure 6). The GS-S13, LO-E30 and LE-G16 cores all show increases in aquatic fatty acid content beginning at 1950, and the LA-S1 core begins to increase at 1930. The aquatic fatty acid content of all cores increases progressively towards the present with the highest value near the surface that is 3-10 times the background value for a given core. The aquatic fatty acid profiles for LO-E30 and LE-G16 could be explained by cultural inputs of nutrients *(31,33)*, but no such mechanism is known for the increases shown for GS-S13 and LA-S1 (Figure 6). Based on phytoplankton biomass Munawar and Munawar *(44)* classified Lake Erie as eutrophic in the early 1970s. Certainly such conditions would imply a greater concentration of aquatic fatty acids in the surface sediments, as is seen in the LE-G16 core. By that measure, the concentrations of aquatic fatty acids in the surficial sediments of the other three lakes correctly suggest that they are all oligotrophic.

Another view of the aquatic fatty acid profiles is that they are influenced by diagenetic processes such as microbial degradation during settling and after deposition. The shape of the profiles from GS-S13, LA-S1 and LE-G16 are all similar and is consistent with decreases over time of an initial high input of aquatic fatty acids. As well the rate of decrease is higher in the upper few cm for each core, the oxidized zone (Table I), where aerobic microbial activity is normally vigorous. The aquatic fatty acid profile of the LO-E30 core appears to result from even more vigorous degradation. This is consistent with the relatively low surface value resulting from selective microbial degradation during the long transport to this site which is by far the deepest of all (Table I), a process which has been measured in sediment trap studies in the Great Lakes *(37)*.

Indicators of Post-Depositional Alteration. Although all fatty acids are susceptible to microbial degradation *(36, 43)*, unsaturated fatty acids are especially labile and are preferentially removed during early diagenesis in lake sediments *(38)*. We have used changes in the ratio of n-$C_{16:1}$ to n-$C_{16:0}$ (US16) as an indicator of the intensity of diagenetic activity. Under conditions that favor microbial degradation (*e.g.* low sedimentation rate, oxidized surface sediments, higher temperatures, longer water column residence time) lower values for US16 would be expected. Conversely, high sedimentation rate would tend to preserve organic matter by quick burial so that it remains in the oxidized zone for less time. Lower temperatures tend to limit microbial activity and a shallow water column exposes sinking particles to oxic microbial action for shorter periods. These latter conditions serve to preserve the ratio closer to the original, sometimes higher, US16 value of the depositing organic matter.

Comparisons of US16 values on cores from different depositional environments must be made cautiously since multiple depositional factors are concerned as mentioned above, and source changes and duration of settling time are also important variables. We are not certain what the original US16 ratio is for any of the lakes studied, but a laboratory culture of *Asterionella formosa*, a common freshwater diatom, yielded a value of 3.8 for

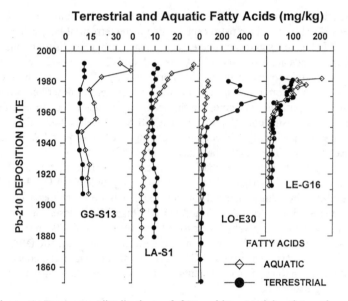

Figure 6. Downcore distributions of fatty acid terrestrial and aquatic biogeochemical markers from sediments of four large North American lakes identified in Figure 1 and Table I. Aquatic = ($C_{12:0}$ + $C_{14:0}$ + $C_{16:0}$), Terrestrial = ($C_{24:0}$ + $C_{26:0}$ + $C_{28:0}$).

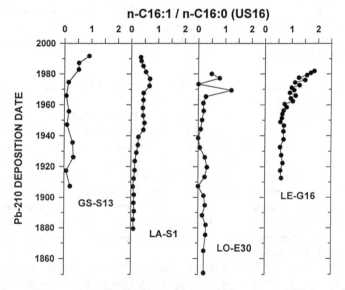

Figure 7. Downcore distributions of fatty acid Unsaturated / Saturated Ratio (n-$C_{16:1}$ / n-$C_{16:0}$), from sediments of four large North American lakes identified in Figure 1 and Table I.

US16 *(45)*. Thus, the high value found for the LE-G16 core (Figure 7) and the gradual nature of the downcore decline in US16 suggests that microbial degradation of deposited fatty acids is inhibited at this site. This could be caused by the high sedimentation rate (Table I) and periodic occurrence of summertime hypolimnetic anoxia *(32)*. The high ratio could also indicate, in this case, a high algal input in line with results discussed above (Figure 6).

Like the LE-G16 core, the GS-S13 core shows a gradual downcore decrease in US16 value. The sedimentation rate is not as high at GS-S13, and hypolimnetic anoxia should not occur at this site. At the depth of this site the year-round temperature is 4° C *(46)*. Such conditions could aid in preservation of fatty acids and slow down degradation rates and could conceivably result in a profile similar to that from LE-G16. The Lake Athabasca coring site is the shallowest of the four sites studied (Table I) and its summer bottom water temperature can reach 9° C *(35)*. The US16 profile (Figure 7) shows evidence of a gradual stepwise decrease at depths which correspond to 1980 and sooner. The most recent values show an increasing trend with depth, which suggests that source changes may have obscured diagenetic effects

The US16 profile for the LO-E30 core is unique (Figure 7). Excluding a few sections near the top of the core, the LO-E30 core shows mostly low US16 values that are among the lowest found. Part of the explanation for this may be that much of the early diagenesis of the lower molecular weight fatty acids occurs during settling of the sediments as they make their way to this deep site.

Conclusions

TOC concentration profiles from the northern lakes record only very small increases in organic matter delivery over the past 100 years in contrast to those from the Great Lakes which show progressively greater delivery of organic matter to the sediments in response to anthropogenic perturbations in the watershed.

The overall CPI indicates natural sources of hydrocarbons predominate in all lakes, but incidents of petroleum contamination may be recorded by the HCPI profiles of the Lake Erie and Great Slave Lake cores. The Lake Athabasca core, and to an even greater extent the Lake Ontario core, show evidence for modern trends in HCPI that suggest persistent contamination by petroleum or its products. Focusing the CPI on lower molecular weight *n*-alkanes (LCPI) shows that the northern lakes are largely unchanged in the past 70-100 years with respect to *n*-alkane sources. Lake Ontario results indicate a slow progression to unity indicating increasing contamination by light petroleum products and the Lake Erie results show that natural sources of *n*-alkanes have increased progressively since the 1940s.

Normal alkane biogeochemical markers show consistent dominance of higher plant sources over the past 100 years for both Lake Athabasca and Great Slave Lake and over the past 70 years for the Lake Erie core. For the Lake Ontario core the predominance of terrestrially sourced *n*-alkanes occurs only over the past 25 years. The Lake Erie and Great Slave Lake cores show cyclic changes in both aquatic and terrestrial *n*-alkanes. Only the Lake Ontario core shows clear modern increases in aquatic sources for *n*-alkanes.

Fatty acid biogeochemical markers show no changes in the amount of terrestrial fatty acids input into Great Slave Lake or Lake Athabasca over the past 100 years. Both Lake Ontario and Lake Erie show modern increases in terrestrial fatty acid inputs and constant values before 1950. The sediment concentrations of fatty acids from aquatic sources appear to be influenced by diagenetic processes, before and after deposition, to varying degrees, for all four lakes. Post-depositional alteration profoundly affects the concentrations of these fatty acids so that they are not reliable as source indicators. The ratio of unsaturated to saturated n-C_{16} acids (US16) appears to be diagnostic of depositional conditions that affect early diagenesis of fatty acids. Such conditions as sedimentation rate, temperature, and settling time appear to explain the downcore patterns of US16 for all four lakes.

Evidence from all of these biogeochemical markers suggests that the northern lakes have not had substantial impact from anthropogenic hydrocarbon sources, nor has the trophic status of these lakes been impacted by human activities. The Great Lakes cores show evidence of petroleum contamination, increases in terrestrial sourced components and increases in organic matter from aquatic sources since the mid 20th century as a result of human activities in the lower Great Lakes basin.

Acknowledgments

Support for coring and biogeochemical marker analyses was provided by Environment Canada, National Water Research Institute and the Northern River Basins Study (Canada-Alberta-N.W.T.). The Great Lakes work was conducted in cooperation with the HI-SED Project of the Great Lakes Environmental Research Laboratory, U.S. National Oceanic and Atmospheric Administration. We thank the Captain and crew of the *CSS Limnos* and members of the NWRI Technical Operations Division for assistance with box coring in the Great Lakes. D. Allen, K. Hill, B. Jackson, J. B. Kemper, J. Kraft and E. Walker assisted with coring on the northern lakes. We thank B. Hilson, K. Lawrynuik, T. Mayer, J. McAndrew, L. O'Conner, D. S. Smith and B. Treen for assistance in the laboratory. Thanks to R. Eganhouse, T. Eglinton and a third reviewer for comments which significantly improved this paper.

Literature Cited

1.   Wakeham, S. G. *J. Wat. Poll. Cont. Fed.* **1977**, *49*, 1680-1687.
2.   Meyers, P. A.; Ishiwatari, R. *Org. Geochem.* **1993**, *20*, 867-900.
3.   Bourbonniere, R. A.; Meyers, P. A. *Limnol. Oceanog.* **1996**, *41*, 352-359.
4.   Bourbonniere, R. A.; Meyers, P. A. *Environ. Geol.* **1996**, *28*, 22-28.
5.   Cranwell, P. A. *Freshw. Biol.* **1973**, *3*, 259-265.
6.   Cranwell, P. A. *Geochim. Cosmochim. Acta* **1978**, *42*, 1523-1532.
7.   Bray, E. E.; Evans, E. D. *Geochim. Cosmochim. Acta* **1961**, *22*, 2-15.
8.   Marzy, R.; Torkelson, B. E.; Olson, R. K. *Org. Geochem.* **1993**, *20*, 1303-1306.
9.   Matsuda, H.; Koyama, T. *Geochim. Cosmochim. Acta* **1977**, *41*, 777-783.
10.   Meyers, P. A.; Maring, H. B.; Bourbonniere, R. A. In *Advances in Organic Geochemistry 1979*; Douglas, A. G.; Maxwell, J. R. Eds.; Pergamon: Oxford, UK, 1980, pp. 365-374.
11.   Bourbonniere, R. A.; Telford, S. L.; Kemper, J. B. Environment Canada, NWRI, **1995**, *Cont. No. 95-76*, 131pp.
12.   Eisenreich, S. J.; Capel, P. D.; Robbins, J. A.; Bourbonniere, R. A. *Environ. Sci. Technol.*, **1989**, *23*, 1116-1126.

13. Turner, L.J. Environment Canada, NWRI, **1994**, *Cont. No. 94-132*, 25pp.

14. Evans, M. S.; Bourbonniere, R. A.; Muir, D. C. G.; Lockhart, W. L.; Wilkinson, P.; Billeck, B. N. Environment Canada, NHRI, **1996,** *NRBS Project Report No. 99,* 171pp.

15. Leenheer, M. J.; Flessland, K. D.; Meyers, P. A. *Org. Geochem.* **1984,** *7,* 141-150.

16. Kemp, A. L. W.; Gray, C. B. J.;. Mudrochova, A. In *Nutrients in Natural Waters;* Allen, H. E.; Kramer, J. R. Eds.; Wiley Interscience: New York, NY 1972, pp. 251-279.

17. Schelske, C.L.; Stoermer, E. F.; Conley, D. J., Robbins, J. A.; Glover, R. M. *Science*, **1983,** *222,* 320-322.

18. Schelske, C. L.; Robbins, J. A.; Gardner, W. S.; Conley, D. J.; Bourbonniere; R. A. *Can. J. Fish. Aquat. Sci.* **1988,** *45,* 1291-1303.

19. *Historical Atlas of Canada;* Gentilcore, R. L., Ed.; Univ. Of Toronto Press: Toronto, ON, 1993; Vol. II.

20. Hunt, J. M. *Petroleum Geochemistry and Geology;* W. H. Freeman: San Francisco, CA, 1979; 617pp.

21. Giger, W.; Schaffner, C; Wakeham, S. C., *Geochim. Cosmochim. Acta*, **1980,** *44,* 119-129.

22. Cranwell; P. A., Eglinton, G.; Robinson, *Org. Geochem.,* **1987,***11,* 513-527.

23. Eglinton, G.; Hamilton, R. J. In *Chemical Plant Taxonomy;* Swaine, T. Ed.; Academic Press: New York, NY, 1963; pp 187-217.

24. Eglinton, G.; Hamilton, R. J. *Science,* **1967,** *156,* 1322-1335.

25. Rieley, G.; Collier, R. J.; Jones, D. M.; Eglinton, G. *Org. Geochem.* **1991,** *17,* 901-912.

26. Cooper, J. E.; Bray, E. E. *Geochim. Cosmochim. Acta* **1963,** *27,* 1113-1127.

27. *Historical Atlas of Canada;* Kerr, D.; Holdsworth, D. W. Eds.; Univ. Of Toronto Press: Toronto, ON, 1990; Vol. III.

28. Kvenvolden, K. A. *Nature* **1966,** *209,* 573-577.

29. Meyers, P. A.; Leenheer, M. J.; Bourbonniere, R. A. *Aquat. Geochem.* **1995,** *1,* 35-52.

30. Fergusen, B. G. *Athabasca Oil Sands - Northern Resource Exploration, 1875 - 1951* Alberta Culture/Canadian Plains Research Centre: Edmonton, AB, 1985, 283pp.

31. Stevens, R. J. J.; Neilson, M.A. *Can. J. Fish. Aquat. Sci.* **1987,** *44,* 2059-2068.

32. Burns, N.M. and C. Ross (eds) (1972) *Project Hypo*, CCIW Paper No. 6 and USEPA Tech, Rpt. TS-05-71-208-24, 182pp.

33. Sly, P. G. *J. Fish. Res. Bd. Can.* **1976,** *33,* 355-370.

34. Kemp, A. L. W.; Thomas, R. L.; Dell, C. I.; Jaquet, J.-M. *J. Fish. Res. Bd. Can.* **1976,** *33,* 440-462.

35. *Atlas of Alberta Lakes;* Mitchell, P; Prepas, E., Eds.; Univ. of Alberta Press: Edmonton, AB, 1990, pp 63-71.

36. Haddad, R. I.; Martens, C. S.; Farrington, J. W. *Org. Geochem.* **1992,** *19,* 205-216.

37. Meyers, P. A.; Eadie, B. J. *Org. Geochem.* **1993,** *20,* 47-56.

38. Matsuda, H.; Koyama, T. *Geochim. Cosmochim. Acta* **1977,** *41,* 777-783.

39. Matsuda, H.; Koyama, T. *Geochim. Cosmochim. Acta* **1977,** *41,* 1825-1834.

40. Cranwell, P.A. (1984) *Org. Geochem.,* **7:** 25-37.

41. Matsuda, H. *Geochim. Cosmochim. Acta* **1978,** *42,* 1027-1034.

42. Ho, E.; Meyers, P. A. *Chem. Geol.* **1994,** *112,* 309-324.

43. Canuel, E. A.; Martens, C. S. *Geochim. Cosmochim. Acta* **1996,** *60,* 1793-1806.

44. Munawar, M; Munawar, I. F. *J. Fish. Res. Bd. Can.* **1976,** *33,* 581-600.

45. Bourbonniere, R. A. *Ph.D. Thesis,* The University of Michigan, Ann Arbor, 1979, p173.

46. Rawson, D. S. *North West Canadian Fisheries Surveys in 1944-1945*; Bulletin No. LXXII; Fisheries Research Board of Canada: Ottawa, ON, 1947; Ch. 6, pp 45-68.

Chapter 10

An Application of Biomarkers in a Preliminary Environmental Study of the Lower Daintree River, North Queensland, Australia

Noel F. Dunlop[1,2], J. David Smith[1], and R. Basil Johns[1,2,3]

[1]School of Chemistry and [2]School of Earth Sciences, University of Melbourne, Parkville, Victoria 3052, Australia

Biomarker distributions found in sediments along a transect of the lower Daintree River follow those recognised in temperate environments for higher plants. In the lower molecular weight range ($<C_{20}$), however, there is much variability which may reflect the local vegetation. In the estuary, anthropogenic, petroliferous hydrocarbons occur as evidenced by low pristane / phytane ratios, n-alkane carbon number distribution patterns from gas chromatographic analyses and typical unresolved complex mixture patterns in the gas chromatograms of the hydrocarbon traces. The preliminary data suggest that the lower Daintree River is being impacted by hydrocarbon contamination.

Amongst the major ecotourist destinations in North Queensland are the Great Barrier Reef Marine Park and the Daintree River region.The drainage basin of the Daintree is 2130 km[2], and the river flows through the Dagmar Range and Cape Tribulation National Parks (cf. Figure 1) (*1*). In its lower reaches it drains sugar cane growing areas and some land now supporting cattle grazing before it flows into the inner waters of the Great Barrier Reef. In recent years the region has come to support a growing tourist industry centered upon boat trips extending from the Daintree Township to the river mouth. In the past three years the number of visitors to North Queensland has totalled over 5.2 million (*2*) raising a potential risk of anthropogenic inputs to the lower Daintree, and because it flows directly into the inner waters of the Great Barrier Reef, there is a possible impact upon the reef itself.

The Daintree River is shallow with a seasonally dependant water flow. The average annual discharge is estimated at 3,560 m^3 x 10^6 (*1*). Sediment release into the inner reef lagoon is marked after periods of high rainfall. The water flow has considerable energy in the upper reaches, but in a wide estuary this energy is dissipated, This favours sedimentation as evidenced by the large sand banks which have been formed. The river enters the reef lagoon through a narrow channel in the sand bar.

[3]Corresponding author

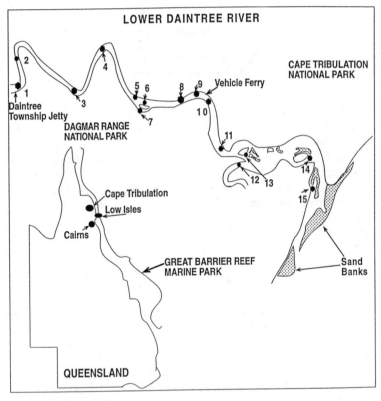

Figure 1. Map of the sampling transect of the lower Daintree River showing sand bank formation in the estuary and at the river mouth.

There is no published record bearing on the geochemistry of the lower Daintree. In order to determine management strategies for the future there is a need for baseline data. This paper describes a preliminary organic geochemical study of a transect down the lower Daintree River using biomarkers (among other parameters) to describe the sedimentary environment.

Sample Collection.

Figure 1 shows the position of the sampling sites. The farthest upstream is located near the Daintree Township Jetty. Sites 9 and 10 were on either side of the vehicular ferry. Site 15 was in the river mouth itself. Table I describes the environments of the sampling sites and the sediments collected over 23 - 24 February, 1995 using a Birge-Ekman grab sampler.

Table I. Sampling Sites

Site	Description of Location
1	Stewart Creek entrance. Seagrass beds heavily vegetated. Grey/black mud. H_2S. Collected mid-stream
2	Daintree Township jetty. Black silt with leaf detritus. Collected near the jetty.
3	Barratt's Creek entrance; fringing mangroves with rain forest. Black silt with leaf detritus. Mid-channel collection.
4	Past Kilkeary Point. Edge of the seagrass beds. Silt with gravel.
5	Off the north bank. Little sediment accumulation; coarse sediment.
6	South bank. Fringing mangroves. Gravel and sand.
7	Backwater on south bank. Fringing mangroves and rain-forest with canefields. Silty mud; brown. Mid-channel collection.
8	Forest Creek entrance; fringing mangroves with grazing land behind. Silty mud; brown. Collected near the north bank.
9	Forest Creek entrance (south). North of the ferry crossing.Fringing mangroves Silty mud; brown. Mid-stream collection.
10	South bank south of the ferry crossing. Mangrove and rainforest. Black silt.
11	East bank of river ~1 km from the ferry. Fringing mangroves and rainforest backing onto cleared land. Silt and coarse sand.
12	Inside South Arm; mangroves. Strong current into the river. Black silt; H_2S. Mid-stream collection.
13	South side of estuary between two sand banks. Open, but protected water. Black silt. H_2S.
14	Between the west bank and a sand-bank island. Mangroves. Silt. H_2S.
15	Within the sand bar at the mouth of the river. Open water to the west of the main channel. Clean white sand with water flow from the river.

Sediment samples were taken from the top 5 cm. The sediments were kept at < $4^{\circ}C$ on collection and stored at < $-4^{\circ}C$ until analyses were performed. The sediment was physically mixed to ensure homogeneity before a sample was removed for analysis.

Table II describes the characteristics of the water column at the sampling sites. The salinity and temperature measurements were taken at the base of the water column using a Yellow Springs salinity/temperature meter calibrated with IAPSO sea water.

Table II. Sampling Site Characteristics

Site	Depth (m)	Salinity ($^o/_{oo}$)	Temperature (^{o}C)	
1	2.8	0.0	27	Zone 1
2	2.0	0.0	30	
3	2.0	0.0	30	
4	1.2	0.0	30	
5	2.8	5.7	30	Zone 2
6	1.8	4.0	28	
7	1.4	4.0	30	
8	1.4	3.1	29	
9	1.5	9.8	30	Zone 3
10	3.5	13.5	30	
11	1.5	20	30	
12	3.5	24	30.6	Zone 4
13	1.3	18	30.2	
14	1.5	26	31	
15	2.0	29.9	30	Zone 5

Experimental.

All solvents were redistilled before use. About 500 g of sediment sample was defrosted and freeze-dried. A sub-sample of the dry sediment (~35 g) was extracted three times in a mixture of chloroform / methanol (2:1/v:v) using ultrasonification for 30 min. The final solvent extract was colourless. The combined extracts were filtered through a glass filter on a sintered glass funnel at reduced pressure; the extracted sediment being finally washed with chloroform. The organic solvent was reduced in volume by rotary evaporation (< 40°C) under reduced pressure to near dryness. The remaining solvent was removed under a stream of N_2 to give a residue - the Total Solvent Extract (TSE) which was then weighed and is reported in Table III as mg/g of dry sediment.

An aliquot of TSE was refluxed in a solution of 5% KOH (w/v) in 80% methanol / water (v/v) at 60°C for 3 hrs. The neutral lipids were obtained by solvent extraction three times using n-heptane/chloroform (4:1/v:v), the extracts were combined and dried over anhydrous sodium sulphate. The decanted solvent, on reduction in volume first by rotary evaporation as above and then under a stream of N_2 became the Total Neutrals fraction and was stored in chloroform at < 4°C. The basic solution was acidified to pH 2 with concentrated HCl, and extracted three times with *n*-heptane/diethylether (4:1/v:v) to give the Total Acids fraction. After drying over anhydrous sodium sulphate and solvent reduction by rotary evaporation, the Total Acids were methylated with boron trifluoride in methanol (14%) on heating at 60°C for 30 min. to give Fatty Acid Methyl Esters (FAME) (*3*). Distilled water was added to destroy excess reagent, and the aqueous layer was extracted three times in *n*-heptane/diethylether (4:1/v:v). The organic layer was dried over anhydrous sodium sulphate, reduced in volume and stored as the fatty acid methyl esters (FAME) fraction.

Separation into hydrocarbon and alcohol/sterol fractions was achieved by column chromatography using an alumina/silica column. One cm of alumina (type 60/e MERK) was placed in a 1 cm diameter glass column followed by 5 cm of silica (Korngrobe 0.040-0.063 mm particle size). The Total Neutrals were transferred to the column in a minimal quantity of chloroform, and the adsorbants eluted with *n*-heptane (100 ml) to give the Hydrocarbon fraction; *n*-heptane/dichloromethane (60:40/v:v; 100 ml) gave the Alcohols fraction which also contained some sterols. The alcohols were treated with bistrimethylsilyltrifluoroacetimide prior to gas chromatographic analysis as the corresponding trimethylsilyl ethers.

Isochratic HPLC using a Waters 6000A system with a 5μm cyanobonded silica column (Alltech 25 cm x 0.46 cm id.) coupled to a Refractive Index Detector (Varian) was used to isolate the monocarboxylic acid esters from the FAME fraction. The solvent system was *n*-heptane / *n*-propanol (98:2) at 1 ml/min.

GC analyses were carried out on a Varian 3700 GC equipped with a glass capillary column (30 m x 0.22 m id; BP-1 phase; SGE Australia) using hydrogen as carrier gas (30 ml/min), nitrogen as make-up gas (30 ml/min) and with flame ionisation detection. Identification of components was by GC retention times, and by GC/Mass Spectrometry using a HP5890 GC interfaced with a JEOL AX505HF mass spectrometer in electron impact mode at 70 ev. Confirmation of structures was achieved by the use of standards and/or literature data.

Results and Discussion.

Characteristics of the Lower Daintree River. At the time of sampling the river was shallow throughout the lower reaches, and the temperature at the base of the water column was close to 30°C (Table II). Salinity measurements shown in Table II suggest the existence of five zones: Zone 1 being fresh water through to Site 15 at the mouth of the river which has a salinity near that of an open marine regime. Salinities at Sites 8 and 13 show significant variations from those at adjacent sites and do not show the expected increasing trend towards a value for a marine environment. These sites are close to tributary entrances and,thus,can be impacted by fresh or less saline waters. Stations 12-14 are located where H_2S was noticeable in the grab sediments. This suggests that these are 'reducing' environments and contrasts with the apparently oxidising and clean white sand occurring at Station 15. The saltwater wedge reached as far upstream as Site 5 at the time of our sampling whilst the estuarine zone appears to start after Site 11.

Near Site 4 there were extensive seagrass beds, and the strength of the current prevented accumulation of sediment in mid-steam. The collection at Site 4 was, therefore, from the eastern side of the channel on the edge of the seagrass bed. The current runs in a narrow channel from Site 11 continuing along the east bank to the river

Table III. Biomarker Ratios

Site	CPI	Pr/Ph	TSE (mg/g)
1	5.4	-	7.1
2	6.0	-	3.4
3	5.5	-	2.3
4	3.8	-	3.7
5	3.5	-	2.7
6	11.7	-	3.4
7	4.6	-	5.2
8	3.5	-	2.6
9	4.5	1.3	6.2
10	3.2	-	8.0
11	3.2	2.2	5.3
12	5.1	1.3	5.5
13	4.8	1.9	6.0
14	3.2	0.8	13.1
15	1.4	0.3	2.7

- = Not calculable; Pr = pristane; Ph = phytane;
TSE = Total Solvent Extract;
CPI = Carbon Preference Index calculated according to reference (15).

mouth, thus, allowing sand bank formation in the broad estuary. Normally, finer sediments are associated with greater adsorption of organic matter, and this is reflected in the increase in the TSE concentrations along the transect until Site 15 is reached where a substantial drop in TSE (Table III) occurs. Here, the sediment is a clean white sand. It is likely that the river out-flow is energetic enough to carry finer particles over the bar at the mouth of the river and into the inner reef lagoon waters, thereby reducing the quantity of organic matter appearing as TSE at Site 15.

Biomarkers in the Sediments. Because the drainage basin of the lower Daintree River is a National Park, the sediment load is expected to carry biomarkers representative of a near pristine, freshwater, tropical environment. Between Stations 5-10 drainage from land being used for agricultural or pastoral purposes might be expected to enter the river. Initially in this discussion, therefore, there is an emphasis on distribution patterns for biomarkers in order to better assess which patterns may be associated with natural inputs from this specialised environment and which biomarkers tend to be anomalous, thus, pointing to anthropogenic inputs or the affects of salinity. Site 15 requires separate consideration.

Alcohol / Sterol Fraction. In these fractions of the TSE β-sitosterol, the common higher plant sterol marker, was by far the major sterol component, emphasizing the strong higher plant input (*4*). Phytol was a major component in the Alcohol fraction for sites other than Site 15, and repesentative distribution patterns are shown in Figure 2. In contrast to Site 15 (open marine environment), Site 13 shows a significant proportion of phytol in the GC distribution pattern in keeping with it being an estuarine sample (Zone 4, Table II), a region where much sediment deposition has occurred as evidenced by sand bank formation. The higher TSE concentrations found in the sediments taken at Stations 13 and 14 (Table III) can be linked with higher plant detritus associated with depositing sediment at these sampling sites. Site 1 has a high relative abundance of phytol in the alcohol fraction of the TSE reflecting the rich photosynthetic vegetation surrounding this sampling site. At sites down the transect from Site 1, the relative abundance of phytol increases. Since phytol is derived from chlorophyll in this environment (*5*) it can be used as a relative measure of photosynthetic biomass, and the observed increase in Zones 3 and 4 would suggest at least a dual input; from higher plants but also from planktonic or benthic algae within salinity Zones 3 and 4.

The *n*-alkanol distributions range from C_{12} to C_{28} and those $> C_{20}$ are taken as indicative of higher plant input (*6*). Apart from Site 15, the patterns reflect an input from higher plants. Site 1 is enriched in higher molecular weight alcohols and exhibits a C_{22} maximum (Figure 2).   Since the salinity at this site is zero , the alcohol distribution pattern may be considered typical of a freshwater tropical environment. Other sites show a strong relative abundance of C_{22} or C_{24} *n*-alkanols. These two strong peaks also appear in sediments taken from a transect off Cairns (cf. Figure 1) (*7*) and occur again in mangrove leaves taken from stands on the Low Isles (directly off the entrance of the Daintree River) (*8*) suggesting that the *n*-alkanols are probably derived from mangroves which line the river banks. Site 9 shows two strong high molecular weight alkanols which are presumably plant derived but for which we cannot yet ascribe a source. A limited series of 2- and 3- methyl alkanols also occur in the alcohol fractions. These have not been observed by earlier workers from this laboratory in mangrove sediments (*8,9*) and may represent other detrital plant material or the products of microbial reworking of organic material or cellular constituents of benthic micro-organisms.

Fatty Acid Distributions. The molecular weight distribution patterns show typical even / odd preferences in the $> C_{20}$ range commonly ascribed to higher plant sources (*10*), and in Figure 3 this is very marked in the pattern for this biomarker fraction from Sites 2 and 5. At Sites 1and 2 acids $< C_{20}$ are low in concentration. Moving down the transect,however, the proportion of fatty acids of carbon number $< C_{20}$ generally increases steadily whilst the higher plant indicators decrease. This trend is broken at Site 9 and this may be associated with drainage from the adjacent farm land entering at this site (cf. Figure 1). At Sites 13, 14 and 15, *n*-C_{16} is a dominant acid. Taken with the presence of the branched acids at C_{15-17}, the patterns are indicative of a change in the microbial community structure (e.g. bacteria) perhaps reflecting changes in the salinity regime or the less oxidizing sedimentary environment (Table I). It should be noted that Site 15 shows a strong pattern of contributions from (marine) microbiota,but contributions from higher plants (i.e., acids $>n$-C_{20}) are minimal. The presence of a

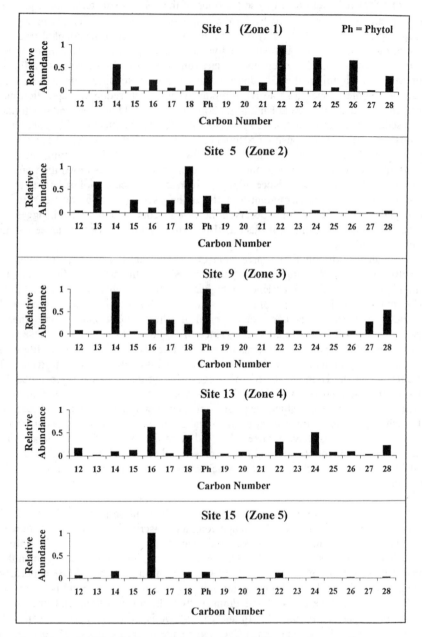

Figure 2. Distribution patterns showing relative abundances of
n-alkanols and phytol from benthic sediments at sites 1, 5, 9, 13 and
15. See Figure 1 for site locations.

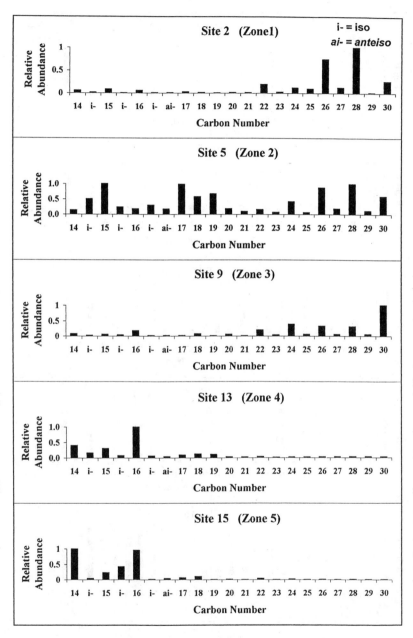

Figure 3. Distribution patterns for fatty acids from the benthic sediments at sites 2, 5, 9, 13, 15. See Figure 1 for site locations.

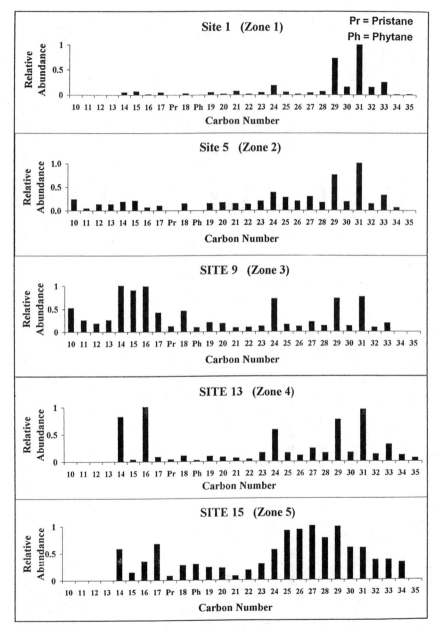

Figure 4. Distribution patterns for *n*-alkanes, pristane and phytane in the benthic sediments from sites 1, 5, 9, 13, 15. See Figure 1 for site locations.

strong n-C_{14} acid at Station 15 comparable in abundance to n-C_{16} is unusual in our experience and suggests a marine planktonic source rich in this cellular constituent. The fatty acid patterns suggest changes in the benthic microbiota and a study of the phospholipid constituents which reflect community structure would be rewarding.

Sedimentary n-Alkanes. The n-alkane patterns in Figure 4 and the CPI values between 3.2-5.4 (Table III) for alkanes with C>20 reflect higher plant input. There are two features, however, which require comment. Excluding Site 15, n-C_{24} occurs at an enhanced relative abundance in most of the alkane patterns. From Sites 9 to 14 n-C_{14} and n-C_{16} also occur at enhanced relative abundances in sediments. It is a reasonable hypothesis that the likely input sources are the mixed vegetation lining the river banks. If this is so, then biomarker patterns for tropical environments such as the Daintree River sediments can be expected to follow similar patterns. Biomarkers peculiar to particular plant species reflecting specific environments, however, can be expected to quantitatively vary the abundances of these organic components in the carbon number distribution patterns.

Biomarker Evidence for Hydrocarbon Impacting Contamination. The alcohol and fatty acid biomarker patterns at Site 15 point to a negligible higher plant input to this sediment. The alkane pattern is anomalous in that although alkanes are clearly present at Site 15, the pattern is not that of higher plants. The low CPI value of 1.4 is consistent with a second contributing source of alkanes to these sediments. The fatty acid pattern is so unlike that for the alkanes, there can be no inter-relationship *via* a decarboxylation step.

When the gas chromatograms of the hydrocarbon fractions were examined a strong baseline rise typical of an Unresolved Complex Mixture (UCM) hump was observed in the hydrocarbons at Site 15 and Sites 7 through 14. Figure 5 illustrates three typical GC traces including that for Site 5 where no UCM hump is observed. Such data have become accepted as strong evidence for a petroleum input to sediments *(11)*.

This anthropogenic input could be derived either from within the river or from shipping in the inner reef lagoon. This last option is the least likely given the flow of the river *into* the reef waters. Figure 4 provides pertinent evidence. Whilst pristane has been found in unpolluted Recent sediments *(12)*, phytane is a component of fossil fuels *(13)*. More recent studies have shown that archaebacteria can be the source of isoprenoid hydrocarbons such as phytane *(14)*, but at Site 15 where the sediment is apparently oxidising (Table I) these are an unlikely source. Phytane can be identified with certainty from Sites 9 to 15 which allows the pristane / phytane ratio to be calculated as shown in Table III. The decrease in this ratio i.e., the increase in abundance of phytane in the estuary Zone 4 and at Site 15 requires an allochthonous source for phytane which we believe to be a fossil fuel.

Conclusions.

The biomarker distribution patterns observed along the transect are consistent with higher plant inputs and can be interpreted in terms commonly used for patterns from temperate sediments. Unexplained variations occur, however, in the n-alkane (C_{24}, C_{14} and C_{16} maxima) and some fatty acid distribution patterns (in the C_{14-16} region)

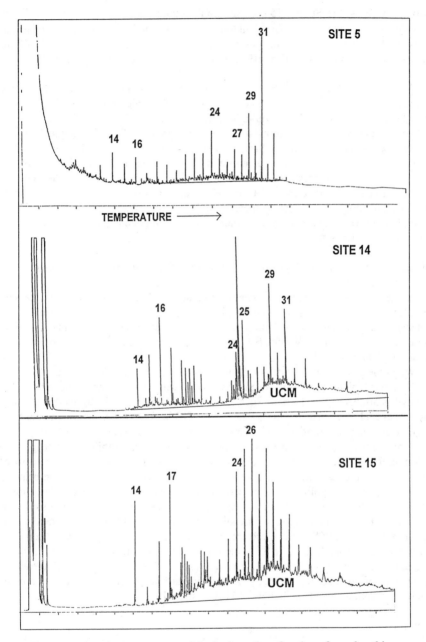

Figure 5. Gas chromatograms of the hydrocarbon fractions from benthic sediments from sites 5, 14, 15 illustrating the UCM humps at Sites 14 and 15.

which appear to reflect input from higher plant species peculiar to this tropical environment.

The UCM, pristane / phytane ratios and the *n*-alkane distribution pattern at Site 15 at the river mouth are consistent with a petroleum contribution to this sediment sample. The biomarker data can be used to track an anthropogenic hydrocarbon input from Site 9 through the estuary to the mouth of the river. This preliminary data suggests that the lower Daintree River system is being impacted by hydrocarbon contamination and that further study is warranted.

Acknowledgments.

The authors acknowledge an ARC Small Grant in part funding of this project and thank the reviewers for constructive contributions.

References

1. Review of Australia's Water Resources and Water Use, 1985. A.W.R.C. Aust. Govt. Publishing Service, Canberra. 1987, Vol. 1, pp 46.

2. Australian and International Visitors to Regions of Queensland. Govt. Report, Bureau of Tourism Research. Canberra, 1995.

3. Metcalf, L.D. and Schmitz, A.A. *Org. Geochem.,* **1961,** *33,* 363.

4. Volkman, J.K. *Org. Geochem.,* **1986,** *9.* 83-99.

5. Volkman, J.K. and Maxwell, J. R. in *Biological Markers in the Sedimentary Record* . Johns, R.B. Ed. Methods in Geochemistry and Geophysics 24; Elsevier, Amsterdam, **1986,** pp 1-42.

6. Cranwell, P.A. *Chem. Geol.,***1988,** *68,* 181-197.

7. Brady, B.A. *Ph.D. Thesis.* University of Melbourne. **1996.**

8. Nichols, P.D. *Ph.D. Thesis.* University of Melbourne. **1983.**

9. Shaw, P.M., and Johns, R.B. *Org. Geochem.,* **1985,** *8,* 147-156.

10. Wannigama, G.P., Volkman, J.K., Gillan, F.T., Nichols, P.D., and Johns, R.B. *Phytochemistry,* **1981,** *20,* 659-666.

11. Venkatesan, M.I., Brenner, S., Ruth, E., Bonilla, J., and Kaplan, I.R. *Geochim. Cosmochim. Acta,* **1980,** *44,* 789-802.

13. Welte, D.H., and Waples, D. *Naturwissenshaften,* **1973,** *60,* 516-517.

12. Blumer, M., and Snyder, W.D. *Science,* **1965,** *150,* 1588-1589.

14. Rowland, S.J. *Org. Geochem.,* **1990,** *15,* 9-16.

15. Bray, E.E., and Evans, E.D. *Geochim. Cosmochim. Acta,* **1961,** *22,* 2-15.

Chapter 11

Preliminary Study of Organic Pollutants in Air of Guangzhou, Hong Kong, and Macao

Jiamo Fu[1], Guoying Sheng[1], Yu Chen[1], Xinming Wang[1], Yushun Min[1],
Ping'an Peng[1], S. C. Lee[2], L. Y. Chan[2], and Zhishi Wang[3]

[1]Guangzhou Institute of Geochemistry, Wushan, Guangzhou 510640,
People's Republic of China
[2]Hong Kong Polytechnic University, Hung Hom, Kowloon, Hong Kong
[3]University of Macao, P.O. Box 3001, Macao

Preliminary investigations were carried out on the extractable organic
matter in aerosols and volatile organic compounds (VOCs) in ambient air
of Guangzhou, Hongkong and Macao. Based on the types and
abundances of biomarkers in the aliphatic and aromatic fractions of the
extracts, as well as contaminant assemblages, a brief discussion has been
made on the sources of organic compounds in air. Our studies reveal
that organic compounds from incomplete combustion of fossil fuels,
especially from vehicular exhaust, predominated both in aerosols and
among VOCs in these samples. However, geological dust and modern
living plants also contributed to the organics in ambient air.

Toxic organic compounds either in the gas phase or in association with suspended
particles in ambient air are of worldwide concern because of their potential impacts on
human health. In this context, molecular markers have been used successfully to identify
the sources of organic air pollutants. The compositions of solvent extractable organic
matter in aerosols have been widely investigated outside China (1-8). The distribution
and sources of polycyclic aromatic hydrocarbons (PAHs) in atmosphere have been the
focus of interest in many studies (9-13). However, studies of organics in aerosols in
China were not reported until 1985 (14-17), and only aerosols from a few large cities
had been investigated. Sheng Guoying et al. attempted to use molecular markers to
identify the sources of hydrocarbons in aerosols from large cities like Beijing, Guiyang
and Guangzhou (16-17). Reports about VOCs in ambient air have also increased in
recent years (18-21). However, because of the difficulty of determining VOCs in air
(22), there are almost no reports yet about VOCs in air from China.

Guangzhou, Hongkong and Macao are among the cities in the Pearl River
Mouth region where air quality has been getting worse in recent years due to rapid
development of industry and increasing population. This paper presents a preliminary
study of the distributions and origins of VOCs in ambient air and extractable organic

matter in aerosols from Guangzhou, Hongkong and Macao in the Pearl River Delta (Figure 1) in order to investigate organic pollution and the possible sources of organic pollutants.

Sampling

Guangzhou Environmental Monitoring Center and Macao Geophysics & Meteorological Service have established environmental monitoring systems in Guangzhou and Macao, respectively, based on the environments of the sampling locations. In this study, we took advantage of these sampling networks. Aerosol samples from the two cities were collected at the monitoring stations. Altogether 70 samples (54 samples from Guangzhou, 4 samples from Hongkong and 12 samples from Macao) for VOCs analysis were collected. Besides air samples from the same locations (aerosols), other urban and suburban samples were collected in Guangzhou and Macao for this comprehensive investigation. In Guangzhou, samples from locations such as a landfill, automobile exhaust, a petrochemical plant, waste water treatment plant, etc., which were considered point sources, were also collected.

Aerosols. Aerosol samples were collected with high volume Total Suspended Particles (TSP) samplers by filtration of air through a 8" ×11" quartz fiber filter (2500 QAT, UP model, Pallflex Products Corp.) at a rate of 1.05 m^3/min for 24 hours. Before sampling, the quartz fiber filters were baked in a muffle furnace at 450°C for 4 hours and then wrapped with aluminum foil. After sampling was completed the quartz fiber filters containing aerosol particles were placed separately into gastight glass bottles that had previously been rinsed with methylene chloride. The air volumes sampled were automatically measured by integration of flow rate over time, and the data, including average flow rate, average temperature and total volume, were stored in the control system of the samplers. The aerosol samples analyzed are listed in Table I.

VOCs. Samples from Guangzhou were collected at 10:00pm on July 10, 1995. Samples from Macao were collected on November 20, 1995. Using a TMP-1500 Timing Minipump with flow-meter (Jianhu Electric Instrument Factory, China), ambient air was drawn through a 7"×1/4" multi-bed stainless steel tube (Tekmar Company, USA) packed with silica gel, Carbonsieve and charcoal. Thus, VOCs in the air were trapped onto solid adsorbents in the tube. The sampling tubes were then sealed in glass containers and transported back to the laboratory for analysis.

Laboratory Analysis

Aerosols. Detailed procedures of extraction and separation of organics in aerosols can be found elsewhere (*15*). Aliphatic and aromatic hydrocarbons were analyzed with a Hewlett Packard 5890 Series II gas chromatograph (GC) equipped with a flame ionization detector and a Hewlett Packard 5972 gas chromatograph/mass selective detector (GC/MSD). For determination of PAHs in aerosols, a standard mixture of 16 polycyclic aromatic hydrocarbons (naphthalene, acenaphthylene, dihydroacenaphthylene, fluorene, phenanthrene, anthracene, fluroranthene, pyrene, benzo(a)anthracene,

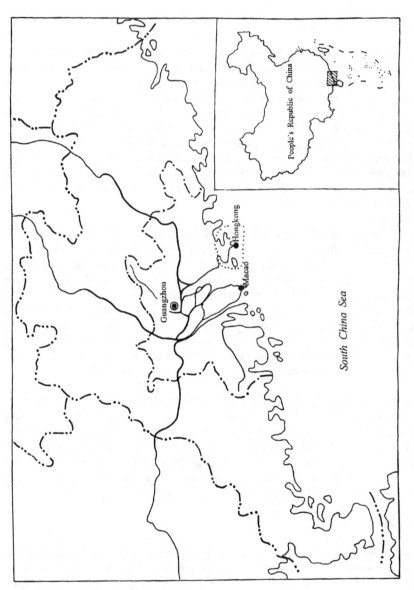

Figure 1. Locations map of the air sampling sites in China.

chrysene, benzo(b)fluroranthene, benzo(k)flouroranthene, benzo(a)pyrene, indeno(1,2,3- cd)pyrene, dibenzo(ghi)pyrene) was used. The purities of individual PAHs were above 98%. The recovery ranged from 80.7 to 110%. The results of blank experiment showed that traces of phthalate esters including diethyl and dibutyl phthalates were detected in the aromatic hydrocarbon eluate. These contaminants did not interfere with the recognition or quantification of the compounds of interest.

Table I Sample location, environmental data and data for extracts of aerosols from Guangzhou, Hongkong and Macao

sample designation*	Elevation above ground(m)	sampling time	Extractable organic matter(EOM) ($\mu g/m^3$)	Normal alkanes		
				Carbon number range	C^{**}_{max}	CPI***
G-S-1	14	4/1994	6.28	14-35	29	1.22
G-S-2	9	4/1994	1.50	n.d.	n.d.	n.d.
G-U-3	12	4/1994	10.47	n.d.	n.d.	n.d.
G-U-4	20	4/1994	7.61	14-34	29	1.21
G-U-5	1.3	4/1994	8.26	14-36	25, 29	1.18
G-U-6	12	4/1994	11.00	14-35	27, 29	1.13
G-S-7	14	4/1994	3.74	n.d.	n.d.	n.d.
G-S-8	9	7/1994	8.97	14-36	29, 31	1.98
G-U-9	20	7/1994	8.61	14-36	29, 31	1.59
G-U-10	12	7/1994	9.93	14-36	29,31	1.39
G-U-11	1.3	7/1994	12.38	n.d.	n.d.	n.d.
G-S-12	14	11/1994	9.32	14-36	25,27	1.20
G-S-13	9	11/1994	35.31	14-36	25	1.26
G-U-14	20	11/1994	51.70	14-36	25,29	1.31
G-U-15	1.3	11/1994	35.5	15-34	22	0.91
G-U-16	12	11/1994	61.04	14-37	27,31	1.49
G-S-17	16	11/1994	28.56	n.d.	n.d.	n.d.
H-U-18	32	1/1995	5.83	17-33	23,24	1.11
H-U-19	32	1/1995	9.21	18-33	23,24	1.07
H-S-20	20	3/1995	3.09	17-35	27,29	1.13
H-S-21	20	3/1995	2.32	17-34	27,29	1.22
M-U-22	15	11/1995	50.13	12-38	24	1.14
M-U-23	1.3	11/1995	36.57	15-39	24	1.16
M-U-24	20	11/1995	19.38	12-38	29	1.39
M-U-25	12	11/1995	19.32	12-38	25	1.26
M-S-26	1.3	11/1995	5.75	13-38	29,31	1.88
M-S-27	1.3	11/1995	19.18	14-38	29	1.55

n.d.=Not determined.
* G=Guangzhou, H=Hongkong, M=Macao, U=Urban, S=Suburb.
** Carbon chain maximum in homologous series.
*** Carbon Preference Index : it is expressed as a summation of the odd carbon number homologs over a range (C_{24}-C_{34}) divided by a summation of the even carbon number homologs over the same range. CPI=A/2[1/A+1/B], A=$\Sigma C_{25}+C_{27}+C_{29}+C_{31}+C_{33}$, B=$\Sigma C_{24}+C_{26}+C_{28}+C_{30}+C_{32}$, C=$\Sigma C_{26}+C_{28}+C_{30}+C_{32}+C_{34}$

VOCs. The determination of VOCs was conducted with a Tekmar 6032 Aerotrap and Tekmar 3000 Purge & Trap Concentrator interfaced to a Hewlett Packard 5972 GC/MSD. The sampling tubes were inserted into the Tekmar 6032, and volatile substances were thermally desorbed and carried in a stream of helium to the GC/MSD for identification and quantitative analysis. Blank sampling tubes were transported and analyzed the same way. For quantitative analysis of VOCs, standards with 60 compounds were prepared in a stainless steel canister (10 liters in volume with initial pressure of 10 mPa) at a concentration of 100 $\mu g/m^3$ in helium for each compound. The recoveries of standard compounds ranged from 47 to 138%. Standard gas mixtures were analyzed using identical conditions to those used for the field samples. Multipoint calibration of the GC/MSD was performed, and quantitation was done by the external standard method. HP 5972 GC/MSD conditions: capillary column: HP 5MS 30m×0.25mm×0.25µm, Initial Temp: 35°C, hold 4min to 200°C at a rate 8°C/min, hold 5min. ionization: EI; EM Volts: 1988 volts; Emission: 50ev; Mass Range: 35-280 amu.

Results and Discussion

Part I Aerosols

Concentrations of extractable organic matter (EOM). Concentrations of EOM in Guangzhou aerosols ranged from 1.5 to 61.0 $\mu g/m^3$, and in two samples from heavy traffic sites (namely, G-U-16 and G-U-14) concentrations were more than 50 $\mu g/m^3$ (Table I). With regard to three groups of samples collected in different seasons, EOM concentrations collected in November were higher than those in samples collected in April and July. Average EOM concentrations in samples collected in April, July and November were 7.52, 8.73 and 36.9 $\mu g/m^3$, respectively, so EOM concentrations in November were approximately five times those collected in April and July. In Guangzhou, the air temperature ranges from 0° to 40° C, with an annual average of 23-25° C. The spring and autumn samples were collected in April and July respectively, times during which heavy rainfall occurs in Guangzhou. The typhoon season also occurs mostly in July. Consequently, precipitation probably leads to removal of suspended particles as well as volatile organic pollutants in the air. However, in winter Guangzhou is characterized by clean skies, less rain and reduced winds. During this period it is more difficult for aerosols and air pollutants to precipitate or be removed, possibly explaining the higher EOM concentrations in winter. In a given season, EOM concentrations varied depending on the character of the sampling locations. For example, heavy traffic areas (G-U-6 and G-U-16) tend to exhibit higher EOM concentrations (Table I).
EOM concentrations in aerosols from Macao and Hongkong are 5.8~50.1 mg/m^3 and 2.3~9.2 mg/m^3 (Table I), and averaged 25.2 mg/m^3 and 5.1 mg/m^3, respectively. In aerosols from the three cities, EOM concentrations are highest in Macao and lowest in Hongkong.

Saturated Hydrocarbons. Three normal alkane distribution patterns were observed in aerosols from Guangzhou, Hongkong and Macao. Type I is characterized by C_{max} (n-alkane carbon chain maximum in homologous series) at n-C_{22} to n-C_{24} and weak or no odd to even carbon number predominance (i.e. CPI of approximately 1.0). Gas chromatograms for samples of this type have an unresolved complex mixture (UCM), characteristic of the residue from incomplete combustion of fossil fuels. Type II has a bimodal distribution (n-C_{25} and n-C_{27} or n-C_{29}). N-alkanes greater than n-C_{25} exhibit measurable odd to even carbon number predominance, with CPI values between 1.1 to

1.3 (Table I). This distribution reflects two sources: plant waxes and fossil fuel pollution from human activities. Samples collected in industrial areas and other low traffic sites often show this kind of distribution. The type III distribution is characterized by *n*-alkanes with longer chains (C_{max} at *n*-C_{29} or *n*-C_{31}) and CPI values near 2.0. This indicates a significant input from higher plant waxes. Type III distributions mostly occurred in samples from green belt sites with areas having less traffic and population (e.g. G-S-8 and M-S-26, Table I). However, we also observed UCMs in gas chromatograms of this type, indicating the presence of background pollution.

Pristane (I in Appendix I) and phytane (II) were widely detected in aerosols, and the ratios of Pr/Ph range from 0.7 to 0.9. The probable source of these two compounds is fossil fuel utilization (i.e. vehicular traffic) and geological dust.

Hopanes (III) and steranes (IV), observed commonly in crude oil, coal and sediments, were found in all aerosol samples analyzed. The extended $17\alpha(H),21\beta(H)$-hopane series appeared to be thermally mature with approximate equal abundances of S- and R-C_{32} hopanes. C_{29} neohopane (V) was also found in all samples. However, gammacerane (VI) found in Guangzhou samples was not detected in Hongkong samples. C_{27}-C_{29} $5\beta(H)$-steranes were not detected in any samples, but the abundance of C_{27}-C_{29} $5\alpha(H),14\beta(H),17\beta(H)$-steranes is higher than that of C_{27}-C_{29} $5\alpha(H),14\alpha(H),17\alpha(H)$-steranes. The presence of the geologically mature polycyclic hydrocarbons indicates that some portion of the organics in these aerosols were derived from fossil fuels or dust originating from diagenetically mature sediments.

Sesquiterpanes and diterpanes found in aerosols, such as cuparene (VII), cadalene (VIII), tetrahydrocadalene (IX), α-cedrene (X) and 19-norabietane (XI), are related to higher plant waxes or resins. Moreover, α-cedrene and dehydroabietane (XII) have been suggested as markers of inputs from conifer plants. The presence of sesquiterpanes and diterpanes in aerosols reflects a contribution from modern plants to the organic matter in aerosols (*23*).

Polycyclic Aromatic Hydrocarbons. More than seventy PAHs were detected in the aerosols, and among them 16, 10 and 15 of the US EPA priority pollutants were found in aerosols from Guangzhou, Hongkong and Macao, respectively (Table II). The priority pollutants that occurs most commonly are naphthalene, acenaphthylene, dihydroacenaphthylene, fluorene, phenanthrene, anthracene, fluoranthene, pyrene, benzo[a]anthracene, chrysene, benzo[a]fluoranthene, benzo[k]fluoranthene, benzo[a]pyrene, indeno(1, 2, 3-cd)pyrene, dibenzo(a, h)anthracene and benzo(ghi)perylene. The concentrations of PAHs in aerosols varied between locations. In Guangzhou, the highest concentrations occurred in commercial regions with heavy traffic (e.g. G-U-14 and G-U-16, Table II). Benzo(a)pyrene, considered to be a product of incomplete combustion of fossil fuels, was found in more than 90% of the samples.

The Ratio of *m*-Quaterphenyl to Fluoranthene. *M*-quaterphenyl has been detected in dust from solid waste incinerators and in exhaust from combustion of polymers. It is considered to be a marker of high temperature polymer combustion (*25, 26*). *M*-quaterphenyl was found in aerosols from Guangzhou, Hongkong and Macao, especially in one sample from the Hongkong seaside (H-U-18). In that sample, *m*-quaterphenyl was found to be most abundant compound in the aromatic fraction (Figure 2). Because

Table II The Composition of Priority Pollutant PAHs in Aerosols of Guangzhou, Hongkong and Macao (ng/m^3)

Samples	Nap	Ace	Dih	Flu	Phe	Ant	Flua	Pyr	Ben(a)	Chr	Ben(b)	Ben(k)	Ben(a)p	Ind	Dib	Ben(ghi)	∑Priority PAHs
G-S-1	n.d.	n.d.	n.d.	n.d.	0.23	n.d.	0.25	0.24	tr.	0.39		0.41*	0.54	0.67	tr.	0.78	3.51
G-S-2	n.d.	n.d.	tr.	tr.	0.06	n.d.	0.12	0.10	0.08	0.15		0.29	0.08	0.19	tr.	0.22	1.29
G-U-3	n.d.	n.d.	n.d.	n.d.	1.27	n.d.	2.45	2.15	4.32	4.16		13.32	7.07	9.03	1.64	10.08	55.49
G-U-4	n.d.	n.d.	n.d.	n.d.	tr.	n.d.	0.27	0.20	0.29	0.37		0.85	0.40	0.58	tr.	0.78	3.74
G-U-5	n.d.	n.d.	n.d.	n.d.	tr.	n.d.	0.81	0.47	0.95	2.08		2.89	1.58	1.90	tr.	2.17	12.85
G-U-6	n.d.	n.d.	n.d.	n.d.	0.17	n.d.	0.26	0.21	0.40	0.46		1.19	0.84	0.86	tr.	1.12	5.51
G-S-7	n.d.	0.17	0.18	n.d.	0.24	n.d.	1.32	0.26	0.14	0.19		0.23	0.17	0.16	tr.	0.18	3.24
G-S-8	n.d.	n.d.	n.d.	n.d.	0.48	n.d.	0.74	0.66	0.74	1.14		4.94	1.94	3.83	0.72	4.38	19.57
G-U-9	n.d.	tr.	tr.	tr.	0.51	n.d.	1.19	0.54	0.76	1.10		2.86	0.88	1.79	tr.	2.23	11.86
G-U-10	n.d.	tr.	tr.	tr.	0.38	n.d.	0.59	0.31	0.40	0.51		2.12	0.85	1.77	0.40	2.25	9.58
G-U-11	tr.	tr.	tr.	tr.	0.87	n.d.	0.97	0.83	0.95	1.12		4.22	1.73	2.90	tr.	3.78	17.37
G-S-12	n.d.	tr.	tr.	tr.	0.99	n.d.	2.42	1.83	2.02	4.01		6.57	2.34	3.36	tr.	3.38	26.92
G-S-13	n.d.	n.d.	n.d.	n.d.	2.35	n.d.	3.63	3.51	6.35	10.67		26.32	12.06	16.15	4.46	18.14	103.64
G-U-14	n.d.	n.d.	n.d.	n.d.	4.94	n.d.	11.54	12.31	16.29	27.07		52.01	25.03	22.16	6.12	23.43	200.90
G-U-15	0.23	0.14	n.d.	0.14	2.25	tr.	5.03	5.46	4.63	5.78		8.68	5.16	6.27	0.63	8.24	52.64
G-U-16	n.d.	n.d.	n.d.	n.d.	4.42	n.d.	7.22	7.85	11.97	33.53		58.97	21.69	21.70	2.54	25.75	193.10
G-S-17	n.d.	n.d.	n.d.	n.d.	1.70	n.d.	3.66	3.68	6.99	12.87		22.00	10.78	9.66	2.54	9.07	82.95
H-U-18	0.41	n.d.	n.d.	n.d.	n.d.	n.d.	1.04	1.25	1.46	1.49	n.d.	4.39	1.53	1.43	0.13	1.75	14.88
H-U-19	0.27	n.d.	n.d.	n.d.	n.d.	n.d.	0.49	0.47	1.61	1.33	n.d.	3.29	1.91	0.90	tr.	1.05	11.32
H-S-20	0.02	n.d.	n.d.	n.d.	n.d.	n.d.	0.05	n.d.	n.d.	n.d.	n.d.	0.05	tr.	n.d.	n.d.	n.d.	0.12
H-S-21	0.03	n.d.	n.d.	n.d.	n.d.	n.d.	0.04	0.02	n.d.	n.d.	n.d.	0.09	n.d.	n.d.	n.d.	n.d.	0.18
M-U-22	0.75	n.d.	n.d.	0.06	2.05	0.21	3.01	4.34	2.31	3.39	3.82	4.17	2.61	4.21	0.46	7.24	38.63
M-U-23	n.d.	n.d.	n.d.	n.d.	1.03	tr.	1.19	2.19	2.42	2.80	2.49	n.d.	0.94	2.80	0.51	4.10	20.47
M-U-24	n.d.	tr.	n.d.	0.04	0.83	tr.	1.12	1.19	0.67	1.24	1.91	1.51	0.80	2.01	tr.	2.67	13.99
M-U-25	n.d.	n.d.	n.d.	n.d.	0.91	tr.	1.34	1.50	0.48	0.86	n.d.	n.d.	n.d.	1.09	tr.	2.83	9.01
M-S-26	n.d.	tr.	n.d.	0.40	1.15	tr.	1.25	1.26	0.52	0.83	1.04	0.76	0.48	0.83	tr.	1.04	9.56
M-S-27	n.d.	n.d.	n.d.	n.d.	0.89	tr.	1.25	1.15	0.40	1.94	1.48	2.13	0.52	1.29	tr.	1.62	12.67

Nap=Naphthalene; Ace=Acenaphthylene; Dih=Dihydroacenaphthylene; Flu=Fluorene; Phe=Phenanthrene; Ant=Anthracene; Flua=Fluoranthene;
Pyr=Pyrene; Ben(a)=Benzo(a)anthracene; Chy=Chrysene; Ben(b)=Benzo(b)fluoranthene; Ben(k)=Benzo(k)fluoranthene; Ben(a)p=Benzo(a)pyrene;
Ind=Indeno(1,2,3.-cd)pyrene; Dib=Dibenzo(a,h)anthene; Ben(ghi)=Benzo(ghi)perylene.
tr.=trace; n.d.=not determined   * Total concentration of Ben(b) and Ben(k).

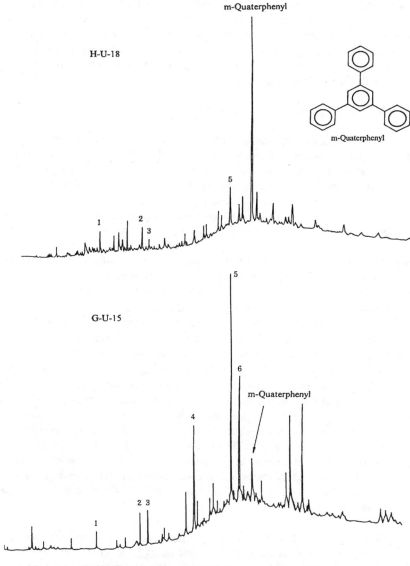

Figure 2. Gas chromatograms of PAHs in aerosols from Hongkong and Guangzhou. 1. Phenanthrene; 2. Fluoranthene; 3. Pyrene; 4. Chrysene; 5. Benzo(k)fluoranthene; 6. Benzo(a)pyrene.

fluoranthene can be found in all samples, the ratio of m-quaterphenyl to fluoranthene (m-Q/Fl) was used to show the relative abundance of m-quaterphenyl in different samples. In two samples from a Hongkong suburb (H-S-20 and H-S-21), the m-Q/Fl ratios were 12.2 and 22.4 respectively, and in Guangzhou aerosols the m-Q/Fl ratios were 0.05~2.42, whereas for Macao aerosols, ratios were 0.49~0.81. In Hongkong, solid wastes are disposed of by incinerator, but in Guangzhou they are dumped into landfills. Thus, the higher ratios seen in Hongkong may reflect the introduction of dust from solid waste incinerators to the atmosphere.

Part II Volatile Organic Compounds in Ambient Air

Aromatic and aliphatic hydrocarbons (AAHs), chlorinated hydrocarbons (CHs) and terpenes together contributed more than 90 % of total VOCs (T_{VOCs}) in most samples. In automobile exhaust, AAHs totaled nearly 100 percent of T_{VOCs}. In air samples collected near streets and roads with heavy traffic high percentages (>80%) of AAHs were also observed, and the contributions of terpenes to T_{VOCs} were less than 5% (1.2~4.6% in Macao). However, the percentage of terpenes increased in suburban air or near greenbelts (9.8~22% in Guangzhou), indicating a greater input of biogenic compounds.

Aromatic hydrocarbons. Aromatic hydrocarbons dominated the VOCs in ambient air from Guangzhou and Macao. Benzene, toluene, ethylbenzene, xylene (BTEX) and styrene were the most abundant aromatic hydrocarbons in most samples, with toluene ranging from 5.4 to 190 $\mu g/m^3$ in Guangzhou, and from 18.5 to 315 $\mu g/m^3$ in Macao. In locations with heavy traffic, not only BTEX, but also C_3- and C_4-benzenes were found in high concentrations (totaled 210-815 $\mu g/m^3$ in Macao and 169-672 $\mu g/m^3$ in Guangzhou) compared with samples collected in greenbelts or in suburban areas (43.8-119 $\mu g/m^3$ in Macao and 22.6-106 $\mu g/m^3$ in Guangzhou) with less traffic. In Hongkong samples, C_3-benzenes and C_4-benzenes concentrations were lower (21.1-154.3 $\mu g/m^3$), probably because of the better control of automobile exhaust in Hongkong. We noticed that the ratio of styrene to o-xylene varied with samples taken near landfills and petrochemical plants, with higher ratios (1.51-2.53) than those in automobile exhaust (averaged 0.52), and in heavy traffic locations (avg.= 0.78 in Macao and 0.91 in Guangzhou). This might reflect the degradation of polystyrene plastics or solvents dumped in landfills, and synthesis and use of styrene at petrochemical plants. An explanation for the low ratios in air from heavy traffic locations is that VOCs came mainly from automobile exhaust.

Cyclic hydrocarbons. D-Limonene (XIII), α-pinene (XIV), β-pinene (XV), camphene (XVI) and myrene, which are considered compounds of plant origin, were found in 1.2-22 % of the samples tested. Limonene, which contributes to the odor in landfills, and originates from the microbial degradation of plants (e.g. vegetables; 22), was the most abundant compound in some air samples from the landfill, indicating that landfills are a possible source of terpenes in air. Air samples from the landfill and greenbelt sites had higher concentrations of such compounds (110-330 $\mu g/m^3$ in landfill air and 19.8-236 $\mu g/m^3$ in greenbelts), but their compositions were distinct (Figure 3). For example, in air

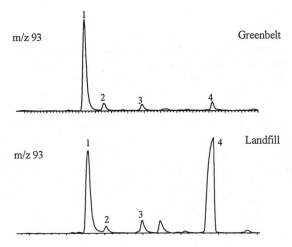

Figure 3 Typical distribution patterns of cyclic hydrocarbons in greenbelt and landfill air samples. 1. α-pinene; 2. β-pinene; 3. camphene; 4. D-limonene.

samples from the greenbelt α-pinene predominated over limonene, whereas in landfill air the reverse was true.

Chlorinated hydrocarbons. Chlorinated hydrocarbons were detected in most samples. These include: chloroform ($CHCl_3$), carbon tetrachloride (CCl_4), trichloroethylene (C_2HCl_3), tetrachloroethylene (C_2Cl_4), chlorobenzene (C_6H_5Cl) and dichlorobenzene ($C_6H_4Cl_2$). Generally, percentages of CHs in T_{VOCs} ranged from 2.8 to 26.8% in samples tested. In air samples from industrial areas, CHs have higher percentages (12.1-26.8%). This might be connected with their use as solvents or raw materials in industry. At the landfill site, tetrachloroethylene, ethylene chloride and trichloroethylene existed in relatively higher concentrations in the air (totaled 34.1-75.9 $\mu g/m^3$). Chloroform was found in a sample collected near the reaction pool of a tap water plant at an unusually high concentration (15.8 $\mu g/m^3$). Its presence probably results from the formation of chloroform during disinfection of source water by liquid chlorine in the treatment plant.

Conclusions

Toxic organic pollutants (PAHs, BTEX, chlorinated hydrocarbons) were detected in ambient air from Guangzhou, Hongkong and Macao. Organics from incomplete combustion of fossil fuels, especially from vehicular exhaust, predominated both in aerosols and among VOCs in these samples. However, geological dust and modern living plants also contribute to the organics in ambient air in the area studied. Our preliminary study shows that the ratio of m-quaterphenyl to fluoranthene, the ratio of styrene to o-xylene, etc., may be useful indicators of different pollution sources connected with human activity. Among the three cities, either from the study of aerosol pollutants or from the analysis of VOCs in the ambient air, Macao shows the most serious organic pollution in its ambient air. This is probably because of the heavy automobile traffic, and the narrow streets and high buildings which impede flushing of pollutants. In Hongkong, pollutants from vehicular exhaust appear to play a less significant role in ambient air than that in Guangzhou and Macao.

Acknowledgments

We thank Guangzhou Environmental Monitoring Center and Macao Geophysics & Meteorological Service for their great help in sample collection. We thank MS Lab in State Key Laboratory of Organic Geochemistry for their help in GC/MS analysis. We are indebted to Mr. Liu Dehan, Mr. Lei Jianquan, Mr. Chen Jianxin and Mr. Liu Zuhua for their help in field sampling. Many thanks to Ms. Hu Yunxia for typing the manuscript. We thank Dr. Robert P. Eganhouse and two anonymous reviewers for comments and suggestions which greatly improved this paper. Partial financial support from China Natural Sciences Foundation (project No. 49375249 and No. 49632060) is gratefully acknowledged.

APPENDIX I: Chemical structures cited

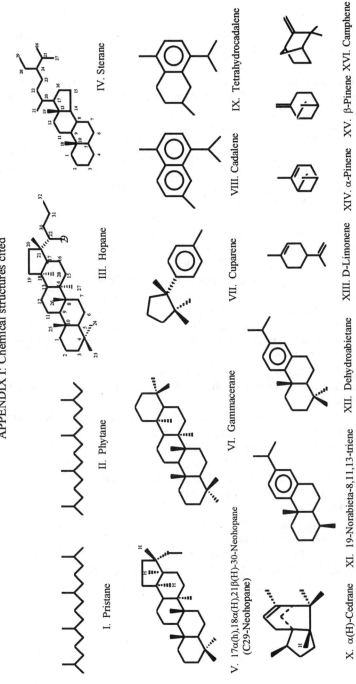

I. Pristane

II. Phytane

III. Hopane

IV. Sterane

V. 17α(h),18α(H),21β(H)-30-Neohopane (C29-Neohopane)

VI. Gammacerane

VII. Cuparene

VIII. Cadalene

IX. Tetrahydrocadalene

X. α(H)-Cedrane

XI. 19-Norabieta-8,11,13-triene

XII. Dehydroabietane

XIII. D-Limonene

XIV. α-Pinene   XV. β-Pinene   XVI. Camphene

Reference

1. Ketseridis, G.; Hahn, J.; Jaenicke, R.; Junge, C. *Atmos. Environ.* **1976**, *10*, 603-610.
2. Broddin, G.; Cautreels, W.; Van Cauwenberghe, K. *Atmos. Environ.* **1980**, *14*, 895-910.
3. Cox R.E.; Mazurek, M.A.; Simoneit, B.R.T. *Nature* **1982**, *296*, 895-910.
4. Mazurek, M.A.; Simoneit, B.R.T. In *Identification and Analysis of Organic Pollutants in Air, ACS Symp.*(edited by Keith, L. H.), Ann Arbor Science/Butterworth, MA. **1984**, pp 353.
5. Simoneit, B.RT. *Atmos. Environ.* **1984**, *18*, 51-67.
6. Simoneit, B R.T. *J. Atmos. Chem.* **1989**, *8*, 251-275.
7. Jaffrezo, J.L.; Clain, M.P.; Masclet, P. *Atmos. Environ.* **1994**, *28*, 1139-1145.
8. Standley, L.J.; Simoneit, B.R.T. *J. Atmos. Chem.* **1994**, *18*, 1-5.
9. Gschwend, P.W.; Hites, R. A. *Geochim. Cosmochim Acta.* **1981**, *45*, 2359-2367.
10. Prahl, F.G.; Crecelius, E.; Carpenter, R. *Environ. Sci. & Tech.* **1984**, *18*, 687-693.
11. Aceres, M.; Grimalt, J.O. *Environ. Sci. & Tech.* **1993**, *27*, 2896-2908.
12. Halsall, C.J. *Environ. Sci. & Tech.* **1994**, *28*, 2380-2386.
13. Chuang, J.C. *Environ. Sci. & Tech.* **1995**, *29*, 494-500.
14. Zhang Anan; Liu Yongxing *Environ. Chem.(China)* **1985**, *4*, 21-27.
15. Simoneit, B.R.T.; Guoying Sheng; Xiaojing Chen; Jiamo Fu; Jian Zhang; Yuping Xu *Atmos. Environ* **1991**, *25A*, 2111-2129.
16. Sheng Guoying; Fu Jiamo; Zhang Jian; Xu Yuping; Simoneit, B.R.T. In *Environmental Biogeochemistry* (edited by Berthelin, J. and Banld, J.), Elsevier Science **1991**, *6*, pp. 77.
17. Sheng Guoying; Feng Jialiang; Fu Jiamo; Wang Xinmin; Min Yushun *The Second Asia Symposium on Academic Activity for Wastereatment and Resources (AAWTR)*, **1995**, pp. 245.
18. Kissel, J. C.; Henry, C.L.; Harrison, R. B. *Biomass Bioenergy* **1992**, *3*, 181-189.
19. Exberg, Lars E. *Atmos. Enviro.*, **1994**, *28*, 3571-3575.
20. Eitzer, B. D. *Environ. Sci. & Tech.* **1995**, *29*, 896-902.
21. Miguel, A. H.; De Aquino Neto, F.R.; Cardoso, J.N.; De C. Vasconcellos, P.; Pereira, A.S.; Marquez, K.S.G. *Environ. Sci. Tech.* **1995**, *29*, 338-345.
22. Mukund, R.; Kelly, T.J.; Gordon, S.M.; Hays, M.J.; Mcclenny, W.A. *Environ. Sci. Tech.* **1995**, *29*, 183a.
23. Kawamura, K.; Kaplan, I.R. *Atmos. Environ.* **1986**, *20*, 115-124.
24. Tancell, P. J. *Sci. Total Environ.* **1995**, *162*, 179-186.
25. Tong, H.Y.; Shore, D.L.; Karasek, F.W. *J. Chromatogr.* **1984**, *285*, 423-441.
26. Karasek, F.W.; Tong, H.Y. *J. Chromatogr.* **1985**, *332*, 169-179.

ANTHROPOGENIC MARKERS

Chapter 12

Anthropogenic Molecular Markers: Tools To Identify the Sources and Transport Pathways of Pollutants

Hideshige Takada[1], Futoshi Satoh[1], Michael H. Bothner[2], Bruce W. Tripp[3], Carl G. Johnson[3], and John W. Farrington[3]

[1]Faculty of Agriculture, Tokyo University of Agriculture and Technology, Fuchu, Tokyo 183, Japan
[2]U.S. Geological Survey, Woods Hole, MA 02543
[3]Woods Hole Oceanographic Institution, Woods Hole, MA 02543

The activities of modern civilization have released to the oceans a wide variety of both mobilized natural compounds and synthetic compounds not found prior to modern times. Many of these compounds provide a means of identifying sources of inputs and pathways of movement of chemicals through oceanic ecosystems and serve as molecular markers of human activities. A coastal ocean (Tokyo Bay) and a deep ocean (Deep Water Dump Site 106 in the Western North Atlantic Ocean) example are presented. In the deep ocean study, the correlation between potential sewage marker, i.e. linear alkylbenzenes (LABs), and polychlorinated biphenyls (PCBs) concentrations indicates a contribution of sewage sludge PCBs to the dump site sediments.

The analysis of organic molecular markers to acquire information about environmental processes has been an important concept in organic geochemistry during the past three to four decades. The absolute and relative concentrations of organic compounds in an environmental sample reflect both the original sources of the organic matter and the alteration processes which have occurred in the environment. Significant advances in analytical methodology have been accompanied by an increase in the number of organic compounds yielding important information about environmental samples of all types (1, 2).

Human activity continues to add an increasing variety of organic compounds into the environment or has changed the ratios and amounts of naturally occurring compounds. Both anthropogenic and naturally occurring compounds are found mixed together in recent environmental samples (Figure 1), and several of these compounds may be used as tracers to study natural processes affecting the fate and effects of chemical contaminants in the ocean. Analyses of anthropogenic compounds and their post-discharge degradation products in the environment concurrent with analyses of natural compounds provide information which can be used to more fully determine the history of pollutant release and the transport pathways and fates of anthropogenic compounds in the ocean. The time and space scales of environmental processes that can be studied using molecular marker compounds is dependent on the number of sources, magnitude of each source, and persistence in the environment.

Classification and Properties of Anthropogenic Markers. Many anthropogenic markers have been investigated in recent decades. Table I shows one possible scheme for classifying markers based on properties such as hydrophobicity and on the markers' major sources. Chemical structures of many of these markers are shown in the Appendix. The criteria for selecting a compound to be used as an anthropogenic marker include source specificity, amount used and released to the environment, and persistence in the environment. An ideal anthropogenic marker should be contributed from specific sources so that an unambiguous pathway from input to deposition can be demonstrated. A marker should be used and released in quantities sufficient to permit detection after dilution in the environment. Finally, a prospective marker should be resistant over time to environmental alterations which might result in loss to unidentifiable or common compounds.

Table I. Classification of Anthropogenic Markers

Source		Hydrophobic markers	Water-soluble markers
SEWAGE	Natural products	(I) Coprostanol (II) α-Tocopheryl acetate	(III) Urobilin (IV) Caffeine (V) Aminopropanone
	Synthetic detergents	(VI) Linear alkylbenzenes (LABs) (VII) Tetrapropylene-based alkylbenzenes (TABs) (VIII) Trialkylamines (TAMs) (IX) Nonylphenols (NPs)	(X) Linear alkylbenzene-sulfonates (LAS) (XI) Tetrapropylene-based-alkylbenzenesulfonates (ABS) (XII) Dialkyltetralin-sulfonates (DATS) (XIII) 4,4'-bis(2-sulfostyryl)biphenyl (DSBP)
Street Runoff		(XIV) 2-(4-morpholinyl)benzothiazole (24MoBT)	
Multiple Sources		Polychlorinated biphenyls (PCBs) Polycyclic aromatic hydrocarbons (PAHs) Polyorganosiloxanes (silicones)	

*Roman numbers in parentheses correspond to numbers in Appendix, which can be found on page 192.

Compounds with multiple sources, widespread industrial application, and resistance to degradation have been used to study local, regional and global processes. These include polycyclic aromatic hydrocarbons (PAHs; e.g. ref. *3*) and polychlorinated biphenyls (PCBs; e.g. ref. *4*). Polyorganosiloxanes have also been used as indicator of inputs from modern human activities (*5*). Sometimes, labile compounds have been utilized in combination with persistent markers from the same source where comparison between labile and stable compounds provides data on the degree

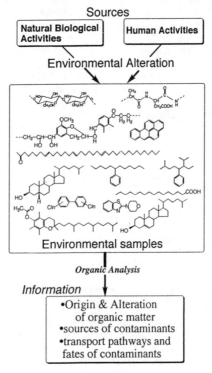

Figure 1. Concept of anthropogenic marker approach.

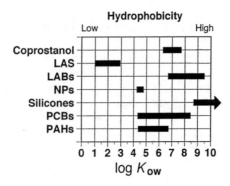

Figure 2. Octanol-water partition coefficient (K_{ow}) of the molecular markers. References for the data source are as follows, coprostanol: (*13*), LAS: (*14*), LABs: (*9*), NPs: (*15*), silicones: (*16*), PCBs: (*17*), PAHs: (*17*). K_{ow} for coprostanol is estimated from water solubility data of cholesterol (*13*) using an equation in ref. *17* p. 139 (Table 7.2)

of environmental alteration or degradation, or about the residence time of the contaminants (e.g. ref. *6*).

We discuss two examples of the use of linear alkylbenzenes (LABs; VI in Appendix) as molecular markers for sewage later in this paper. LABs having C_{10} - C_{14} normal alkyl chains are sulfonated in the industrial production of linear alkylbenzenesulfonates (LAS; X in Appendix), which are widely used anionic surfactants. However, the sulfonation of LABs into LAS is not complete, and the unsulfonated residue is present in LAS-type synthetic detergents (*6, 7*). Use of LAS-detergents and subsequent disposal, thus, brings LABs into aquatic environments (*6, 8*); in most instances by way of sewage discharges. Since LABs are highly hydrophobic (*9*) they sorb to particles and become useful for tracking sewage particles and sewage-derived hydrophobic pollutants (*10, 11*).

Differences in hydrophobicity between "hydrophobic" markers (e.g. comparing PCBs and LABs) and within a class of hydrophobic markers (e.g. comparing PCB congeners) are important to consider when using these compounds as tracers of transport processes. Measurements of differences in environmental distributions correlated with measures or estimates of hydrophobicity provide clues to processes influencing the fate of the compounds. For example, less hydrophobic markers such as nonylphenols (IX in Appendix) could selectively desorb from sediment particles relative to more hydrophobic markers such as LABs (*12*). The octanol-water partition coefficient (K_{ow}), is a surrogate measure of the hydrophobicity of a compound. Figure 2 presents K_{ow}s for several groups of markers.

The use of markers as tools for contaminant source discrimination also requires consideration of the similarity and/or differences in hydrophobicity and sorption mechanism between the markers and particular media. Sorption processes have been described using the concept of compounds partitioning between organic matter in particles and water depending on the hydrophobicity of the compound. During the 1980's, it was demonstrated that the sorption of organic compounds to particles is more complex than simple two-phase particle-water equilibrium (*17, 18* and references therein). The association of contaminants with organic colloids is seen to enhance the mobility of hydrophobic compounds in sediments (*18, 19*). Combustion-derived PAHs are strongly associated with soot particles and are not easily available for equilibrium partitioning (*20-22*). Each of these complex mechanisms requires careful consideration in the quantitative application of anthropogenic markers.

Field Application of The Molecular Marker Approach

We provide two examples, studies in Tokyo Bay and at Deep Water Dump Site 106, from our own research as illustrations of the more general approach of using molecular markers by researchers in the field.

Tokyo Bay Sewage Effluent Contributions to Organic Contaminants in Sediments.

There are considerable horizontal gradients of LAB concentrations in surface sediments of Tokyo Bay as shown in Figure 3 (*10, 23*). In Tokyo, some untreated wastewater is discharged directly into streams and rivers which ultimately flow into Tokyo Bay. The highest concentrations of LABs are clearly seen 5-10 km off the mouths of the major inflowing rivers. The embedded graph in Figure 3 further documents that the deposition of river discharged hydrophobic contaminants are concentrated in the northern Bay. Knowledge of this depositional pattern based on molecular markers is beneficial to other studies such as a harbor-wide monitoring program. For example, if we were concerned about inputs of chemicals of environmental concern such as PAHs or PCBs, inputs from the major rivers accumulating in the sediments and also contaminating the animals in the benthic ecosystem, we could use the LAB data to guide our selection of stations and samples for detailed analyses.

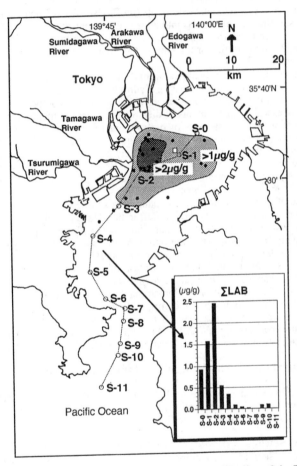

Figure 3. Horizontal distribution of LABs in surface sediments of the Tokyo Bay. Contour lines based on data for 24 stations in the northern bay (10). Bar graph showing ΣLAB concentrations along with the transect from S-0 to S-1 I (23).

Anthropogenic compounds have been deposited since the onset of industrial synthesis and/or the increase in human activities and can, thus, serve as recent (usually annual to decadal scale resolution) geochronometers in sediments. Data from a core in Tokyo Bay illustrates a simple use to estimate the date of a depth interval in the sediment. For this core, excess lead-210 shows constant values throughout the upper 20 cm (Katoh, Y., Tokai University, personal communication, 1994), probably because of strong sediment mixing due to physical or biogeochemical processes. Thus, the excess lead-210 procedure for estimating sediment accumulation rates (24) does not provide a reliable sediment accumulation rate. In this type of situation, anthropogenic markers with a known production record may provide time reference points in sediment cores.

We can use the historic record of alkylbenzene production in Japan (Figure 4a) to provide a time reference point for Tokyo Bay sediments. Tetrapropylene-based alkylbenzenes (TABs; VII in Appendix) have branched alkyl chains which result in poor biodegradability of the sulfonated surfactants (XI in Appendix) and have caused some environmental problems. As a result, LABs were introduced and ultimately replaced TABs by the late 1960s. Figure 4b shows the vertical distribution of LABs and TABs in a sediment core collected from Tokyo Bay. Depth profiles in this core provide an approximate time for the depth at which compounds initially entered the ecosystem. The TAB and LAB concentration *vs.* depth profiles would also be influenced by sediment mixing processes. The decreasing TAB concentration *vs.* sediment depth from a maximum at about 30 cm to the lower concentrations at the sediment-water interface can be explained by a reduction of environmental loading accompanied by regular dilution through mixing with more recently deposited sediments. A date of no older than approximately 1958 can be assigned for sediment at the bottom of the "mixed zone" - approximately 35 cm since TABs were not produced prior to 1958. By similar reasoning, using the depth profiles of LAB concentrations, the sediment at 30 cm depth is no older than approximately 1965.

If there had not been a disturbance of the sediment by either physical or biological mixing, then the depth profile could have provided us with a more detailed historical record of TAB and LAB releases to Tokyo Bay in the manner that historical records of other contaminants such as PAHs, PCBs, DDT have been measured in sediment cores from other aquatic ecosystems (25). Use of the historic record approach of measuring molecular markers in undisturbed sediments to document the release of LABs to coastal areas has been reported by Eganhouse and Kaplan (26).

Deep Water Dumpsite (DWDS) 106 Study. DWDS 106 is located 106 miles off the coast of New Jersey, USA, in 2300-2900 m water depth (Figure 5) and, while operational, was the world's largest deep water sewage sludge dump site (27, 28). Approximately 42 million wet tons (metric) of sewage sludge were dumped between 1986 and 1992 when all dumping stopped (28). Sewage sludge, sediment trap and bottom sediment samples obtained from the site in 1989 and 1990 were analyzed for molecular markers in order to obtain evidence of transport and accumulation of dumped sludge particles to the deep sea floor (11, 27). We summarize our results reported in detail earlier for LABs and coprostanol (11) and use these to constrain our results for PCBs reported in this paper.

Sewage sludge contains PCBs, and the dumped sludge could contribute to PCBs in DWDS-106 sediments. PCBs are known to have diverse sources of input (e.g. atmospheric transport or lateral transport from coastal zones) in addition to sludge disposal. We have attempted to estimate a specific contribution of sewage-derived PCBs to DWDS-106 sediments. This estimation is based on the assumptions that LABs are sewage-specific, the K_{ow}s of LABs and PCBs overlap significantly (9), and that the sorption mechanisms of the two compound classes are similar.

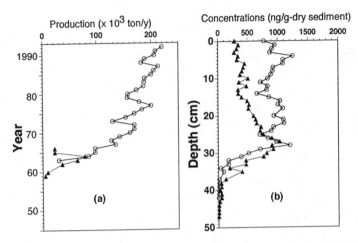

Figure 4. (a) Historic record of long chain alkylbenzene production in Japan; open circle: LABs; closed triangle :TABs; (b) Vertical distribution of LABs and TABs in a sediment core collected from Tokyo Bay in 1993 (Satoh & Takada, unpublished data); open circle: LABs excluding peaks coeluting with TABs (i.e. 6-, 4-, 3- and 2-C_{11}); closed triangle: TABs excluding peaks coeluting with LABs. Sampling location is indicated in Figure 3 as open square.

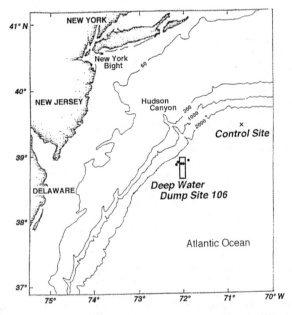

Figure 5. Sampling location for the Deep Water Dump Site 106 Study. Closed circles indicate surface sediment locations in and around the dump site.

Sampling and Analytical Methods. Surface (0-1 cm) sediment samples were collected from the dump site and the control site by DSRV *Alvin* in September 1989 and August 1990 (Figure 5). The control site is ca. 100 km northeast (i.e. up-current) of the dump site. Sinking particles were also collected by deploying sediment traps from September 1989 to August 1990 in the lower 110 m of water column at two locations along the western edge of the dump site. The details of sampling were described previously (*27*). Two sewage sludge samples from New York City were also analyzed.

The previously described analysis of LABs and coprostanol (*11*) is summarized in Figure 6. The sum of all 26 secondary LAB congeners is represented as "ΣLAB". The detection limit of ΣLAB was 0.2 ng/g-dry sediment. Coprostanol, epicoprostanol and cholesterol were separated from each other and quantified by capillary GC-FID after acetylation. The detection limit of coprostanol was 1 ng/g-dry sediment. Procedural blanks for LABs and coprostanol were below the limit of detection.

The extraction and isolation method for PCBs is outlined in Figure 6. Individual concentrations of PCBs were determined on a Hewlett Packard 5890 gas chromatograph equipped with a 30-m DB-5 capillary column (0.25 mm i.d.; 0.25 μm film thickness; J&W Scientific) and an electron capture detector. Seventy nine chromatographic peaks were identified and quantified using a standard PCB mixture (1: 1: 1: 1 mixture of Kanechlor 300:400:500:600, kindly supplied by Dr. Tanabe, Ehime University). The peak identification of the standard was performed by GC-MS, retention index (*29*), and previously reported identification of Kanechlor (*30*). The congener composition of the standard mixture was determined by GC-FID analyses. The sum of the 79 chromatographic peaks is expressed as ΣPCB. Replicate analyses (n=4) of a sediment sample showed less than 5% relative standard deviation for individual PCB concentrations. Although PCB concentrations were not corrected for recovery, analyses of sediment samples spiked with the PCB standard showed a recovery of greater than 90%. Procedural blanks for PCBs through the entire procedure correspond to less than 0.06 ng ΣPCB/g-dry sediment.

Transport of Sewage Particles to the Deep Sea Floor. LABs were detected in the dump site sediments and sediment trap samples, whereas no LABs were detected in the control site sediments. This fact clearly indicates that despite horizontal dispersion and dilution immediately after dumping (*28*) the dumped sludge particles were transported vertically through the water column of over 2300 m, and accumulated in the deep sea sediments. The distribution pattern of LABs is consistent with the sewage sludge deposition model of Fry and Butman (*31*) and is well correlated with those for other anthropogenic markers: coprostanol (I in Appendix), silver, and *Clostridium perfringens*, a bacterium found in human fecal material (*27*). The coprostanol detected at the control site at a concentration of 12 ng/g (significantly above the analytical blank) is attributed to contributions from the feces of birds (*32*), marine mammals such as whales (*33*), or *in situ* microbial or biogeochemical alternation of cholesterol to coprostanol (*34*).

The isomeric distribution of LABs indicates the absence of significant microbial attack during transport. Microbiological incubation experiments (*35, 36*) and environmental observations (*8, 37*) have demonstrated the selective depletion of isomers having phenyl substitution positions near the end of alkyl chains (i.e. external isomers) relative to isomers having phenyl substitution positions near the center of alkyl chains (i.e. internal isomers). To quantitatively express the ratio of the internal to external isomers, an I/E ratio ([6-C_{12} AB + 5-C_{12} AB]/[4-C_{12} AB + 3-C_{12} AB + 2-C_{12} AB]) is defined; m-C_n; m : phenyl substitution position, n : alkyl carbon number (*36*). I/E ratios in untreated wastewater are approximately 0.7, a higher I/E ratio indicates a greater depletion of external isomers and, thus, more extensive degradation (*36*). As is obvious from Figure 7b, within the dump site samples, no increasing trend in I/E

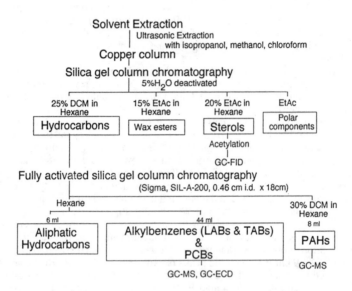

Figure 6. Analytical scheme of the molecular markers.

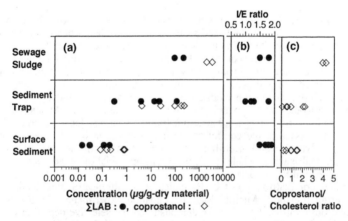

Figure 7. (a) LAB and coprostanol concentrations, (b) I/E ratio of LABs, and (c) a ratio of coprostanol relative to cholesterol in the sewage sludge, sediment trap, and sediment samples collected from the DWDS 106 site (*11*).

ratio was observed from sludge (1.7 ± 0.2), to sediment trap (1.3 ± 0.3), to sediment samples (1.7 ± 0.2). This suggests that no significant biodegradation of LABs occurs in the water column or at the sediment-water interface. This lack of LAB degradation in the deep sea is probably caused by rapid sinking of dumped sludge particles, or incorporation of sewage particles into rapidly sinking fecal material. Low water temperature and (36) and high pressure (38) in the deep sea might also suppress microbial activity leading to LAB degradation.

Dilution Factors of Sewage Particles. LAB concentrations in the sediment trap samples are about two orders of magnitude lower than those in sludge (Figure 7a). In the sediment trap, sludge particles are probably diluted by naturally produced particles and less contaminated resuspended sediment. This interpretation is supported by a lower ratio of coprostanol to cholesterol in sediment trap samples than in sludge (Figure 7c). Because phytoplankton and zooplankton produce cholesterol but not coprostanol, the addition of naturally produced sterols to sewage-derived sterols decreases the ratio. Using an average LAB concentration of sewage sludge of 152×10^{-6} g/g dry weight, and assuming conservative behavior for LABs during transport through the water column, we estimate a sewage sludge dilution factor of between 1:1 and 1:500 from our sludge and sediment trap data (11). Hunt et al. (28) have reported on analyses of seven sludge samples from Sewage Authorities contributing sludge to the material dumped at DWDS-106. The range of LAB concentrations was 39.8 to 150×10^{-6} g/g dry weight sludge with a mean of 123×10^{-6} g/g dry weight sludge.

In addition to the absence of significant microbial degradation of the LABs as evidenced by the I/E ratio discussed above, dissolution of LABs from the particles into the aqueous phase should be negligible due to the extremely high hydrophobicity of LABs (log K_{ow}: 6.90 - 9.29; ref. 9). Using the coprostanol concentrations, dilution factors in sediment trap samples ranging between 1:11 and 1:700 can be estimated. (Figure 7a). These dilution factors are somewhat higher than those calculated using LABs perhaps due to faster degradation and/or dissolution of coprostanol relative to LABs.

LAB concentrations in the sediments from the dump site are roughly two orders of magnitude lower than those for the sediment trap samples (Figure 7a). This decrease is probably due to mixing with sediments containing no LABs which had been deposited before the onset of sludge dumping. The mixing hypothesis is supported by the vertical distribution of LABs in a sediment core collected from the dump site in which LABs were present to depths of greater than 5 to 6 cm (11). Because the sediment accumulation rate for the continental slope is reported to range from 0.7 cm/100 yr. to 2.2 cm/100 yr. (39), and the sewage disposal to this site began about 4 years before the sampling, deeper penetration suggests rapid sediment mixing in the dump site. As noted above, the sewage sludge dilution factors are estimated at between 1:800 and 1:10000 based on LABs data and 1:3200 to 1:32000 based on coprostanol data. The differences are probably due to lower stability of coprostanol as compared to the LABs.

Contribution of Sewage Sludge PCBs to Dump Site Sediments. We summarize here a more detailed presentation of our PCB data for the 1989 and 1990 samples from DWDS-106 (Takada et al., in preparation). ΣPCB concentrations were 11 to 21 ng/g dry sediment for the dump site, similar to concentrations reported for sediments sampled in the DWDS-106 area in 1992 (40). These values are significantly higher than the value of 3 ng/g for the control site, suggesting that the dump site sediments were contaminated by PCBs derived from ocean dumping or other processes. There is a good correlation ($r^2 = 0.93$) between ΣLABs and ΣPCBs with a positive intercept (Figure 8a). A similar correlation was observed for sediment sam-

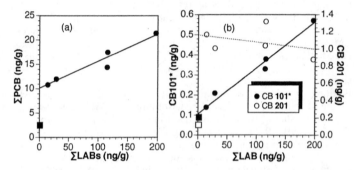

Figure 8. Relationships of PCBs with ∑LABs in the deep-sea sediments. (a): ∑PCBs, (b): CB101 (filled symbols) and CB201 (open symbols). Circle symbols: the dump site; square symbols: the control site. The solid and broken lines represent least squares linear regressions for the dumpsite data. The equations and the regression coefficients are as follows, ∑PCB = 0.056∑LAB + 10.0 (r^2=0.93), CB101 = 0.0022∑LAB + 0.118 (r^2=0.97), CB201 = -0.00085∑LAB + 1.164 (r^2=0.14). *CB101 may also contain CB90.

ples collected from the same area in 1992, although the slope for the correlation is different (*40*). Figure 8a indicates that PCBs are contributed by sludge dumping but that additional sources must exist as well. The positive intercept, ~10 ng/g, is a somewhat higher concentration than PCB concentrations at the control site (~3 ng/g) but not much different than the 5.1 and 7.5 ng/g dry sediment reported for the 1992 control/reference stations (*40*). Both the control site and the dump site are approximately at the same water depth and distance from shore and would be expected to receive similar inputs of PCBs from atmospheric and non-point coastal sources. A possible additional source of PCBs at DWDS 106 could be the chemical waste dumping which occurred prior to sewage sludge disposal (*31*). This may be consistent with the unusual PCB congener distribution at the dump site. Compared to the control site sediment and the sewage sludge, higher chlorinated congeners (especially octa- and nona-chlorinated biphenyls) are enriched in the dump site sediments as seen in Figure 9. This difference in relative distribution of PCB congeners suggests different sources. Within the dump site we observed different spatial distributions of PCBs. The stations which are more heavily affected by sludge show a higher concentration of LABs and higher proportion of lower-chlorinated biphenyls relative to the less-affected dump site stations. Lower-chlorinated PCB congeners were abundant in the sewage sludge samples we analyzed as well as the samples analyzed by Hunt *et al* (*28*). These data are consistent with the scenario that dump site sediments initially acquired higher-chlorinated biphenyls from an independent source, followed by sewage sludge-associated PCBs rich in lower-chlorinated congeners.

The relative contribution of sludge-derived PCBs to the dump site sedimentary environment can be estimated by two approaches (Estimation-1 and Estimation-2). Estimation-1 is based on the correlation between LABs and PCBs shown in Figure 8. Assuming a constant contribution rate for non-sludge-derived PCBs, the increment in PCB concentrations in the highest-sludge-impacted station *vs*. y-intercept in the LABs-PCBs diagrams can be considered as sludge-derived PCBs. For $\sum$PCBs, the contribution of sludge-derived PCBs is estimated at ~ 50% (Figure 8a). The greater contribution is calculated for the lower-chlorinated congeners (e.g. ~ 80 % for CB101) and lesser contribution of the higher-chlorinated congeners (e.g. no contribution for CB201), as shown in Figure 8b. Table II summarizes the estimate of sludge-derived PCBs (as a percentage of the total) for selected congeners.

Table II. Percent contribution of sewage sludge PCBs to the dump site sediment[a]

Degree of Chlorination		Estimation-1[b]	Estimation-2[c]
5	CB101[d]	80	26 ±13
6	CB138[e]	80	14 ± 8
7	CB180	30	8 ± 4
8	CB201	0	2 ± 1
9	CB206	0	0

[a]For a station (Dive#2165) where is most heavily affected by sewage sludge
[b]sludge-PCBs is estimated as increment in PCBs concentrations *vs*. y-intercept in the LABs-PCBs diagrams.
[c]sludge-PCBs is estimated multiplying PCB/LAB ratios for sludge by the LAB concentrations in the sediment sample.
[d]CB90 may also be contained.
[e]CB163 and CB164 may also be contained (*41, 42*).

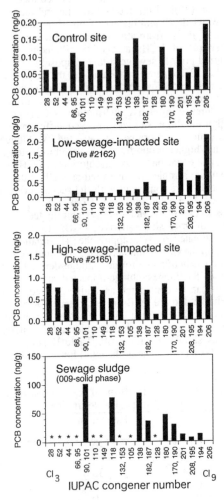

Figure 9. Congener distribution of PCBs in the dump site sediments, the control site sediments, and the sewage sludge. Asterisks for the sewage sludge sample mean no data because of the congeners lost during column fractionation. CB163 and CB164 may coelute with CB138 (41, 42).

Another approach (Estimation-2) uses LAB data and the ratio of PCBs-to-LABs in sewage sludge samples to estimate the sludge contributions of PCBs to the dump site sediments. Averaged PCBs/LABs ratios ($CB101/\Sigma LAB=0.74 \times 10^{-3}$, $CB138/\Sigma LAB=0.60 \times 10^{-3}$, $CB180/\Sigma LAB=0.36 \times 10^{-3}$, $CB201/\Sigma LAB=0.10 \times 10^{-3}$; $CB206/\Sigma LAB=0$; Takada *et al.* in preparation) for sludge are multiplied by the LAB concentrations in dump site sediments. The estimates are summarized in Table II. We find greater contributions of the lower-chlorinated congeners and minimal contribution of the higher-chlorinated congeners. However, the estimates of sludge-contribution between two approaches are significantly different (i.e. higher contributions are estimated by Estimation-1 than Estimation-2). Several factors may cause this difference. We have individual chlorobiphenyl congener data available for only the two sludge samples. Hunt *et al.* (28) in their analyses of six sludge samples reported total PCB concentrations, not individual congener data, in their analyses of seven sludge samples. Uncertainty is introduced by the small number of sludge samples analyzed and the potential for varying compositions among sludge samples for mixtures of congeners. The assumption of constant background PCBs in Estimation-1 may not be valid, and the higher sludge-impacted stations may have received greater contributions of non-sludge (i.e. background) PCBs. This may lead to an overestimation for Estimation-1 using the LAB-PCB correlation. Also, the calculation using PCB/LAB ratio may be underestimated if preferential loss of LABs relative to PCBs occurs. Lamoureux and co-authors (40) observed significantly lower concentrations of LABs (more than a factor of 3) in sediment samples collected in the same area in 1992 compared to our 1989-90 samples. They proposed a hypothesis that LABs are selectively assimilated by deposit-feeders relative to inorganic markers (i.e. Ag). If selective processing occurs between LABs and PCBs, it would lead the Estimation-2 approach to underestimate sewage sludge contributions of PCBs to sediments.

Phase associations of LABs and other molecular markers in sludge and their compound-specific environmental behaviors should be examined in future studies. Our studies of the molecular markers as tracers of sludge particles at DWDS-106, and those of others (28, 40), are hampered by the fact that the compliance monitoring analyses of sludge samples had relatively high detection limits. Thus, we have limited data for concentrations of PCBs, PAHs and LABs in the sludge material prior to dumping. Furthermore, we agree with Hunt *et al.* (28), that studies of the particle size distributions within the sludge and the analyses of molecular markers within size classes are important for future studies of this type. Nevertheless, the molecular markers studies that have been conducted at the DWDS-106 site have provided the basis for longer term studies of biogeochemical processes in the deep sea such as biological and physical mixing of the hydrophobic sludge organic chemicals into the sediments and the resuspension, transport and dispersion of these chemicals in the deep sea epibenthic and benthic environments.

Conclusions and Future Directions for Research

One important application of anthropogenic markers during the past two decades has been the identification of sources of pollutants and their fate in coastal and open ocean sediments. Estimates of the contribution made by a specific source is important, especially in cases where contaminants are derived from multiple sources, because this information can be used in decisions about where to concentrate management efforts for controlling sources of significant input. Analyses of multiple molecular markers have great promise in sorting out contributions from multiple sources. The recent addition of analytical capabilities to measure ratios of stable isotopes of carbon and carbon-14 content of individual molecular markers greatly enhances the discriminatory power of molecular markers as source identifiers (43, 44) and for revealing biogeochemical processes acting on contaminants released to all environments, including the ocean.

Even as discharges of several chemicals of environmental concern are being reduced or eliminated in many areas of the world, we are challenged by the legacy of past human activities. Large amounts of chemicals of environmental concern have accumulated for decades in coastal sediments which are a potential source for continuing contamination of the benthic ecosystems and overlying water after reduction or elimination of discharges of these chemicals. Molecular markers can be used to understand sedimentary processes such as resuspension, deposition, burial, biological mixing, and bioavailability thereby contributing to the knowledge base needed to reduce or eliminate the pollution threat associated with these sediments.

Urban sewage discharges, urban runoff, agricultural runoff or atmospheric transport to the ocean continue to be significant sources of chemicals which are of environmental concern either because of potential human health impacts as a result of transfer through the food web back to humans, or because of potential adverse impacts on valuable living natural resources. These inputs may increase because of human population growth and development in coastal areas around the world. Studies to date have provided valuable guidance for management of contaminant and pollutant inputs. Studies of a wider variety of molecular markers specific to sewage discharges, urban runoff, agriculture related runoff, and atmospheric inputs in varying ocean regimes, particularly different types of coastal ecosystems, are urgently needed to improve our quantitative knowledge of how biogeochemical processes influence the fate of chemicals of environmental concern. As demonstrated by the research reported in the references cited and discussed in this paper, sufficient documentation exists for using several molecular markers in monitoring programs to meet environmental management needs.

APPENDIX. Structures and Sources of the Anthropogenic Molecular Markers

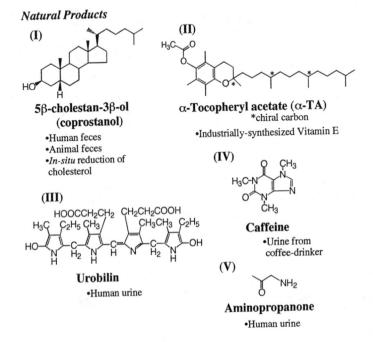

Natural Products

(I)

5β-cholestan-3β-ol
(coprostanol)
•Human feces
•Animal feces
•*In-situ* reduction of
  cholesterol

(II)

α-Tocopheryl acetate (α-TA)
*chiral carbon
•Industrially-synthesized Vitamin E

(III)

Urobilin
•Human urine

(IV)

Caffeine
•Urine from
  coffee-drinker

(V)

Aminopropanone
•Human urine

Synthetic products

(VI)

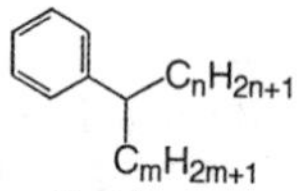

C_nH_{2n+1}
C_mH_{2m+1}
n+m = 9 - 13

Linear alkylbenzenes (LABs)
•Anionic surfactants

(VII)

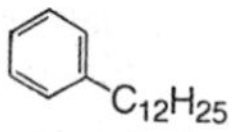

$C_{12}H_{25}$

Tetrapropylene-based alkylbenzenes (TABs)
•Anionic surfactants

(VIII)

$C_nH_{2n+1}C_mH_{2m+1}NCH_3$
n = 1, 14, 16 or 18
m = 14, 16, or 18

Trialkylamines (TAMs)
•Cationic surfactants

(IX)

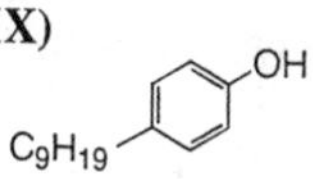

C_9H_{19} — OH

Nonylphenols (NPs)
•Nonionic surfactants

(X)

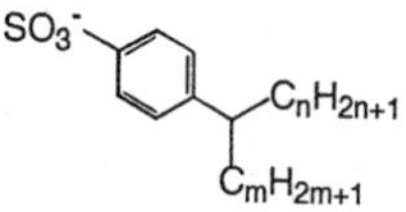

SO_3^-

C_nH_{2n+1}
C_mH_{2m+1}
n+m = 9 - 13

Linear alkylbenzenesulfonates (LAS)
•Anionic surfactants

(XI)

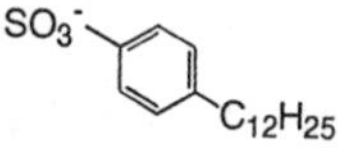

SO_3^-

$C_{12}H_{25}$

Tetrapropylene-based alkylbenzenesulfonates (ABS)
•Anionic surfactants

(XII)

$C_nH_{2n\mp1}$ — C_mH_{2m+1}

SO_3^-

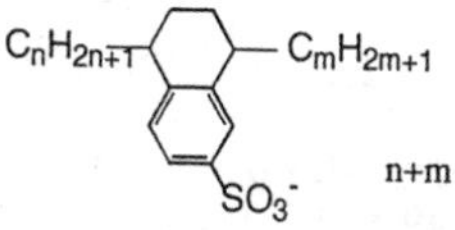

n+m = 6 - 10

Dialkyltetralinsulfonates (DATS)
•Anionic surfactants

(XIII)

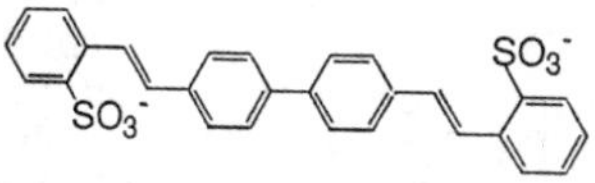

SO_3^- ... SO_3^-

4,4'-bis(2-sulfostyryl)biphenyl (DSBP)
•Fluorescent Whitening Agents (FWAs)

(XIV)

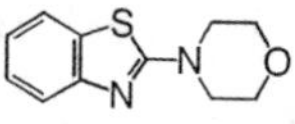

2-(4-morpholinyl)benzothiazole (24MoBT)
•Tire rubber debris

Acknowledgments

We appreciate the stimulating discussions with Drs. Robert P. Eganhouse and Paul Sherblom about molecular markers. We thank Dr. J. Frederick Grassle for organizing the research project of DWDS 106; Mr. Hovey Clifford and Ms. Rose Petracca for collecting samples; Dr. Damian Shea for kindly supplying the sludge samples; Dr. Susan McGroddy for stimulating discussions and cooperation in the laboratory. The officers and crews of RV *Oceanus*, RV *Atlantis II*, and DSRV *Alvin* provided essential sampling assistance. Financial support was provided by U.S. NOAA National Undersea Research Program and by Ministry of Education of Japan. This is Contribution Number 9404 of Woods Hole Oceanographic Institution.

Literature Cited

1. Blumer, M. *Angewandte Chemie Intern. Edition in English,* **1975**, *14*, 507-514.
2. Peters, K.E. ; Moldowan, J.M. *The biomarker guide: Interpreting molecular fossils in petroleum and ancient sediments*; Prentice-Hall: Englewood Cliffs, NJ, 1993.
3. Laflamme, R.E. ; Hites, R.A. *Geochim. Cosmochim. Acta,* **1978**, *42*, 289-303.
4. Iwata, H.; Tanabe, S.; Sakai, N.; Tatsukawa, R. *Environ. Sci. Technol.,* **1993**, *27*, 1080-1098.
5. Pellenbarg, R. *Mar. Pollut. Bull.,* **1979**, *10*, 267-269.
6. Takada, H. ; Ishiwatari, R. *Environ. Sci. Technol.,* **1987**, *21*, 875-883.
7. Eganhouse, R.P.; Ruth, E.C.; Kaplan, I.R. *Anal. Chem.,* **1983**, *55*, 2120-2126.
8. Eganhouse, R.P.; Blumfield, D.L.; Kaplan, I.R. *Environ. Sci. Technol.,* **1983**, *17*, 523-530.
9. Sherblom, P.M.; Gschwend, P.M.; Eganhouse, R.P. *J. Chem. Eng. Data,* **1992**, *37*, 394-399.
10. Takada, H.; Ogura, N.; Ishiwatari, R. *Estuarine Coastal Shelf Sci.,* **1992**, *35*, 141-156.
11. Takada, H.; Farrington, J.W.; Bothner, M.H.; Johnson, C.H.; Tripp, B.W. *Environ. Sci. Technol.,* **1994**, *28*, 1062-1072.
12. Chalaux, N.; Bayona, J.M.; Venkatesan, M.I.; Albaigés, J. *Mar. Pollut. Bull.,* **1992**, *24*, 403-407.
13. Saad, H.Y. ; Higuchi, W.I. *J. Pharm. Sci.,* **1965**, *54*, 1205-1206.
14. Hand, V.C. ; Williams, G.K. *Environ. Sci. Technol.,* **1987**, *21*, 370-373.
15. Ahel, M. ; Giger, W. *Chemosphere,* **1993**, *26*, 1471-1478.
16. Bruggeman, W.A.; Weber-Fung, D.; Opperhuizen, A.; Van Der Steen, J.; Wijbenga, A.; Hutzinger, O. *Environ. Toxicol. Chem.,* **1984**, *7*, 287-296.
17. Schwarzenbach, R. P.; Gschwend, P.M.; Imboden, D.M. *Environmental Organic Chemistry*; John Wiley & Sons: New York, NY, 1993.
18. Brownawell, B.J. ; Farrington, J.W. *Geochim. Cosmochim. Acta,* **1986**, *50*, 157-169.
19. Baker, J.E.; Capel, P.D.; Eisenreich, S.J. *Environ. Sci. Technol.,* **1986**, *20*, 1136-1143.
20. McGroddy, S.E. ; Farrington, J.W. *Environ. Sci. Technol.,* **1995**, *29*, 1542-1550.
21. Maruya, K.A.; Risebrough, R.W.; Horne, A.J. *Environ. Sci. Technol.,* **1996**, *30*, 2942 - 2947.
22. Gustafsson, Ö.; Haghseta, F.; Chan, C.; MacFarlane, J.; Gschwend, P.M. *Environ. Sci. Technol.,* **1997**, *31*, 203-209.
23. Chalaux, N.; Takada, H.; Bayona, J.M. *Marine Environ. Res.,* **1995**, *40*, 77-92.

24. Goldberg, E.D.; Hodge, V.; Koide, M.; Griffin, J.; Gamble, E.; Bricker, O.P.; Matisoff, G.; Holden, G.R., Jr. *Geochim. Cosmochim. Acta,* **1978**, *42*, 1413-1425.
25. Charles, M.J. ; Hites, R.A., In *Sources and Fates of Aquatic Pollutants;* Hites, R.A.; Eisenreich, S.J. EDs; Advances in Chemistry Series 216; American Chemical Society: Washington, D.C., 1987, pp. 364-389.
26. Eganhouse, R.P. ; Kaplan, I.R. *Mar. Chem.,* **1988**, *24*, 163-191.
27. Bothner, M.H.; Takada, H.; Knight, I.T.; Hill, R.T.; Butman, B.; Farrington, J.W.; Colwell, R.R.; Grassle, J.F. *Marine Environ. Res.,* **1994**, *38*, 43-59.
28. Hunt, C.D.; Peven, C.S.; Pabst, D. *J. Marine Environ. Eng.,* **1996**, *2*, 259-283.
29. Mullin, M.D.; Pochini, C.M.; MaCrindle, S.; Romkes, M.; Safe, S.H.; Safe, L.M. *Environ. Sci. Technol.,* **1984**, *18*, 468-476.
30. Kannan, N.; Schulz-bull, D.E.; Petrick, G.; Duinker, J.C. *Intern. J. Environ. Anal. Chem.,* **1992**, *47*, 201-215.
31. Fry, V.A. ; Butman, B. *Marine Environ. Res.,* **1991**, *31*, 145-160.
32. Leeming, R.; Ball, A.; Ashbolt, N.; Nichols, P. *Water Res.,* **1996**, *30*, 2893-2900.
33. Venkatesan, M.I. ; Santiago, C.A. *Mar. Biol.,* **1989**, *102*, 431-437.
34. Mackenzie, A.S.; Brassell, S.C.; Eglinton, G.; Maxwell, J.R. *Science,* **1982**, *217*, 491-504.
35. Bayona, J.M.; Albaigés, J.; Solanas, A.M.; Grifoll,, M. *Chemosphere,* **1986**, *15*, 595-598.
36. Takada, H. ; Ishiwatari, R. *Environ. Sci. Technol.,* **1990**, *24*, 86-91.
37. Takada, H.; Ogura , N.; Ishiwatari, R. *Environ. Sci. Technol.,* **1992**, *26*, 2517-2523.
38. Jannasch, H.W. ; Wirsen, C.O. *Science,* **1973**, *180*, 641-643.
39. Anderson, R.F.; Bopp, R.F.; Buesseler, K.O.; Biscaye, P.E. *Cont. Shelf. Res.,* **1988**, *8*, 925-946.
40. Lamoureux, E.M.; Brownawell, B.J.; Bothner, M.H. *J. Marine Environ. Eng.,* **1996**, *2*, 325-342.
41. Hillary, B.R.; Girard, J.E.; Schantz, M.M.; Wise, S.A. *J. High Resolut. Chromatogr.,* **1995**, *18*, 89-96.
42. Larsen, B.R. *J. High Resolut. Chromatogr,* **1995**, *18*, 141-151.
43. Lichtfouse, E. ; Eglinton, T.I. *Org. Geochem.,* **1995**, *23*, 969-973.
44. Eglinton, T.I.; Aluwihare, L.I.; Bauer, J.; Druffel, E.R.M.; McNichol, A.P. *Anal. Chem.,* **1996**, *68*, 904-912.

Chapter 13

Alkylbenzenesulfonates in Recent Lake Sediments as Molecular Markers for the Environmental Behavior of Detergent-Derived Chemicals

René Reiser[1], Heidi Toljander[2], Achim Albrecht[1], and Walter Giger[1,3]

[1]Swiss Federal Institute for Environmental Science and Technology (EAWAG) and Swiss Federal Institute of Technology Zurich (ETH), CH–8600 Dübendorf, Switzerland
[2]Helsinki University of Technology, SF–02150 Espoo, Finland

Alkylbenzenesulfonates together with soap are the most widely used anionic surfactants. Linear alkylbenzenesulfonates (LAS) were introduced in the mid 1960's as substitutes for the poorly biodegradable tetrapropylenebenzenesulfonates (TPS). In this work, residual amounts of alkylbenzenesulfonates have been determined in lake sediments. The historical release of these anthropogenic chemicals into surface waters has been preserved in dated sediments as cores from two Swiss lakes have demonstrated. Concentrations of LAS and TPS in the sediment cores exhibit subsurface concentration maxima of 2.4 and 3.4 mg per kg dry sediment, respectively. Based on depth profiles of LAS-homologs and isomers it can be inferred that the postdepositional fate of LAS and TPS in recent sediments is determined by both the lack of biodegradation under reducing conditions and by diffusion in the pore waters. The latter is governed by partitioning between sediment particles and porewater. Evidence is provided for postdepositional diffusion by comparing measured data to results of a mathematical model.

Detergents for household and institutional use are very complex formulations containing a variety of substances including (i) surfactants, (ii) builders, (iii) bleaching agents, and (iv) auxiliary agents (1). Detergents are employed in large amounts and are discarded with the wastewater after use. For this reason, the environmental fate and effects of detergent-derived chemicals, including their behavior during sewage treatment, distribution in aquatic ecosystems, and toxic effects on organisms, have been the subject of much research. The aim of these studies was to estimate their potential ecotoxicological risk (2, 3). Some of the detergent-related chemicals (or their by-products) have the potential of being source-specific molecular markers (4, 5) for domestic wastewater. This is based on their widespread use, their resistance to chemical and biochemical alteration in the environment, and the physicochemical properties that control their transport. Environmental chemicals that can be used as molecular markers are: linear alkylbenzenes LAB (6, 7) trialkylamines (8) and fluorescent whitening agents (9).

[3]Corresponding author

Surfactants constitute the most important group of detergent components and are contained to 10 to 20 % in commercial detergent products (10). Fatty acids (soap) and linear alkylbenzenesulfonates (LAS) are the two most widely used anionic surfactants. As of 1993 LAS had an annual global production of approximately 2.4 x 10^6 tons (11). In many countries LAS were introduced in the mid 1960's as a replacement for the tetrapropylenebenzenesulfonates (TPS), which were the first synthetic surfactants that were produced using petrochemicals. Both LAS and TPS are complex mixtures of homologs and isomers. LAS-homologs have linear alkyl chains that range from 10 to 14 carbon atoms. In addition, each homolog group comprises a series of isomers that vary in the position of substitution of the benzenesulfonate moiety. The symbol "m-C_n-LAS" is used in this work for referring to individual LAS isomers, where "m" and "n" are numbers that indicate the position of the benzenesulfonate moiety and the length of the linear alkyl chain, respectively. One decylbenzenesulfonate (C_{10}-LAS) isomer and one possible isomer of TPS are shown in Figure 1. Isomers having positions of substitution close to the terminal carbon atom (e.g. 2-C_{10}-LAS in Figure 1) are referred to as *"external"* isomers, whereas isomers having positions of substitution close to the center of the alkyl chain (e.g. 5-C_{10}-LAS) are referred to as *"internal"* isomers. Some detergent products used from 1965 to the late 1970's contained LAS with alkyl chains of up to 16 carbon atoms; however, they were gradually phased-out after about 1980 (personal communication with E. Stähli, Lever, Zug, Switzerland). LAS are efficiently eliminated (98 to 99 %) in mechanical-biological sewage treatment plants (3, 12-15). They are readily biodegradable under oxic conditions. However, under anoxic conditions, LAS are very slowly degraded, if at all (16, 17). Thus, LAS that are transported to the sludge digester, accumulate in the anaerobically digested sewage sludge. Sewage sludge has been found to contain LAS at concentrations between 3 and 30 g/kg dry matter (3, 13, 14). In contrast to LAS, the alkyl chains of TPS are highly branched, so that TPS are only slowly biodegraded under oxic and anoxic conditions. In the USA and in Europe prior to 1965 and in developing countries where TPS are still used, high residual concentrations of TPS in sewage receiving rivers cause extensive foaming problems.

The objective of this study was to investigate recent sediment cores of lakes in order to document the historical record of alkylbenzenesulfonates input and the potential use of these compounds as molecular markers. Both LAS and TPS adsorb to suspended particles that are contained in sewage receiving rivers and lakes. Partition coefficients of LAS in sediments determined by Hand and Williams (18) varied between 3 and 26,000 L/kg, as sediment type, chain length, and phenyl positions were varied. In natural waters, particle-associated LAS are incorporated into sediments (19-25). Since TPS resist biodegradation under both oxic and anoxic conditions and LAS are persistent under anoxic conditions, both are expected to remain unaltered once they become buried in anoxic sediments. In 1969 Ambe (25) has determined TPS in a sediment core from the Tokyo Bay. He has found TPS in concentrations above the detection limit only in the uppermost 30 cm of the sediment. This probably indicated the increasing use of TPS during the 1950's and 1960's. Amano (26) determined LAS in sediments of Lake Teganuma (Japan). He reported an LAS maximum at about 300 mg/kg at a sediment depth of about 4 to 6 cm. Even though the susceptibility of LAS to aerobic biodegradation limits their potential use as molecular markers, it was likely that the sedimentary record of LAS traces the fate of sewage contaminants during sewage treatment and subsequent transport in receiving waters as well as after deposition in sediments.

In contrast to LAB, which are present as residual intermediates in concentrations of 0.04 to 0.2 % in technical LAS mixtures (27), LAS and TPS constitute the bulk of the LAS and TPS surfactant mixture. Consequently, LAS are released in amounts up to 2,500 times larger than that of LAB. In addition, LAS are less hydrophobic as indicated by their log K_{ow} value (1.2 - 2.7) compared to that of LAB (6.9 - 9.2) (28). As a consequence LAS are molecular markers for a more hydrophilic category of sewage contaminants.

Analogous to LAB (7, 27, 29), depletion of the external LAS isomers indicates that biological degradation has occurred (30). Thus, the information on aerobic biological processes that might have taken place during sewage treatment and transport to the sediment may be recorded in the isomeric patterns of LAS in the sediments. In summary, alkylbenzenesulfonates have the potential of being used as molecular markers in sediments, tracing both the changes in input (over time) and the postdepositional behavior of sewage contaminants.

We report here the results obtained in an investigation involving two Swiss lakes (Lake Wohlen and Lake Biel). This study is part of a comprehensive research program, that includes also a study of fluvial sediments and a lake sediment core precisely dated by varvecounting.

Sampling of the Sediment Cores

Sediment cores were collected at two sites within the watershed of the Aare river near the city of Bern, Switzerland (Figure 2). Sampling site A is located in Lake Wohlen, which is a manmade lake used for generating electric power. The hydroelectric dam was built between 1917 and 1920, allowing continuous sediment deposition since then. Site A is close to the dam, and about 10 km downstream from the outfalls of two sewage treatment plants (total population equivalent of 333,000) near Bern. Sampling site B in Lake Biel is 20 km farther downstream. The input of wastewater effluents is dominated by the city of Bern at both sampling sites.

The geology of the Aare catchment has a controlling influence on the suspended particle load. It is made of crystalline basement rocks in the Alps and molasse in the midlands. Part of the particles settle out in Lake Thun and Lake Brienz approximately 40 and 60 km, respectively, upstream of Lake Wohlen. The Aare is therefore naturally filtered and carries mostly clay-size particles. The tributary Saane, which joins the Aare 4 km downstream of Lake Wohlen, drains an area of active erosion and carries substantial amounts of sand and silt in its suspended load (31). The yearly average run-off for the Aare is about 120 m^3/s below Lake Wohlen and 240 m^3/s below Lake Biel.

Sediment cores were obtained with a gravity corer. The coring device was lowered to within ca. 1m of the lake bottom and allowed to free fall the remaining distance. Once recovered, the cores (in PVC liners of 6 cm inner diameter) were subsampled. Layers of 0.5 or 1 cm thickness (corresponding approximately to 0.25 to 0.5 year of sediment deposition) were extruded using a piston. Each extruded layer was cut off with a spatula and filled in a Plexiglas box. The samples were frozen solid with dry ice before being transported to the laboratory. In order to avoid contamination, all equipment was carefully rinsed with water prior to sampling. Surface contamination occurred in Lake Biel, because the material inserted into the liner for stabilizing the core during transport on the boat contained LAS. As a consequence, the top section of the core from Lake Biel could not be used for the interpretation of the LAS concentration profile.

Dating

The cores were dated on the basis of the mass-related activities of the radionuclides Cs-134, C-137 and Co-60 (32, 33). Approximately 10 g of dry sediment were used to measure radionuclides by γ-spectroscopy using the lines 1173 and 1332 keV for Co-60, 661.7 keV for Cs-137, 604.7 and 795.5 keV for Cs-134 and 477 keV for Be-7 on a high purity Ge well detector. The presence of Be-7 in top sediments of both cores

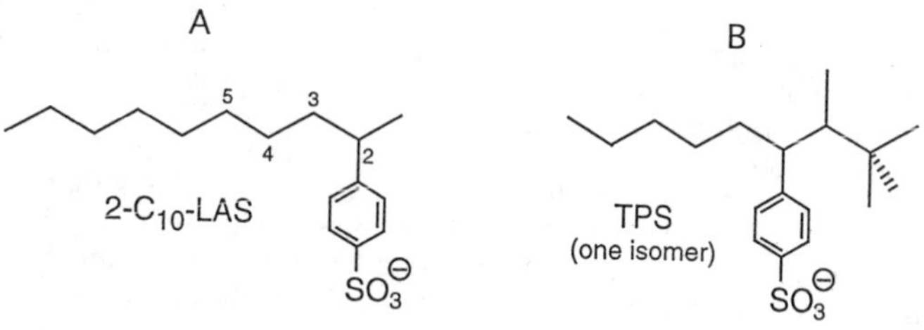

A

B

2-C$_{10}$-LAS

SO$_3^{\ominus}$

TPS
(one isomer)

SO$_3^{\ominus}$

Figure 1. Structures of alkylbenzenesulfonates.
A) Linear alkylbenzenesulfonates, m-C$_n$-LAS. Homologs: $10 \leq n \leq 14$, isomers: $2 \leq m \leq 7$.
B) One possible isomer of tetrapropylenebenzenesulfonates (TPS).

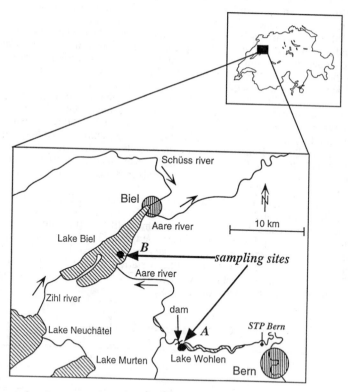

Schüss river

Biel

Aare river

Lake Biel

N

10 km

B

sampling sites

Aare river

Zihl river

dam

Lake Neuchâtel

A

STP Bern

Lake Murten

Lake Wohlen

Bern

Figure 2. Geographic map showing the sampling sites A and B in Lake Wohlen and Lake Biel, respectively. STP: sewage treatment plant.

indicate no significant sediment loss near the core-top. The depth profiles of Cs-137 activities in both Lake Wohlen and Lake Biel indicated two peaks at 16 cm and 57 cm and at 9 cm and 27 cm, respectively (Figure 3). The upper peak, which is also characterized by the presence of Cs-134, is attributed to fallout from the accident at the nuclear power plant of Chernobyl in 1986. The lower peak contained only Cs-137 and is the result of atmospheric nuclear weapons testing in 1963. The additional peaks of Cs-137 and Co-60 activities in the sediment of Lake Biel related to discharges of the local nuclear power plant which had maxima in the mid 1970's for Cs-137 and in 1982 for Co-60, respectively (34). This time markers were used for sediment dating. The assumption of constant sedimentation rate between the markers is reasonable, because there is no evidence for discontinuous erosion in the catchment.

Sediment characteristics

The dark grey to black colour of the sediments indicated anoxic conditions. The sediments were not varved. However, the cores showed rhythmical successions of dark/light laminae which provide evidence for non disrupted lamination and, hence, for absent or negligible mixing processes such as bioturbation. Further evidence for the negligibility of mixing processes is provided by the relatively thin horizons of elevated radioisotope concentrations (Figure 3).

The water contents were determined as the differences between sample weights before and after freeze drying. In the top 5 cm of the core from Lake Biel the water content dropped from about 60 to 55 % and subsequently varied between 51 and 56 %. In the top 5 cm of the core from Lake Wohlen the water content dropped from about 80 % to 60 % and subsequently varied between 55 and 60 %.

Total organic carbon content (TOC) was calculated as the difference between total carbon and carbonate carbon. Total carbon was measured using a Carlo-Erba CNS-Analyzer. The carbonate carbon was determined by acidic digestion of the sediment sample and subsequent titration of the released carbon dioxide using a CO_2-Coulometer 5011 (Coulometrics Inc., Wheat Ridge, CO). The relative standard deviation of 6 duplicate TOC determinations in Lake Wohlen was 0.9%. Particle size distributions were determined using a Mastersizer X laser particle sizer (Malvern Instruments Ltd., England).

Averages of both TOC and grain size distributions are listed in Table I. The intervals indicated quantify the overall standard deviations along the sediment cores. Both, TOC and grain size distributions were remarkably constant along the sediment cores. However, TOC in Lake Wohlen decreased from 2.9 % at the sediment/water interface to 1.9 % at a depth of 5 cm. The data set from Lake Biel lacks values near the sediment water interface and thus, interpretation on TOC in the near surface section is not possible. The relatively large TOC corroborates the assumption of anoxic conditions in the sediments of both lakes.

Table I. Characteristics of the sediments

	TOC %	clay %	silt %	sand %
Lake Wohlen	2.1 ±0.4	11.3 ±1.6	87.3 ±1.5	1.4 ±0.3
Lake Biel	2.2 ±0.2	11.8 ±1.6	85.5 ±1.0	2.8 ±0.9

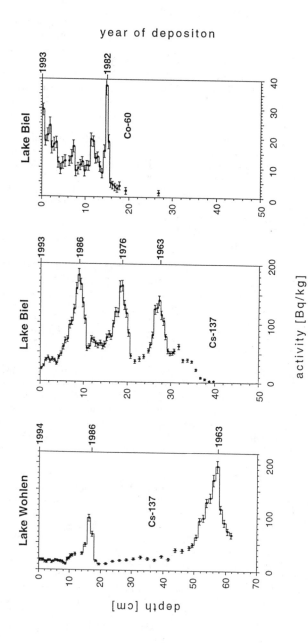

Figure 3. Depth profiles of mass related activities of Cs-137 and Co-60 in the sediment cores of Lake Wohlen and Lake Biel.

Analytical Method for the Quantitative Determination of LAS and TPS

Approximately 10 g of the freeze-dried sediment sample were extracted with 80 mL of methanol in a Soxhlet apparatus. The anionic components were separated from the methanolic extract by passing it through a strong anionic exchange column (SAX, Macherey-Nagel, Düren, Germany) (35). Subsequently, the SAX column was rinsed with methanol and then eluted with acidic methanol (0.4 mL HCl 32% in 100 mL methanol). The acidic eluate was evaporated to dryness by a stream of nitrogen while heated to about 60 °C. The alkylbenzenesulfonates contained in the dry residue were converted to their corresponding trifluoroethylesters in order to make them suitable for gas chromatography. We applied a two-step procedure converting the sulfonic acids to the sulfonyl chloride forms that were then converted to their trifluoroethylesters (20, 36). The dry sample was redissolved in 0.2 mL $SOCl_2$ and heated to 95 °C during 1h in a Reacti vial which was capped with a Teflon lined septum. After cooling to room temperature, $SOCl_2$ was removed under a gentle stream of nitrogen. Subsequently, 0.2 mL of pyridine and 0.2 mL of trifluoroethanol were added. Then, the vial was recapped and heated to 95 °C during 10 min. Reaction by-products and matrix components were removed on a column containing 1.5 g alumina in hexane/ethylester 1:1. The first 2 mL of the eluate were collected and evaporated to dryness under a gentle stream of nitrogen. The residue was redissolved in 0.2 to 0.5 mL of chloroform and analyzed using GC/MS (gas chromatography/mass spectrometry).

GC/MS analysis was performed on an Hewlett-Packard 5890 gas chromatograph coupled to a Hewlett-Packard 5971A mass spectrometer. An HP-5 column (20 m x 0.2 mm i.d., 0.33 μm film thickness) connected to a retention gap (2 m x 0.53 mm i.d. deactivated fused silica) was used for chromatographic separations. Volumes of 1 or 2 μL were injected on-column. The GC program started at an initial temperature of 60 °C and an initial time of 1 min. The temperature was then raised from 60 °C at a rate of 10°/min to 200 °C followed by a second ramp at 6 °/min to 300 °C, which was held at 3 min.

The mass spectrometer was operated in the selected ion monitoring (SIM) mode in order to enhance both selectivity and sensitivity. The yield of homolog-specific molecular ions was improved by running the ion source under positive chemical ionization conditions with methane as reagent-gas. The masses used to monitor the molecular ions of LAS are listed in Table II. The branched alkylbenzenesulfonates were determined in separate runs, during which the protonated molecular ion of the C_{12} homolog of TPS (m/z = 409), the major homolog (>50% by mass) (7), was monitored.

In general, the different isomers of derivatized linear alkylbenzenesulfonates were well separated by gas chromatography. Yet, complete chromatographic separation was not achieved for the LAS isomers 5- and 6-C_{11}-LAS, 6 and 7-C_{13}-LAS, 7- and 8-C_{15}-LAS, and 7- and 8-C_{16}-LAS. Likewise, some of the 2-phenyl isomers coeluted with the internal isomers of the LAS-homolog group which eluted next. However, coeluting isomers of different homologs were distinguished in the mass spectrometer due to their different molecular masses.

Concentrations of LAS and TPS were determined on the basis of the peak areas obtained from the SIM chromatograms. The responses of the individual isomers were calibrated separately. External calibration was applied because the concentration-response relations were non-linear under CI conditions. For calibration a concentration series of a mixture of the C_8-LAS internal standard and the C_{10}- to C_{13}-LAS-homologs was used. Likewise, a separate series of the internal standard and TPS was prepared. Calibration curves were constructed for each of the LAS peaks and seven selected TPS peaks which did not overlap with homolog LAS peaks. For the quantitation of C_{14}-, C_{15}- and C_{16}-LAS-homologs, which were not commercially available, the calibration parameters of the C_{13}-isomers were used. The total concentration of TPS in sediment samples was calculated using the seven selected peaks as representatives of the total TPS. Concentrations of total TPS were calculated as the average of the seven individually determined total concentrations. Each sediment sample was spiked with the C_8-LAS internal standard prior to extraction. The concentrations of LAS and TPS determined in the sediment with external calibration were corrected for losses during sample preparation by the individual recovery rates of the C_8-LAS standard.

A quality assurance and quality control (QA/QC) program was established. The QA/QC protocol comprised data on calibration and calibration checks, spike recoveries, glassware blanks, detection limits, and reproducibility. The recovery of total LAS added to six replicates of fresh wet sediment samples was 94.5±6.8 %. The standard deviation of the determination of total LAS in nine replicates of a homogenized sediment slurry was 5.3%. The limit of quantification of LAS was calculated as ten times the standard deviation of the blank values (37) which were measured within every run of six samples. For both LAS isomers of the C_{14}- to C_{16}-homologs and TPS, which had no detectable blank concentrations, the instrumental detection limit (peak height ten times higher than baseline noise) was assumed. The resulting detection limit for single LAS isomers ranged from 1.6 µg/kg to 4.5 µg/kg and that of the total of TPS was at about 200 µg/kg.

Results of GC/MS analysis of sediments from Lake Wohlen are displayed in Figure 4. Chromatogram A is from a sediment horizon (at depth 46 cm) which was deposited in approximately 1970. It displays the distribution of LAS-homologs and isomers containing up to 16 carbon atoms in the linear alkyl chain. Chromatogram B is from a sediment horizon deposited in the mid 1960's (depth 55 cm), showing the complexity of the TPS mixture. For comparison, a chromatogram of a standard solution of TPS is shown in Figure 4C.

Table II. Molecular Ions for Trifluoroethylesters of the LAS and TPS

LAS-homolog	$(M+1)^+$ [m/z]
C_8 (internal standard)	353
C_{10}	381
C_{11}	395
C_{12}	409
C_{13}	423
C_{14}	437
C_{15}	451
C_{16}	465

Concentrations of LAS and TPS in the Sediment Cores

Using the dated sediment cores, we acquired time-correlated depth profiles of total LAS and TPS concentrations (Figure 5). The profiles reflect the input history of the two anthropogenic markers, LAS and TPS, into the sediment, and, thus, the Aare river. An increasing consumption of TPS in the late 1950's resulted in increasing concentrations in the sediment between 65 cm and 55 cm in Lake Wohlen (Figure 5A). In a sediment layer corresponding to about 1965 (depth 56 cm), TPS

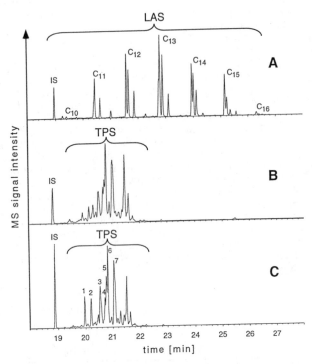

Figure 4. Results of GC/MS analyses of sediments from Lake Wohlen. Chromatograms recorded in the single ion monitoring mode using molecular ions. Details see text and Table II.
A) Sediment layer deposited in about 1970 containing LAS at 2.1 mg/kg dry sediment.
B) Sediment layer deposited in about 1965 containing TPS at 3.4 mg/kg dry sediment.
C) Standard solution of 100 µg/mL of TPS. 1, 2, ..., 7: Peaks used for quantitation.

IS: C_8-LAS internal standard, C_{10} - C_{16}: LAS-homologs.

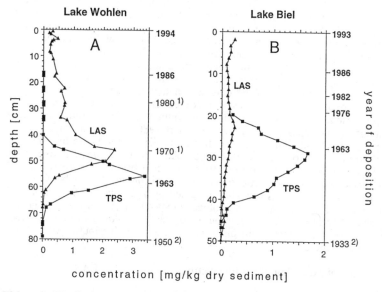

Figure 5. Vertical concentration profiles of total LAS and TPS in sediment cores of A) Lake Wohlen and B) Lake Biel. [1] linearly interpolated, [2] linearly extrapolated.

concentrations exhibit a maximum at 3.4 mg/kg dry sediment. This maximum correlates well with the time when TPS were replaced by LAS. Subsequently, concentrations of LAS increase as TPS decrease in the overlying layers. A maximum in LAS concentrations at 2.4 mg/kg occurs in sediments deposited in about 1970 (depth 45 cm), which roughly coincides with the time when the sewage treatment plant of the city of Bern was put into operation. This decline with approach to the sediment-water interface clearly indicates the effectiveness of the sewage treatment plant in removing LAS.

A similar profile pattern was found in the sediment core of Lake Biel (Figure 5B). However, the difference between the maximum concentrations of LAS and TPS is considerably larger in Lake Biel. This finding can be explained by the greater biodegradability of LAS. The residence time of sewage markers in the oxic surface waters of the Aare river before sedimentation is longer at the sampling site in Lake Biel than in Lake Wohlen. Thus, the majority of LAS had been degraded during transport in the river before they were buried in the anoxic sediment of Lake Biel. At the same time TPS had persisted and showed a much more conservative behavior.

In Figure 6 concentration profiles of LAS from Lake Wohlen are plotted for each LAS-homolog group revealing both input changes and postdepositional fate of LAS. Input changes are evident in the near surface sections of the profiles. Long-chain homologs (C_{14} - C_{16}) were present only in sediment layers corresponding to years prior to 1980 and could not be detected in the most recent (>1980) layers. The phasing out of long-chain LAS is recorded clearly in the sedimentary record. Postdepositional fate is documented by the width of the concentration peak present in sediment layers deposited between 1965 and 1980. This peak width is inversely correlated with the length of the linear alkyl chain. Assuming the homolog distribution of the originally deposited LAS remained constant over time, this correlation must be attributed to different postdepositional processes controlled by differences in partitioning properties. Since partitioning of LAS between sediment particles and pore water is probably governed by hydrophobic interactions (18, 38), the mobility of the short-chain homologs is expected to be enhanced relative to the long-chain homologs. The assumption of constant homolog distribution, however, is difficult to verify, as data on the history of the homolog distribution of LAS products are rare in the literature.

Mathematical Modeling of Postdepositional Transport of LAS

A mathematical model was applied in order to simulate postdepositional transport of LAS. As a first approach, a simple one-dimensional model was used. The sediment/water interface was chosen as the origin of the depth coordinate. The diagenetic equation for solute transport in the sediment pore water was derived from the general diagenetic equation developed by Berner (39)

$$\phi\left(\frac{dC}{dt}\right)_x = \frac{\partial\left(\phi D_b \frac{\partial C}{\partial x}\right)}{\partial x} - \frac{\partial(\phi\omega C)}{\partial x}$$

C: concentration of the solute in mass per volume pore water [μg/L]

D_b: molecular diffusion coefficient in the bulk sediment [cm²/y,]

t: time [y]

x: depth [cm]

ω: velocity of burial of particles below the sediment/water interface; average from 1963 to 1994: 1.82 cm/y

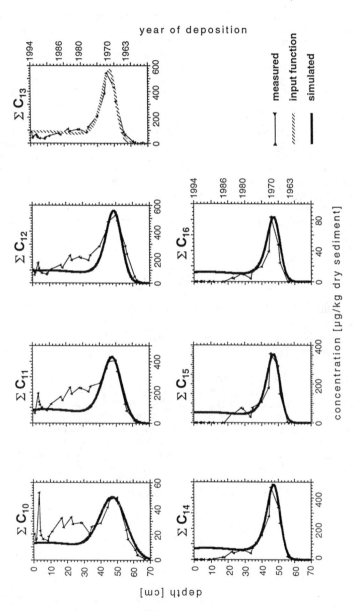

Figure 6. Mathematically simulated and measured concentration profiles of LAS homolog groups (Σ m-C_n-LAS) in the sediment of Lake Wohlen.

ϕ: porosity, i.e. volume fraction of interconnected pores; set to an estimated average value of 0.76

The isomeric composition of LAS was constant with depth between 5 and 60 cm indicating negligible biodegradation in the subsurface sediment (not shown). However, significant changes in favour of the internal isomers occurred in the top 5 cm of the core, possibly indicating biodegradation and, thus, oxic conditions in the near surface sediment section. Yet, no significant decrease in LAS concentrations with depth can be inferred from Figure 5A. Therefore it is evident that biodegradation which occurs in the top section of the sediment has a minor impact on the subsurface concentration profile. Consequently, diagenetic loss reactions were neglected in the diagenetic equation.

The velocity of burial ω was set constant at 1.82 cm/y, based on sediment deposited between 1963 and 1994. In fact, ω is greater (2 cm/y) for layers deposited between 1986 and 1994 and smaller (1.6 cm/y) for layers deposited between 1963 and 1986. Presumably this difference is a result of compaction as indicated by the drop in water content in the near surface section. However, the sampling and dating technique used does not provide sediment parameters which allow the inclusion of effects of compaction in the model.

Porosity ϕ was estimated using the content of water and the volume of dry sediment.

$$\phi = \frac{V_w}{V_w + M_s / \rho_s}$$

V_w: volume of pore water, i.e. mass of pore water [cm^3]
M_s: mass of dry sediment [g]
ρ_s: density of the solid sediment phase; 2.5 g/cm^3 mean value from Greifensee (33)

The velocity of burial of pore water was set equal to the velocity of particle burial, i.e. advective flow of pore water relative to sediment particles was neglected. Evidence for absent porewater flow is provided by the coincidence of the two peak maxima of LAS and TPS with the time of the advent of the sewage treatment plant of the city of Bern and the TPS replacement, respectively. On the contrary, significant porewater flow (i.e. groundwater discharge) would shift these peak maxima.

As there was no evidence for mixing processes, the diffusion coefficient of only molecular diffusion was assumed for D_b. The molecular diffusion coefficient, however, was corrected for partitioning of analytes onto sediment particles (39) and, thus, became the governing parameter in this approach of modeling LAS transport in the sediment.

$$D_b = \frac{D_s}{1 + \rho K_p \left(\frac{1 - \phi}{\phi} \right)} \qquad \text{(39) p. 76}$$

D_s: molecular diffusion coefficient in pore water; 92 cm^2/y, estimated for C_{12}-LAS using acetic acid as reference compound (40) and effects of tortuosity (39)
K_p: partition coefficient of the individual LAS-homologs, see Table III

Partition coefficients, used for this modeling exercice were taken from the literature (18), and were determined using sediments from Rapid Creek, SD, and radio labelled LAS-homologs. Coefficients determined for a sediment with a grain size distribution similar to that of Lake Wohlen were selected and corrected for the difference in TOC (Table III).

Table III. Partition coefficients K_p for LAS-homologs

LAS-homolog	K_p [L/kg]
C_{10}-LAS	140
C_{11}-LAS	390
C_{12}-LAS	1,100
C_{13}-LAS	3,500
C_{14}-LAS	8,700
C_{15}-LAS	26,000*
C_{16}-LAS	73,000*

Based on values published by Hand and Williams (18).
* estimated by exponential extrapolation of the C_{10}- to C_{13}- homologs

In Figure 6 the simulated profiles are superimposed on the measured profiles. The profiles were simulated on the basis of relative calibration. This means that the profile of the C_{13}-homolog was fitted, in that input functions at the sediment/water interface were defined and calibrated with the measured concentrations of the C_{13}-homolog, which is the most conservative of the currently manufactured homologs. The input functions were

i) a Gaussian term for the years 1950 - 1972 which describes the rapidly increasing LAS consumption as a result of TPS replacement

$$C_{(0)} = C_{max} \cdot \frac{\exp-\frac{(t-a)^2}{2b^2}}{\sqrt{2\Pi b}}$$

a: time of peak maximum
b: width of the peak
C_{max}:   peak height

ii) an exponential term for the years 1970 - 1994 which describes the decreasing LAS release as a result of the opening of the sewage treatment plant. The linear and quadratic terms correct for the steadily increasing LAS consumption

$$C_{(0)} = c \cdot (t-a) + d \cdot (t-a)^2 + C_{max} \cdot \frac{\exp-(t-a)}{\sqrt{2\Pi b}}.$$

Parameters *a* to *d* were qualitatively fitted by manual iteration using the partition coefficient of the C_{13}-homolog in Table III. Using these calibrated input functions, we computed the profiles for the other homologs by substituting the corresponding partition coefficients.

In general, the measured profiles were reasonably well predicted by only varying the partition coefficients, specifically in the sediment section which was deposited during the time when LAS were introduced to the market. There is disagreement for concentrations in sediment layers deposited during the 1980's, the time when the long-chain types were phased out. Here the assumption of constant homolog distribution was invalid.

In summary, partition coefficients seem to be key parameters in the process of postdepositional transport of LAS. Particularly, molecular diffusion is nearly inhibited, if the linear carbon chains are longer than 14 carbon units, or K_p's exceed 10,000 L/kg.

Conclusions and Outlook

The input history of LAS and TPS is recorded in recent lacustrine sediments. The replacement of slowly biodegradable TPS by the readily biodegradable LAS as well as the phasing out of the long-chain components of LAS are reflected in the concentration depth profiles. Based on the fact that the input history of LAS was faithfully recorded and showed no evidence of postdepositional biodegradation in the subsurface sediment, it can be inferred that LAS are persistent under the anoxic conditions in these sediments. This is analogous to the persistent behavior of LAS under methanogenic conditions in sewage sludge digesters (41).

Although, amphiphilic surfactants are more hydrophilic than other molecular markers such as linear alkylbenzenes, trialkylamines, and fluorescent whitening agents (4, 6-9) they can be used to trace relatively hydrophilic contaminants. Therefore, alkylbenzenesulfonates can be used as molecular markers for domestic waste water, although some limitations must be considered. The susceptibility of LAS to aerobic biodegradation diminishes to some extent their usefulness as molecular markers. With respect to its persistence under oxidizing conditions, TPS appear to be more suitable molecular markers than LAS particularly with respect to sewage inputs prior to 1965. On the other hand, the lower concentrations of LAS in Lake Biel relative to TPS indicate the importance of aerobic biological activity during transport to Lake Biel.

The concentration peak of LAS in the sediment of Lake Wohlen (Figure 5A), was caused by the rapid introduction of LAS to the market and by the advent of the sewage treatment plant of Bern. Moreover, the range of partition coefficients K_p exhibited by LAS offer a means of assessing the potential for remobilization of these contaminants in sediments. Our studies have shown that contaminants having partition coefficients larger than that of the C_{14}-LAS-homolog ($K_p > 10,000$), are immobilized in the sediment due to sorption onto sediment particles.

Acknowledgments

We gratefully acknowledge VISTA Chemical Company and the Committee for Linear Alkylbenzene Environmental Research (CLER) which supported this work financially and by their scientific expertise. As partners in the Rhine Basin Program we acknowledge Hewlett Packard Company for the donation of the GC/MS equipment. We thank Marc Suter and Christian Schaffner for advising us in MS and GC techniques. We acknowledge Mike Sturm, and Alfred Lück of EAWAG's department of environmental physics for their highly instructional help during sample collection and sedimentological characterization.

Literature Cited

1. Jakobi, G.; Löhr, A. *Detergents and Textile Washing;* VCH Verlagsgesellschaft mbH: Weinheim, 1987.
2. de Oude, N.T. *Detergents;* Handbook of Environmental Chemistry, Springer-Verlag: Berlin, 1992.
3. Giger, W.; Alder, A.C.; Brunner, P.H.; Marcomini, A.; Siegrist, H. *Tenside Surfactants Detergents,* **1989,** *26,* 95 - 100.
4. Takada, H.; Eganhouse, R.P. In *The Encyclopedia of Environmental Analysis and Remediation;* John Wiley & Sons, Inc.: New York, 1997 (in review).
5. Vivian, C.M.G. *Total Environ.,* **1986,** *53,* 5 - 40.
6. Ishiwatari, R.; Takada, H. *Nature,* **1983,** *301,* 599 - 600.
7. Eganhouse, R.P.; Blumfield, D.L.; Kaplan, I.R. *Environ. Sci. Technol.,* **1983,** *17,* 523- 530.
8. Fernandez, P.; Valls, M.; Bayona, J.M.; Albaigés, J. *Environ. Sci. Technol.,* **1991,** *25,* 547 - 550.
9. Stoll, J.-M.A.; Poiger, T.F.; Lotter A.F.; Sturm M.; Giger, W. In *Molecular Markers in Environmental Geochemistry;* Eganhouse R. P., Ed.; Amer. Chem. Soc. Symposium Series; 1997.
10. Upadek, H.; Krings, P. In *Chemiefaser-Tagung;* 1991. Dornbirn, Germay.
11. deAlmeida, J.L.G.; Dufaux, M.; Ben-Taarit, Y.; Naccache, C. *J. Am. Oil Chem. Soc.,* **1994,** *71,* 675 - 694.
12. Prats, D.; Ruitz, F.; Vazquez, B.; Zarzo, D.; Berna, J.L.; Moreno, A. *Environ. Toxicol. Chem.,* **1993,** *12,* 1599 - 1608.
13. Brunner, P.H.; Capri, S.; Marcomini, A.; Giger, W. *Wat. Res.,* **1988,** *22,* 1465-1472.
14. Berna, J.L.; Ferrer, J.; Moreno, A.; Prats, D.; Ruiz Bevia, F. *Tenside Surfactants Deterg.,* **1989,** *26,* 101-107.
15. Waters, J.; Feijtel, T.C.J. *Chemosphere,* **1995,** *30,* 1939 - 1956.
16. Federle, T.W.; Schwab, B.S. *Wat. Res.,* **1992,** *26,* 123 - 127.
17. Wagner, S.; Schink, B. *Wat. Res.,* **1987,** *21,* 615-622.
18. Hand, V.C.; Williams, G.K. *Environ. Sci. Technol.,* **1987,** *21,* 370-373.
19. Takada, H.; Ishiwatari, R.; Ogura, N. *Estuarine, Coastal and Shelf Science,* **1992,** *35,* 141 - 156.
20. Trehy, M.L.; Gledhill, W.E.; Orth, R.G. *Anal. Chem.,* **1990,** *62,* 2581-2586.
21. Rapaport, R.A.; Eckhoff, W.S. *Environ. Toxicol. Chem.,* **1990,** *9,* 1245-1257.
22. DeHenau, H.; Matthijs, E. *Intern. J. Environ. Anal. Chem.,* **1986,** *26,* 279 -293.
23. Hon-Nami, H.; Hanya, T. *Jap. J. Limnol.,* **1980,** *41,* 1 - 4.
24. Tabor, C.F.; Barber, L.B. *Environ. Sci. Technol.,* **1996,** *30,* 161 - 171.
25. Ambe, Y. *Environ. Sci.Technol.,* **1973,** *7,* 542 - 545.
26. Amano, K.; Fukushima, T.; Nakasugi, O. *Hydrobiologya,* **1992,** *235/236,* 491 - 499.
27. Takada, H.; Ishiwatari, R. *Environ. Sci. Technol.,* **1987,** *21,* 875 -883.
28. Takada, H.; Satoh, F.; Bothner, M.; Tripp, B.; Farrington, J. In *Molecular Markers in Environmental Geochemistry;* Eganhouse R. P., Ed.; Amer. Chem. Soc. Symposium Series; 1997.
29. Bayona, J.M.; Albaigés, J. *Chemosphere,* **1986,** *15,* 595 - 598.
30. Swisher, R.D. *Surfactant Biodegradation;* Dekker: New York, 1987.
31. Albrecht, A.; Reichert, P.; Beer, J.; Lück, A. *J. Environ. Radioactivity,* **1995,** *28,* 239 - 269.
32. Cundy, A.B.; Croudace, I.W. *Estuarine, Costal and Shelf Science,* **1996,** *43,* 449 - 467.
33. Wan, G.J.; Santschi, P.H.; Sturm, M.; Farrenkothen, K.; Lück, A.; Werth, E.; Schuler, C. *Chemical Geology,* **1987,** *63,* 181 - 196.

34. Albrecht, A.; Lück, A.; Weidmann, Y., Submitted to *Applied Geochemistry,* **1997**.
35. Matthijs, E.; DeHenau, H. *Tenside Surfactants Deterg.,* **1987,** *24,4,* 193 -199.
36. Kataoka, H.; Muroi, N.; Makita, M. *Analytical Sciences,* **1991,** *7,* 585 - 588.
37. Keith, L.H. *Environmental Sampling and Analysis;* Lewis Publishers: Chelsea, Michigan, 1991.
38. Brownawell, J.B.; Chen, H.; Zhang, W.; Westall, J.C. In *Organic Substances and Sediments in Water;* Baker, R.A., Ed.; Lewis Publishers: Chelsea Michigan, 1991, Vol. 2; pp 127 - 147.
39. Berner, R.A. *Early Diagenesis: A Theoretical Approach;* University Press: Princeton, N.Y., 1980.
40. Schwarzenbach, R.P.; Gschwend, P.M.; Imboden, D.M. *Environmental Organic Chemistry;* John Whily & Sons, Inc.: New York, 1993.
41. McEvoy, J.; Giger, W. *Environ. Sci.Technol.,* **1986,** *20,* 376 - 383.

Chapter 14

Linear Alkylbenzenesulfonates and Polychlorinated Biphenyls as Indicators of Accumulation, Biodegradation, and Transport Processes in Sewage Farm Soils

Thorsten Reemtsma, Irena Savric, Claudia Hartig, and Martin Jekel

Department of Water Quality Control, Technical University of Berlin, Secretariate KF 4, Strasse des 17. Juni 135, D–10623 Berlin, Germany

A comparative study was conducted on the concentrations and compositions of polychlorinated biphenyls (PCB) and linear alkyl-benzenesulfonates (LAS) in two former sewage farms to obtain insights into the history of operation and post-depositional processes having affected the sewage-derived organics accumulated in the soils over the last decades. Total concentrations in the surface soils are in the range of 1 - 6 μg g^{-1} for PCB and 0.1 - 4 μg g^{-1} for LAS. Differences in the intensity of sewage discharge between the two sites are recorded in the total contents of PCB and LAS, while different qualities of wastewater treatment prior to disposal are visible in the PCB composition and the LAS homologs distribution of the surface soils. Alterations in the homolog composition of LAS in depth profiles reflect desorption and sorption processes during downward transport through the unsaturated zone, while the largely unaltered isomeric composition might suggest that LAS were not degraded in the soil column.

Sewage farms were installed as an alternative to wastewater treatment plants by the City of Berlin in the second half of the 19th century. The infiltration basins were flooded with settled wastewater up to six times a year (1), and the water itself as well as the organic matter and dissolved nutrients of the wastewater were used for tillage. Sewage farming reached its largest expansion in the early 1920s with up to 110 x 10^6 m^2 of farm land and an average amount of 700,000 m^3 sewage d^{-1} treated at the farms. Since the 1920s the sewage farms were successively closed down and only a very few sewage farms are still in use.

During their time of operation the sewage farm soils accumulated nutrients and heavy metals (2) as well as organic wastewater constituents such as polycyclic aromatic hydrocarbons (PAH) and polychlorinated biphenyls (PCB) (3) and large amounts of unknown extractable organohalogens (4). Due to the high infiltration rates of some meters per year applied over a period of approximately a hundred years, organic wastewater constituents originally deposited in the surface soils might have entered deeper soil layers and the surficial aquifer. Breakthrough of wastewater constituents into the surficial groundwater has been investigated in rapid infiltration systems by field (5-6) and column studies (7). However, detailed knowledge of the organic xenobiotics deposited in the sewage farms as well as of their long term dynamics and controlling factors is very limited.

We selected two classes of substances, PCB and linear alkylbenzenesulfonates (LAS), to obtain information on the processes determining the fate of organic wastewater constituents in the sewage farms. Based on differences in their biogeochemical profiles, we used these compound classes to provide complementary information on sewage sources, and on post-burial migration and degradation processes in the sewage farm soils.

While PCBs are among the most intensively investigated xenobiotic compounds (8), present knowledge on the biogeochemistry of LAS in soils is still fragmentary. Although several investigations on the behaviour of LAS in sludge amended soils (9-13) and in groundwater influenced by secondary treated wastewater (14-16) or septic tanks (17-20) have been performed, LAS have apparently not been used as molecular markers for transport and degradation processes in soils.

Materials and Methods

Sampling Sites. The sampling sites are displayed in Figure 1. Both sewage farms were established in the late 19th century and were used for the treatment of raw wastewater. However, the history of operation at these sites differed significantly. Site I (SF-I, Karolinenhöhe) was primarily used for the treatment of municipial wastewater, with average infiltration rates below 4 m yr^{-1}. The application of raw wastewater ceased in the 1960s, after which secondary effluent of a sewage treatment plant was discharged. Municipal and industrial wastewater was treated on site II (SF-II, Buch). Situated outside of the City boundaries, this farm was operated by the water authorities of East Berlin (German Democratic Republic) after World War II. Sewage discharge was intensified in the 1960s, and the sewage farm was closed in 1985.

Soils from both sites are loamy, fine to medium grain sands with low organic carbon (< 0.1 %) and clay (< 1 %, often < 0.5 %) contents in the C-horizons. The Yah-horizons (0 - 20 cm; humic layer) exhibit organic carbon contents of 2 - 5%. The water table is 6 m and 9 m below land surface at sites II and I, respectively.

Sampling. Soil samples were collected from each of the sewage farms in spring of 1994 (SF-I; SF-II/1 - II/3); samples for additional LAS analyses were taken in spring 1996 at site II (SF-II/4). All samples were scraped from the vertical walls of freshly excavated holes after careful removal of the surficial material. Sludge aggregates could be isolated from the surface soil of sewage farm II. The soils were sieved over 5 mm (removal of small stones and plant litter) and stored frozen until analyses.

Frozen samples were put in an oven at 45 °C and dried within 3 - 4 days until they reached constant weight. They were then homogenized in a stainless steel disk vibratory mill, and an aliquot was subjected to Soxhlet extraction.

PCB analyses. Sample preparation, work-up and analysis followed the procedure of Nolte et. al. (*21*). Briefly, the soils were Soxhlet-extracted with hexane/toluene (9/1) overnight, and the extracts were cleaned by shaking with concentrated sulfuric acid and by passing the extracts through silica columns (500 mg Bakerbond SPE; Baker, Philipsburg, USA). A PE 8420 gas chromatograph with electron capture detection (GC-ECD) and equipped with an AS 8300 autosampler (all from Perkin-Elmer, Überlingen, Germany) was employed for PCB analyses. Separation was performed on a CP-SIL 8cb column 50 m x 0.25 mm i.d. (Chrompack, Frankfurt, Germany) with helium as carrier gas (170 kPa, column head pressure). Injector temperature was 250°C, detector temperature 350°C; the column was held at 120°C for 1 min, heated at 20°C min^{-1} to 180°C and at 1°C min^{-1} to 240°C; after 3 min of isothermal operation the column was heated at 10°C min^{-1} to 280°C and was finally kept constant for 10 min. 1 μl of sample was injected in the splitless mode.

About 100 congeners could be identified based on coelution with individual reference congeners (obtained from Promochem, Wesel, Germany). Quantitation of all detected peaks was based on relative response factors determined by a two point calibration with two technical PCB mixtures (Clophen A30, Clophen A60; Promochem, Wesel, Germany) of known relative composition (*22*).

Recovery from spiked samples was between 75% and 90% for all but three detected peaks, and the standard deviation of 24 replicates was 5 - 10% (*21*). Detection limits range from 0.2 to 1 ng g^{-1} depending on the relative response factors of the individual congeners. Blank analyses were performed with every series of PCB analyses, and their PCB contents were below the detection limit.

LAS notation.   All LAS used in detergent formulations are 1-alkyl-4-sulfonyl benzenes. The length of the alkyl chain and the position of the sulfophenyl-substituent in the alkyl chain, however, vary from C10 to C13 and from the 2- to the mid-chain position in recent detergent formulations. An individual LAS compound is, therefore, clearly defined by the position of the sulfophenyl substituent at the alkyl chain and by the alkyl chain length: 3-LAS12 denotes dodecyl-3-benzenesulfonate.

LAS analyses.   Extraction and clean-up were adopted from the procedure developed by Reiser for fluvial sediments (*23; see also Reiser and Giger, this volume*). Briefly, the dried sample was Soxhlet-extracted with methanol, and octyl-1-phenylsulfonate (1-LAS8) was added as internal standard. Nonionic and cationic compounds were removed by passing the methanolic extract through a strong anion exchange cartridge, from which LAS were eluted with 1% (v/v) concentrated HCl in methanol. LAS in the dried eluates were derivatized by a two step procedure: i) formation of sulfonic acid chlorides by thionylchloride and ii) formation of trifluoroethylesters by the reaction with trifluoroethanol in triethylamine. The derivatives were then separated from more polar impurities by aluminium oxide (Alox N) chromatography on a Pasteur-pipette minicolumn.

LAS trifluoroethylesters were analyzed with a GC-quadrupole-MS system (MD 800; Fisons, Manchester, Great Britain) with 5 μl of the extract splitless injected at an injector temperature of 220°C. Separation was performed on a SPB5 30 m x 0.25 mm i.D. column (Supelco, Bellefonte, USA) with helium as carrier gas at 80 kPa column head pressure; the column temperature was held at 80°C for 1 min, then heated at 10°C min^{-1} to 200°C and at 6°C min^{-1} to 300°C, followed by 10 min isothermal operation. The mass spectrometer was operated in the electron-impact mode (EI), and the trifluoroethylesters of LAS were detected by monitoring their molecular ions. Since the elution times of the isomers of two adjacent homologs overlap, two traces were recorded simultaneously at a scan rate of 0.7 sec^{-1}. All isomers from LAS10 to LAS13 (total of 20 components) were quantified on a mass basis assuming identical relative response factors.

The average recovery of LAS from soil samples spiked with 0.22 μg g^{-1} total LAS was 94% (85% - 112%). The standard deviation of triplicate samples was 8%. The detection limit of individual LAS components ($S/N > 10$) is in the range of 0.1 ng injected onto the column, corresponding to a concentration of 0.1 - 0.2 ng g^{-1} of an individual component and 2 - 4 ng g^{-1} total LAS detectable from 20 g of sample. A blank sample was run in each analytical sequence. If care is taken to avoid contamination by LAS during extraction and work-up, blank values below the detection limit are obtainable.

Results and Discussion

Factors Determining the Organic Matter of Sewage Farm Soils. The major processes determining the fate of organic substances in the sewage farms are displayed in Figure 2. The amount of organics deposited in the farms is dependent on the amount of wastewater discharged and the kind of wastewater treatment applied prior to the sewage disposal. The primary removal process of organic compounds in the sewage farms is mineralization in the soil surface layer, which is accompanied by humification. Less degradable or persistent substances accumulate in the surface soil. A part of the organics deposited at the soil surface might be removed by wind erosion, volatilization and photolysis; the importance of these processes depends on the degree of soil covering by plants and may, therefore, strongly vary in time and space. Dissolved and colloidal substances delivered with the wastewater or desorbed from the soil surface layer penetrate the unsaturated zone of the soil column, where further microbial degradation and multiple sorption/desorption processes determine the amounts remaining in the soil column or entering the surficial groundwater.

Total Contents of LAS and PCB. Total PCB and LAS contents of the surface soils are given in Table I. The level of contamination is generally one order of magnitude higher in sewage farm II than in sewage farm I. This is consistent with prolonged and more extensive discharge of untreated wastewater on sewage farm II. Sewage farm soils are known to exhibit strong spatial variations in their inorganic and organic load, with contaminant contents decreasing from the location of wastewater discharge onto the infiltration basins (Figure 2, left) towards the opposite side (26).

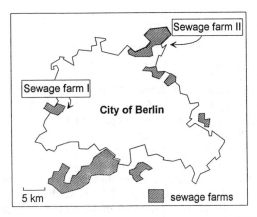

Figure 1. Sewage farms of the City of Berlin (around 1979) and the two sampling sites.

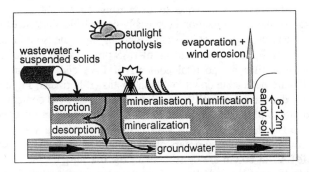

Figure 2. Major processes governing the organic matter pool of sewage farms.

Table I. Contents of PCB and LAS in sewage farm surface soils

	depth (cm)	PCB (μg g^{-1})	LAS (μg g^{-1})
surface soil I	0 - 12	0.9	0.1
surface soil II	0 - 10	3 - 6	0.4 - 4
sludge aggregates II	0 - 10	14 - 86	10 - 11
sewage sludges		0.1 - 20[a]	2000 - 12000[b]

Data from [a]) Ref. *24, 25* [b]) Ref. *11, 24.*

The elevated concentrations of PCB and LAS in the sludge aggregates isolated from the soil surface layers of sewage farm II reflect the importance of suspended wastewater particles for the input of these contaminants.

While the PCB-contents of the soils are in the range of those reported for sewage sludges during the 1980s (*24-25*), the LAS contents are significantly lower and are comparable to sewage sludge amended soils (*9, 12*). This reflects the purification potential of the surface layers towards easily degradable wastewater constituents such as LAS. It is notable that LAS were determined at all. Half lives of LAS in sludge amended, well aerated soils have been calculated to range from 3 - 30 days (*9; 12*). The nine years between the end of sewage application (1985) on sewage farm II and the time of our sampling, thus, correspond to at least one hundred half lives. Assuming first order kinetics, LAS should be completely degraded after this space of time, regardless of their initial concentration. This is, however, not the case. Either due to decreasing concentrations or due to the increasing age of the contamination even easily degradable substances such as LAS are being preserved in the soil system. Binding onto the soil matrix has been suggested to be an important mechanism of LAS preservation in soils (*10*).

The depth profiles of the total contents of LAS and PCB are displayed in Figure 3. PCB are detectable down to 40 cm depths (B-horizons, Figure 3a), but the concentrations decrease drastically. At 40 cm depth total PCB contents are between one and two orders of magnitude lower (0.01 - 1 μg g^{-1}) than in the uppermost layer (1 - 6 μg g^{-1}). The depth profiles of LAS (Figure 3b) show a similar pattern with decreases between one and two orders of magnitude. In all but one case, concentrations of total LAS at 60 cm are close to or below 0.1 μg g^{-1}.

These depth profiles indicate that wastewater constituents deposited in the surface layers of the sewage farms have been transported to deeper soil layers. In the case of LAS, elevated contents at depths of 30 - 40 cm in sewage farm II (0.1 - 1 μg g^{-1}) compared to sewage farm I (0.02 μg g^{-1}) correspond to higher surface soil contents of LAS and to elevated discharge rates.

PCB Composition in Surface Soils. The PCB composition in the surface layers of the two farms as determined by gas chromatography differs strongly (Figure 4): only late eluting highly chlorinated biphenyls are found in sewage farm I soil, while

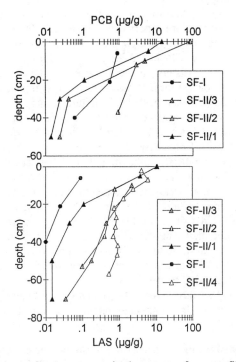

Figure 3. Total contaminants contents in the sewage farm profiles; (top): PCB (μg g^{-1}), (bottom): LAS (μg g^{-1}); depth 0 cm refers to sludge aggregates.

mixtures with a wide range of chlorination are found in the surface soils of farm II. Quantitative data are given in Table II. The PCB distribution appears to reflect the quality of wastewater treatment prior to the sewage disposal onto the sewage farm land.

Sewage farm I soil resembles the Clophen A60 distribution (Table II) with an additional minor contribution of lower chlorinated PCBs. Untreated wastewater was discharged onto sewage farm I until the mid 1960s together with the highly chlorinated PCBs such as Clophen A60, which have been used in open systems until 1972 in Western Germany (27). The secondary treated wastewater discharged since the mid 1960s did not transport appreciable amounts of PCB, since these compounds are effectively eliminated during wastewater treatment. Elimination rates of up to 99% by activated sludge treatment are reported for PCBs (28-30) with sorption onto the activated sludge being the major removal process (29). Therefore, the replacement of highly chlorinated PCBs (Clophen A60) by low chlorinated mixtures (Clophen A30) in the early 1970s (27) is not visible in the PCB pattern of sewage farm I.

The distribution in sewage farm II, however, reflects input from high (Clophen A60) to low chlorinated (Clophen A30) PCB mixtures (Table II). At sewage farm II untreated wastewater was discharged over the whole period of operation (until 1985), and the soil PCB distributions are, thus, likely to record the qualitative changes in PCB use and discharge. Unfortunately, data on the use of PCB and its restriction in the former German Democratic Republic are not available. The data presented here, however, suggest that highly chlorinated PCBs have been replaced by less chlorinated mixtures in the German Democratic Republic, too. This view is supported by comparing the PCB composition of the surface soils of sewage farm II and the sludge aggregates isolated from it (Table II). The latter are representative of the latest particulate organic matter deposited in the sewage farm prior to its shut-down in 1985. The proportions of di- to tetrachlorinated biphenyls in the sludge aggregates is significantly higher (73 %) than in the surrounding soil (54 %), suggesting that the most recently discharged PCBs were less chlorinated than the PCBs previously accumulated in the soil surface layers. Alternatively, the higher percentage of less chlorinated PCBs in the sludge aggregates might be due to their shorter residence times in the soils.

Table II. PCB distribution (%) according to the degree of chlorination in the sewage farm surface soils and in Clophen A30, Clophen A60

chlorine cont.	2	3	4	5	6	7	8	9
soil I	5	3	6	12	42	27	5	< 1
soil II[a]	2	20	32	20	20	5	1	< 1
sludge II[a]	5	29	39	16	9	3	< 1	< 1
Clophen A30[b]	20	48	25	6	1	-	-	-
Clophen A60[b]	-	-	1	17	52	30	6	< 1

[a] average of three samples; [b] data from Ref. 22.

PCB Composition in the Depth Profiles. The octanol-water partition coefficients of PCBs (log K_{ow}) range from 4.5 for monochloro- up to 8.1 for octachloro-biphenyls (*31*) and are inversely related to their water solubilities (*32*). Correspondingly, laboratory experiments have shown a distinct inverse relation between the degree of chlorination and mobility in sediments (*33*) and soil (*34*). An additional retardation of the more hydrophobic congeners might be attributed to the partly irreversible nature of their sorption onto soil material (*35*).

Therefore, downward transport of PCBs *via* the dissolved phase in soils should be recorded in a relative enrichment of the more soluble less chlorinated congeners. Transport of PCBs as colloids or sorbed onto particles might not alter their composition or would even result in a relative increase of the more hydrophobic congeners, since those are suggested to form more stable aggregates. Colloidal transport, also regarded as a partitioning of hydrophobic compounds into dissolved organic matter (DOM), has gained much attention (*36*) and it has been shown that DOM can substantially enhance the mobility of PCBs in soil columns (*37*). In other cases, however, aggregate formation of DOM with hydrophobic contaminants might decrease the mobility of the latter (*38*). It can, thus, hardly be predicted which mechanisms underlie the contaminant transport at a certain site. Changes in the homolog or isomeric composition of a class of compounds with depth might, however, allow one to identify the mechanisms that have promoted the transport of the respective compounds.

In the sewage farms less chlorinated biphenyls are relatively enriched with increasing depth as displayed in Figure 5; this corresponds to the transport of PCBs by the aqueous phase. However, even hexachlorinated PCBs are detected at depths of 50 cm, suggesting that PCB have also been displaced as colloids or on particles.

It has to be noted, that the most water soluble fraction of PCBs is also most susceptible to aerobic degradation (*39*), although the extent of PCB degradation in the field can vary substantially (*40*), and might even be negligible for aged contaminations (*41*). The less chlorinated PCB are also preferentially lost by volatilization (*42*). In fact, volatilization has been shown to be a major removal process for PCBs (Aroclor 1254) in sludge amended soils (*43*). Accordingly, one might look upon this PCB pattern (Figure 5) as an impoverishment of less chlorinated biphenyls in the soil surface, rather than as a relative enrichment with increasing depth.

It is a feature common to most biogenic and xenobiotic compound classes that the isomers or homologs of highest water solubility are also most susceptible to biodegradation and volatilization. This substantially limits their use as molecular markers in many environmental systems. Beside PCB this applies especially to PAHs (*44*).

Use of LAS as Molecular Markers. The biogeochemical profile of LAS is governed by the biodegradability and sorption behaviour of its homologs and isomers. Biodegradation of LAS is initiated by the ω-oxidation of the alkyl chain, and the rate of primary degradation of homologs and isomers has been shown to be related to the distance between the ω-end of the alkyl chain and the xenobiotic sulphophenyl moiety ("distance principle", *45*). The rate of complete degradation, however, is reported to be independent of structural influences (*46*), indicating that the rate determining step of mineralization is the ring opening rather than the initial oxidation

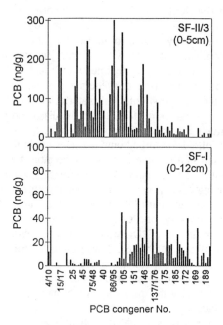

Figure 4. Contents of individual PCBs (ng g⁻¹) (in the order of elution from GC-column) in the surface soils of sewage farm I (bottom) and sewage farm II (top); every fifth congener is denoted.

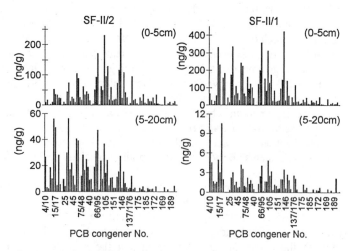

Figure 5. Contents of individual PCBs (ng g⁻¹) (in the order of elution from GC-column) at two depths of the profiles SF-II/1 and SF-II/2; every fifth congener is denoted.

of the alkyl chain. The sorption of LAS onto soil and sediments appears to be governed by hydrophobic interactions *(47)* and is, therefore, subjected to the same structural influences as primary degradation. LAS are, thus, one of the rare examples, where primary biodegradation and leaching result in different compositions of homologs and isomers. Due to their ionic nature, losses of LAS by volatilization can be neglected *(48)*.

The LAS-homolog composition can be characterized by the LAS13/LAS11 ratio:

$$LAS13/LAS11 = \frac{\sum_{j=2}^{7} [\phi_j - LAS13]}{\sum_{j=2}^{6} [\phi_j - LAS11]} \tag{1}$$

$[\phi_j - LAS]$: *concentration of the j-phenyl isomer*

We reviewed literature data on LAS homolog compositions of various environmental compartments and calculated the corresponding LAS13/LAS11 ratios (Figure 6). This ratio provides good insight into the mechanisms governing the LAS composition: all aqueous compartments such as raw sewage, biologically treated sewage and technical mixtures exhibit low LAS13/LAS11 ratios, due to the higher water solubility of the lower homologs *(47)*. The LAS13/LAS11 ratios of particulate matter are generally higher and vary from 0.3 in fluvial sediments up to 1.5 in sewage sludges (Figure 6). The preferential loss of lower homologs by leaching increases the LAS13/LAS11 ratio of the remaining LAS of particulate matter, while this ratio would be expected to decrease as a result of biodegradation.

Comparatively little use has been made of examining the isomeric composition of LAS so far *(50; 53)*, though the influence of the position of the sulfophenyl group onto LAS partitioning in a water/sediment system is nearly as strong as the effect of the alkyl chain length. Hand and Williams *(47)* observed an increase in the distribution coefficients (K_d) of one homolog by a factor of 2 for moving the sulfophenyl group from the 5- to the 2-position, while it increased by a factor of 2.8 between two homologs. Terzic et al. *(53)* pointed out, that primary degradation kinetics in water differ more strongly between the isomers of one homolog than between different homologs: 90% removal of 2-LAS12 was accomplished within one day, while approximately 10 days were required for 90% removal of 5-LAS12 to 7-LAS12. Thus, the isomeric composition might be equally useful as the homolog distribution to follow biogeochemical processes in soils, and it might provide complementary information.

The isomeric composition of LAS mixtures can be described by the 2-LAS/5-LAS ratio:

$$2-LAS/5-LAS = \frac{\sum_{i=10}^{13} [2-LAS_i]}{\sum_{i=10}^{13} [5-LAS_i]} \tag{2}$$

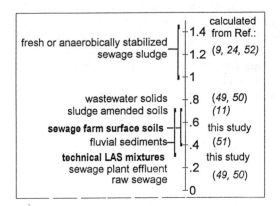

Figure 6. LAS13/LAS11 ratios (equation 1) in various environmental compartments; references according to the text.

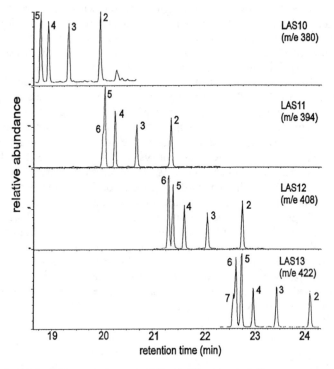

Figure 7. Molecular mass traces of the LAS extracted from a sewage farm surface soil (total LAS contents 0.72 μg g^{-1}); numbers at the peak tops denote the phenyl isomers.

At present, the effects of biogeochemical processes in soils on the homolog and isomeric distribution of LAS can only be surmised, since no investigations considering this aspect have been reported so far.

LAS Composition in Surface Soils.   A typical gas chromatogram of the LAS determined in sewage farm surface soil is displayed in Figure 7. LAS with alkyl chains ranging from 10 to 13 carbons (LAS 10 to LAS13) are regularly determined; LAS14, if detectable at all, remains quantitatively unimportant. Sulfophenyl isomers from the 2- to the mid chain position of all homologs are found; 6-LAS11 and 5-LAS11 as well as 7-LAS13 and 6-LAS13 are, however, only partly resolved.

The LAS13/LAS11 ratios of the sewage farm surface soils range from 0.4 to 0.7 (Table III) and fall between the LAS13/LAS11 ratios of wastewater (0.1 - 0.2) and wastewater solids (0.8) (Figure 6). This might be attributed to the supply of LAS by these two sources, the dissolved and the particulate fraction of raw sewage. Elevated LAS13/LAS11 ratios as reported for wastewater particles (Figure 6) are also observed in the sludge aggregates of sewage farm II (Table III).

The LAS13/LAS11 ratio of sewage farm I (0.4) is slightly lower than that of sewage farm II (0.5 - 0.7) and corresponds to the additional supply of LAS of treated wastewater with a low LAS13/LAS11 ratio (0.2) onto sewage farm I in the last decades. Similar effects are discernible in the 2-LAS/5-LAS ratio, which is also lower in sewage farm I (0.5) than in farm II (0.8 - 1.1).

Table III. LAS13/LAS11 ratios and 2-LAS/5-LAS ratios of the surface soils

	LAS13/LAS11	2-LAS/5-LAS
surface soil I	0.4	0.5
surface soil II	0.5 - 0.7	0.8 - 1.1
sludge aggregates II	0.6 - 1.5	0.8 - 0.9

LAS Composition in Depth Profiles.   The decreases in the LAS contents of the soils with depth are accompanied by significant changes in the LAS13/LAS11 ratios (Figure 8). The ratio decreases from the surface soil towards medium depths (20 - 30 cm) in all soil profiles, clearly reflecting the preferential transport of low molecular weight LAS in the water/soil system. As outlined above, primary degradation of LAS preferentially affects the longer chain homologs. Therefore, biodegradation of the dissolved LAS during downward transport might contribute to the decreasing LAS13/LAS11 ratios, too.

Astonishingly, in three out off the five profiles yet analyzed the LAS13/LAS11 ratio then increases again towards deeper soil layers (30 - 70 cm; B- and C-horizons). Two conditions must be met in order to detect organic matter of the soil surface layers in soil material of greater depth: i) the compounds must be leached from the surface, and ii) they must re-adsorb in the deeper layers. The deeper soil horizons, however, are poor in organic carbon (< 0.1 %). At these low organic

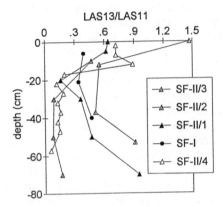

Figure 8. LAS13/LAS11 ratios in the sewage farm soil profiles; depth 0 cm refers to sludge aggregates.

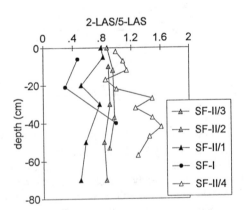

Figure 9. 2-LAS/5-LAS ratios in the sewage farm soil profiles; depth 0 cm refers to sludge aggregates.

carbon levels the interaction of dissolved organics with the particulate phase is more accurately described by sorption onto (a limited number of) specific surface sites than by partitioning processes (54). Correspondingly, McAvoy et al. (20) observed a three- to tenfold decrease in soil K_d values with organic carbon contents decreasing from 0.5% to < 0.1%.

The concentration of LAS in leachates from the sewage farm surface soils with total LAS contents of a few micrograms per gram must be rather small compared to the dissolved organic carbon contents in the percolating water (30 - 60 mg l[-1], *Hoffmann, C., Technical University of Berlin, personal communication, 1996*). Hence, the LAS have to compete with a large amount of dissolved organics for the limited sorption sites in deeper layers. The more hydrophobic LAS13 are preferred over LAS11 in this competition and might, thus, become relatively enriched as observed in most profiles (Figure 8).

Contrary to the homolog composition, the isomeric distribution of LAS is not subjected to systematic alterations with increasing depth (Figure 9). Only one out of the five profiles (SF-II/1) exhibits a decrease in the 2-LAS/5-LAS ratio (from 0.8 at 5 cm to 0.5 at 70 cm), as would have been expected from literature (see above). The 2-LAS/5-LAS ratio of the other profiles stays constant or fluctuates. Obviously, the isomeric composition (Figure 9) of the LAS is governed by other factors than the homolog composition (Figure 8). Presuming that the influence of the isomeric composition on biodegradation (and *vice versa*) is as strong in the water/soil systems as observed in aqueous systems (53), the lack of a significant decrease in the 2-LAS/5-LAS ratio with depth would indicate that primary biodegradation of LAS was negligible during downward transport in the soil.

To our knowledge, LAS degradation in percolating waters have not been investigated in field experiments, and we can, thus, not evaluate the findings outlined above. However, investigations on the potential of soil and groundwater to mineralize LAS have been performed; it was shown in laboratory experiments with LAS spiked to soil and groundwater samples, that the mineralization potential was widely distributed in soils and groundwater influenced by a septic tank system (17-18).

The suggested absence of significant LAS degradation during downward transport in the sewage farms soils might be interpreted assuming that LAS degradation would occur in the dissolved phase only. LAS in percolating waters are supplied by the LAS reservoir of the surface soils. With contents of a few micrograms per gram in the surface soils (Table I), the LAS concentration in the percolating water must be extremely low; it might then be too low to initiate LAS degradation in the water phase. Additionally, LAS degradation is likely to be hampered by their very minor portion of the DOM (see above). Thereby, LAS escapes biodegradation during downward transport and accumulates without alteration of the isomeric composition in deeper soil layers.

Conclusions

The analysis of PCB and LAS in surface soils, sludge aggregates and soil profiles up to 75 cm in depth of two former sewage farms of the City of Berlin has been proven to be useful for evaluating the various processes governing the contaminant budget in the sewage farm soils, such as former sewage input, microbial degradation in the surface soil and the unsaturated zone and transport processes through the soil column.

The PCB composition of the surface soils of the two sites was indicative of the quality of wastewater treatment prior to its dicharge onto the sewage farms. The interpretation of compositional alterations of PCB with increasing depth is hampered by the fact that the most mobile PCB congeners are also most susceptible to biodegradation and volatilization.

LAS, although well degradable substances in water, can be used as molecular markers over some decades in soil systems influenced by municipal wastewater. The LAS13/LAS11 ratio and the 2-LAS/5-LAS ratio are useful to record alterations of the homolog and isomeric composition, respectively. Changes in the homolog composition down the soil profiles can be interpreted in terms of desorption and sorption processes. The factors affecting the isomeric composition of LAS (2-LAS/5-LAS ratio) in the soils remains less clear. This is partly due to the limited attention previously paid to the isomeric composition of LAS.

The following conclusions might be drawn with regard to future work:

i) Investigations of the behaviour of LAS in the environment are recommended to cover both the homolog and isomeric composition of LAS.

ii) The behaviour of LAS in the unsaturated soil column requires further investigations, in order to elucidate the factors governing LAS degradation and sorption.

iii) It might be profitable to employ other degradable and water soluble classes of compounds as molecular markers in biogeochemical research on processes in the soil/water system.

Acknowledgements

We are indebted to René Reiser and Walter Giger for making available the analytical procedure for LAS and providing reference materials. We are grateful to Susanne Dill and Hella Schmeißer and to Heike Podey for laboratory assistance. The advice of Christine Nolte in PCB analyses is gratefully acknowledged. We thank Bob Eganhouse and two anonymous reviewers for their helpful comments and constructive criticism. A part of this work was financially supported by a grant from the Technical University of Berlin (IFP 7/21).

Literatur Cited

(1) Blume, H.-P.; Horn, R.; Alaily, F.; Jayakody, A.N.; Meshref, H. *Soil. Sci.* **1980**, *130*, 186.

(2) Blume, H.-P.; Horn, R. *Z. Kulturtech. Flurbereinig.* **1982**, *23*, 236.

(3) Bechmann, W.; Grunewald, K. *Z. Pflanzenernähr. Bodenk.* **1995**, *158*, 543.

(4) Reemtsma, T.; Jekel, M. *Chemosphere* **1996**, *32*, 815.

(5) Tomson, M.B.; Curran, C.; Hutchins, S.R.; Lee, M.D.; Waggett, G.; West, C.C.; Ward, C.H. In *Ground Water Quality*; Ward, C.H., Giger, W., McCarthy, P.L., Eds.; Wiley: New York, 1985, pp. 188-215.

(6) Wilson, L.G.; Amy, G.L.; Gerba, C.P.; Gordon, H.; Johnson, B.; Miller, J. *Wat. Environ. Res.* **1995**, *67*, 371.

(7) van der Meer, J.R.; Bosma, T.N.P.; de Druin, W.P.; Harms, H.; Holliger, C.; Rijnaarts, H.H.M.; Tros, M.E.; Schraa, G.; Zehnder, A.J.B. *Biodegradation* **1992**, *3*, 265.
(8) *PCBs and the Environment*; Waid, J.S., Ed.; CRC Press: Boca Raton, 1986/1987, Vol. I - III.
(9) Marcomini, A.; Capel, P.D.; Lichtensteiger, T.; Brunner, P.H.; Giger, W. *J. Environ. Qual.* **1989**, *18*, 523.
(10) Marcomini, A.; Capel, P.D.; Lichtensteiger, T.; Brunner, P.H.; Giger, W. In *Organic Contaminants in Wastewater, Sludge and Sediments*; Quaghabeur, D., Temmermann, I., Angeletti, G., Eds.; Elsevier: London, 1989, pp. 105-123.
(11) Holt, M.S.; Matthijs, E.; Waters, J. In *Organic Contaminants in Wastewater, Sludge and Sediments*; Quaghebeur, D., Temmermann, I., Angeletti, G., Eds.; Elsevier: London, 1989, pp. 161-182.
(12) Waters, J.; Holt, M.S.; Matthijs, E. *Tens. Surf. Deterg.* **1989**, *26*, 129.
(13) Figge, K.; Schöberl, P. *Tens. Surf. Deterg.* **1989**, *26*, 122.
(14) Thurman, E.M.; Willoughby, T.; Barber, L.B.; Thorn, K.A. *Anal. Chem.* **1987**, *59*, 1798.
(15) Field, J.A.; Leenheer, J.A.; Thorn, K.A.; Barber, L.B.; Rostad, C.; Macalady, D.L.; Daniel, S.R. *J. Contam. Hydrol.* **1992**, *9*, 55.
(16) Field, J.A.; Barber, L.B.; Thurman, E.M.; Moore, B.L.; Lawrence, D.L.; Peake, D.A. *Environ. Sci. Technol.* **1992**, *26*, 1140.
(17) Larson, R.J.; Federle, T.W.; Shimp, R.J.; Ventullo, R.M. *Tens. Surf. Deterg.* **1989**, *26*, 116.
(18) Shimp, R.J.; Lapsins, E.V.; Ventullo, R.M. *Environ. Toxicol. Chem.* **1994**, *13*, 205.
(19) Federle, T. W.; Ventullo, R.M.; White, D.C. *Microb. Ecol.* **1990**, *20*, 297.
(20) McAvoy, D.C.; White, C.E.; Moore, B.L.; Rapaport, R.A. *Environ. Toxicol. Chem.* **1994**, *13*, 213.
(21) Nolte, C.; Gunkel, G.; Jekel, M. *Vom Wasser* **1996**, *86*, 127.
(22) Schulz, D.E.; Petrick, G.; Duinker, J.C. *Environ. Sci. Technol.* **1989**, *23*, 852.
(23) Reiser, R. *PhD-thesis*; EAWAG/ETH Zürich, 1997, **in prep.**
(24) McEvoy, J.; Giger, W. *Environ. Sci. Technol.* **1986**, *20*, 376.
(25) Tarradellas, J.; Muntau, H.; Beck, H. In *Polychlorinated biphenyls (PCB) - Determination in Sewage Sludge and Related Samples*; Leschber, R. et al., Ed.; COST 681 report, 1985, pp. 11- 42.
(26) Grunewald, K. *Z. Pflanzenernähr. Bodenk.* **1994**, *157*, 125.
(27) Rüffer, H. In *ATV Schriftenreihe*; GFA-Verlag: St. Augustin, 1989, pp. 7-26.
(28) Petrasek, A.C.; Kugelman, I.J.; Austern, B.M.; Pressley, T.A.; Winslow, L.A.; Wise, R.H. *Journal WPCF* **1983**, *55*, 1286.
(29) Shannon, E.E.; Ludwig, F.L.; Valdmanis, I. *Environ. Can. Res. Rep.* **1976**, *49*, 1.
(30) Morris, S.; Lester, J.N. *Water Res.* **1994**, *28*, 1553.
(31) Hawker, D.W.; Connell, D.W. *Environ. Sci. Technol.* **1988**, *22*, 362.
(32) Chou, S.F.J.; Griffin, R.A. In *PCBs and the Environment*; Ward, J.S., Ed.; CRC Press: Boca Raton, 1986, Vol. I, pp. 101- 120.

(33) Wood, L.W.; Rhee, G.-Y.; Bush, B.; Barnard, E. *Water Res.* **1987**, *21*, 875.

(34) Tucker, E.S.; Litschgi, W.J.; Mees, W.M. *Bull. Environ. Contam. Toxicol.* **1975**, *13*, 86.

(35) Di Toro, D.M.; Horzempa, L.M. *Environ. Sci. Technol.* **1982**, *16*, 594.

(36) *Organic substances in soil and water: natural constituents and their influence on contaminant behaviour*; Beck, A.J., Jones, K.C., Hayes, M.H.B., Mingelgrin, U., Eds.; Royal Society of Chemistry: Cambridge, 1993.

(37) Dunnivant, F.M.; Jardine, P.M.; Taylor, D.L.; McCarthy, J.F. *Environ. Sci. Technol.* **1992**, *26*, 360.

(38) Rav-Acha, C.; Rebhun, M. *Water Res.* **1992**, *26*, 1645.

(39) Furukawa, K.; Tomizuka, N; Kamibayashi, A. *Appl. Environ. Microbiol.* **1979**, *38*, 301.

(40) Iwata, Y.; Westlake, W.E.; Gunther, F.A. *Bull. Environ. Contam. Toxicol.* **1973**, *9*, 204.

(41) Alexander, M. *Environ. Sci. Technol.* **1995**, *29*, 2713.

(42) Sawhney, B.L. In *PCBs and the Environment*; Waid, J.S., Ed.; CRC Press: Boca Raton, 1986, Vol. I, pp. 47-64.

(43) Fairbanks, B.C.; O'Connor, G.A.; Smith, S.E. *J. Environ. Qual.* **1987**, *16*, 18.

(44) Wild, S.R.; Berrow, M.L.; Jones, K.C. *Environ. Pollut.* **1991**, *72*, 141.

(45) Swisher, R.D. In *Surfactant Biodegradation*, 2nd ed.; Schick, M.J., Fowkes, F.M., Eds.; Marcel Dekker: New York, 1987, pp. 1-1085.

(46) Larson, R.J. *Environ. Sci. Technol.* **1990**, *24*, 1241.

(47) Hand, V.C.; Williams, G.K. *Environ. Sci. Technol.* **1987**, *21*, 370.

(48) Holysh, M.; Paterson, S.; Mackay, D. *Chemosphere* **1986**, *15*, 3.

(49) Romano, P.; Ranzani, M. *Water Sci. Technol.* **1992**, *26*, 2547.

(50) Cavalli, L.; Gellera, A.; Landone, A. *Environ. Toxicol. Chem.* **1993**, *12*, 1777.

(51) Tabor, C.F.; Barber, L.B. *Environ. Sci. Technol.* **1996**, *30*, 161.

(52) Berna, J.L.; Ferrer, J.; Moreno, A.; Prats, D.; Ruiz Beva, F. *Tens. Surf. Deterg.* **1989**, *26*, 101.

(53) Terzic, S.; Hrsak, D.; Ahel, M. *Water Res.* **1992**, *26*, 585.

(54) Beck, A.J.; Johnston, A.E.; Jones, K.C. *Critic. Rev. Environ. Sci. Technol.* **1993**, *23*, 219.

Chapter 15

Fluorescent Whitening Agents as Molecular Markers for Domestic Wastewater in Recent Sediments of Greifensee, Switzerland

Jean-Marc A. Stoll, Thomas F. Poiger, André F. Lotter, Michael Sturm, and Walter Giger[1]

Swiss Federal Institute for Environmental Science and Technology (EAWAG) and Swiss Federal Institute of Technology Zurich (ETH), CH–8600 Dübendorf, Switzerland

Three different fluorescent whitening agents (FWAs) were examined in sediment cores of Greifensee, a small lake in Switzerland. Two of these FWAs (DAS 1 and DSBP) are currently used in domestic detergents. The third one (BLS) was contained in detergents until some years ago. During sewage treatment, FWAs are only partly eliminated, and, hence, residual amounts reach the aquatic environment. They are partly associated with particles (K_d= 30 - 440 L/kg) and can, therefore, settle to the bottom of surface waters. This behavior and their persistence to biodegradation allow the application of FWAs as molecular markers for domestic waste water. Total DAS 1 and DSBP inventories in the sediments of Greifensee were found to be 18 - 270 mg/m^2 and 7 - 80 mg/m^2, respectively. Inventories of BLS ranged from 0.3 - 11 mg/m^2. With increasing distance from the discharge points, FWA inventories generally decreased. A sediment profile in Greifensee collected by means of a freeze core device shows the input history of FWAs from their first use in the mid 1960s. Because the equilibrium of FWAs in lake sediments between dissolved and particulate fraction lays strongly on the particulate side, remobilization in the sedimentary core is assumed to be negligible. Thus, concentrations in a particular sediment layer can be attributed to the inputs occurring at the time of deposition.

Detergents used for laundry washing and cleaning of surfaces are mixtures of different synthetic chemicals which are used in very large quantities. Worldwide consumption of detergent products was estimated at 31 million tons per year in 1984 (*1*). The most important components, on a weight basis, are surfactants, builders, and bleaching agents. Other components are e.g. enzymes, foam regulators, dyes, and perfumes. To improve the whiteness of textiles, fluorescent whitening agents (FWAs, Figure 1) are added to laundry detergents in relatively small amounts of about 0.15% on a dry weight basis (*2, 3*). Exact quantities of FWA use are unknown, but estimates showed a world wide production of 14,000 t of DAS 1 and 3,000 t of others (predominantly DSBP) in 1989 (*2*). FWAs contained in detergents serve to replace textile FWAs which are photochemically degraded during wearing or are washed out during

[1]Corresponding author

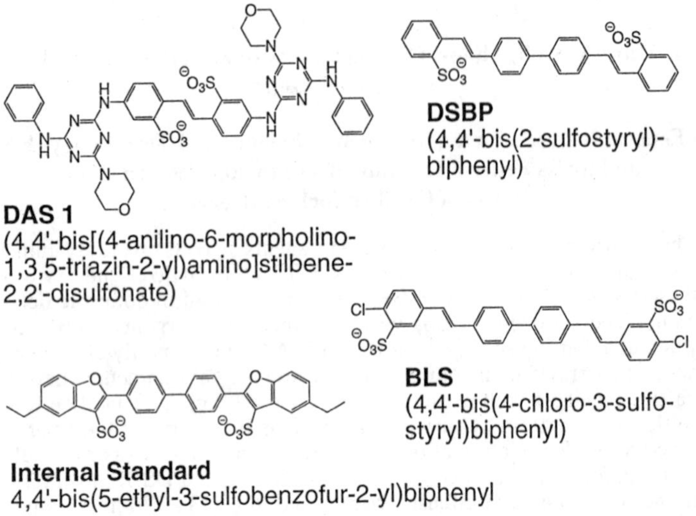

DAS 1
(4,4'-bis[(4-anilino-6-morpholino-
1,3,5-triazin-2-yl)amino]stilbene-
2,2'-disulfonate)

DSBP
(4,4'-bis(2-sulfostyryl)-
biphenyl)

BLS
(4,4'-bis(4-chloro-3-sulfo-
styryl)biphenyl)

Internal Standard
4,4'-bis(5-ethyl-3-sulfobenzofur-2-yl)biphenyl

Figure 1. Structures of the fluorescent whitening agents included in this study. The molecules containing one or two stilbene moieties can occur in two or three isomeric forms, respectively. Only the (E)- and (E,E)- isomers are shown.

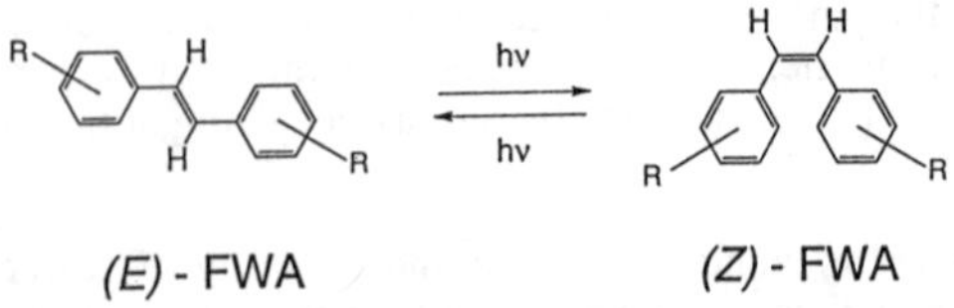

(E) - FWA   *(Z)* - FWA

Figure 2. E-Z-isomerization of stilbene-type FWAs.

the washing process (2). FWAs that are not attached to the fabric during the washing process are discharged with the wash water. In Switzerland, two compounds are currently being used as detergent FWAs: DAS 1 since the 1960s and DSBP since 1972 (4, 5). BLS, the third FWA included in this study, was used in large-scale laundry facilities (e.g. hospitals) until quite recently (6). The internal standard of this study is a fluorescent research compound, that was never used as FWA and, therefore, does not occur in environmental samples. All FWAs included in this study are based on one or two stilbene moieties.

Exposure of dissolved FWAs to sunlight causes reversible E-Z-isomerization of the stilbene moiety (Figure 2, (7)). Hence DAS 1 containing one stilbene moiety occurs in two isomeric forms, herein called (E)-DAS 1 and (Z)-DAS 1. With two stilbene moieties present in FWAs (DSBP and BLS) three isomeric forms are possible, (E,E)-, (E,Z)- and (Z,Z)-FWA. The Z,Z-isomer of DSBP is not formed photochemically. Thus, only amounts of (E,E)- and (E,Z)-DSBP are reported. The three isomers of BLS could not be separated chromatographically, so only total amounts of BLS are reported. FWAs are produced and added to laundry detergents in their fluorescent E- or E,E-forms. Photoisomerization to the corresponding Z-, E,Z- or Z,Z-forms leads to a complete loss of fluorescence (8).

The environmental fate of DAS 1, DSBP, and BLS is summarized in Table I. Concentrations of FWAs in raw influent typically range from 10 to 20 µg/L (6, 9, 10). FWAs are partly retained in municipal waste water treatment plants due to adsorption onto activated sludge yielding sludge concentrations of 10 - 100 mg/kg dry matter.

Table I. Environmental Fate of Fluorescent Whitening Agents

	DAS 1	DSBP	BLS	Reference
Sewage treatment				
- biodegradation	n. o.[a]	n. o.[a]	n. o.[a]	(6, 9, 11)
- photodegradation	n. o.[a]	n. o.[a]	n. o.[a]	(6, 9)
- sorption on sludge	89 %	53 %	98 %	(6, 9)
	41 - 81 %			(10)
- concentration ranges (µg/L)				
- before treatment	7.1 - 11.4	6.9 - 21.3	0.27 - 2.40	(6, 9)
	14 - 57			(10)
- after treatment	2.6 - 4.5	3.3 - 8.9	0.01 - 0.15	(6, 9)
	6 - 27			(10)
- sludge concentrations, dry matter (mg/kg)	55 - 105	27 - 58	3 - 11	(6, 9)
Surface waters				
- biodegradation	n. o.[a]	n. o.[a]	n. o.[a]	(4, 12-14)
- photodegradation, $t_{1/2}$	4.5 h	1.5 h	n. d.[b]	(7)
- sorption (L/kg)				
- K_d (E or E,E-isomer)[c]	444	218	n. d.[b]	(9)
- K_d (Z or E,Z-isomer)[c]	109	32	n. d.[b]	(9)
- concentration ranges (µg/L)				
- rivers in Switzerland	0.005 - 0.9	0.01 - 1.0	< 0.0002	(15)
- rivers in U. S. A.	0.06 - 0.7			(13)
- Greifensee, Switzerland	0.04 - 0.1	0.01 - 0.1	< 0.0002	(15)

[a] n. o. = not observed, [b] n. d. = not determined, [c] in river water and sediment

Neither biological degradation by activated sludge nor photochemical degradation were observed during sewage treatment (6, 9, 11). Both seem to be too slow compared to the residence times in the treatment plants. The concentration of DAS 1 in secondary effluents of four Swiss sewage treatment plants was found to range from 2.6 to 4.5 µg/L, which corresponds to an average elimination of 89% through activated sludge treatment. The corresponding values for DSBP and BLS were 3.3 - 8.9 µg/L (53% eliminated) and 0.01 - 0.15 µg/L (98%), respectively (6, 9). Average concentrations of DAS 1 in thirty-five rivers in the United States have been determined to range from 0.06 µg/L to 0.7 µg/L at locations above and below sewage outfalls, respectively (13). Recent studies in Swiss rivers showed concentrations for DAS 1 and DSBP ranging from 0.005-0.9 µg/L and 0.01-1.0 µg/L, respectively (15). BLS concentrations were too low to be quantified in these studies (< 0.2 ng/L).

In surface waters, FWA concentrations are reduced by photodegradation (7) and sorption/sedimentation (9), whereas biodegradation was not observed over 28 days (4, 12 - 14). Sorption to sediment and little or no biodegradation indicate that FWAs may be present in sediments in detectable quantities.

The objectives of this study were to investigate whether FWAs could serve as molecular markers for domestic wastewater and whether the sedimentary column could be used as a natural archive for the use of detergent chemicals by humans and for changes in loading of natural waters with pollutants. This project is part of a larger research program that includes other components of laundry detergents such as LAS, a widely used synthetic surfactant (16).

Study Site

Greifensee (Figure 3) is a highly eutrophic lake with regular overturn in winter (December - March). It is located 10 km to the east of Zurich, Switzerland, and has an area of 8.5 km², a volume of 0.15 km³, and a maximum depth of 32 m. Water enters the lake mainly through two tributaries (Aa and Aabach) and leaves through the outflowing river Glatt with a mean flow of 4.3 m³/s. The residence time of the water in the lake is 1.1 years. A high population density of 600 inhabitants per km² (total 100,000) in the catchment area (167 km²) produces a significant amount of wastewater (17).

The wastewater is treated by seven sewage treatment plants, five of which deliver their water into the two tributaries, Aa and Aabach. The two others, situated on the east and on the west lakeside, deliver their water directly into the lake, the one on the east side (Uster) being much larger. Summing up, three quarters of the wastewater produced in the catchment area enter the lake from the east side near Uster and one quarter is discharged to the lake through the tributary Aabach, coming from the southeast. The sewage treatment plant located on the western lakeside of Greifensee contributes only a few percent of the total wastewater input. In the catchment area of Greifensee sewage treatment commenced in 1967 - 1971, and direct filtration was added to sewage treatment in 1981 - 1984. These changes resulted in two significant decreases of particle discharge to the lake.

Since the middle of this century anoxic conditions in the hypolimnion have prevailed during summer, preventing higher life forms in contemporary sediments. Consequently, the surficial sediments are not bioturbated. Individual yearly sediment layers (varves) can be distinguished by the changing colors of the sediments (spring: gray, autumn: black), thus, permitting visual dating of sediment cores.

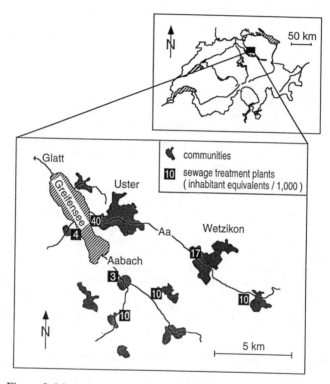

Figure 3. Map of Switzerland and catchment area of Greifensee.

Experimental

Samples. Two methods were used to collect sediments of Greifensee: (i) a conventional gravity coring device and (ii) a recently developed freeze corer (*18*). For the gravity coring method a tube of PVC (polyvinylchloride, diameter: 6 cm) was charged with a weight and allowed to sink into the lake bottom, closed on the top, and pulled out. Onshore the core of 30 - 40 cm was directly divided into 5 cm sections corresponding each to roughly 10 years of sedimentation time. For the freeze core method, a metal sword was lowered 70 cm into the sediments and cooled down to -65 °C with carbon dioxide in methanol. The cold sword froze a layer of 4 cm of sediment around itself and was pulled out. The frozen sediment along the length of the sword was then divided into layers corresponding to periods of two years each. The conventional coring method was used to collect 16 cores from different sites in the lake. The freeze core technique was applied for collecting a sediment core from the center of the lake. The samples were placed in jars, freeze-dried, homogenized with pestle and mortar and stored at 4°C.

Reagents. FWAs, as well as the internal standard (all technical grade), were obtained as sodium salts from Ciba-Geigy AG, with purities of 97% (DAS 1), 90 % (DSBP), and 68 % (BLS). 100 % of the FWAs were present as E- or E,E-isomers. Ammonium acetate (analytical grade) was purchased from Merck ABS AG (Basel, Switzerland). Tetrabutylammonium hydrogen sulfate (TBA) was purchased from Fluka AG (Buchs, Switzerland). All solvents (HPLC grade) were purchased from F.E.R.O.S.A. (Barcelona, Spain) and were used as received.

Liquid Extraction. Samples of 200 mg dry sediment were mixed with 9 ml of 0.03 M TBA in methanol in screw-cap test tubes and briefly shaken, followed by sonication for 30 min. The sample was centrifuged at 2500 rpm for 5 min and decanted into a 50 ml measuring flask. The extraction was repeated twice, and the extracts were combined. The extract was evaporated to dryness under vacuum and re-dissolved in 3 ml of a mixture of water and DMF (dimethylformamide, 1:1). After addition of 30 μl internal standard (1 mg/l 4,4'-bis(5-ethyl-3-sulfobenzofur-2-yl)biphenyl in a mixture of water and DMF (1:1)) the extract was transferred to a vial, centrifuged at 2500 rpm for 5 min and decanted into an autosampler vial. An injection volume of 10 μl was used to analyze the sediment extract by HPLC. To prevent isomerization or photochemical degradation, samples with FWAs in solution were not exposed to UV or blue light.

High-Performance Liquid Chromatography. All analyses were performed using a Hewlett-Packard model 1090L Series HPLC equipped with an autosampler and a ternary solvent delivery system. The FWAs were separated on a reversed phase microbore column (Hypersil ODS, 3 μm, 100 x 2.1 mm i.d. with pre column, Hewlett Packard) operated at room temperature with an eluent flow rate of 0.4 ml/min. The mobile phase solvents were a 2:3 mixture of methanol and acetonitrile (eluent A) and 0.1 M aqueous ammonium acetate buffer of pH 6.5 (eluent B). A 22 min linear gradient from 30% A / 70 % B to 60 % A / 40 % B followed by a 2 min linear gradient to 90 % A / 10 % B was used for analyses. The outlet of the HPLC column was connected to a postcolumn UV irradiation apparatus (Beam Boost, ict AG, Basel, Switzerland) equipped with a UV lamp with a maximum intensity at 254 nm. The eluate was irradiated in a 0.3 mm i.d. x 0.5 m Teflon capillary for 5 seconds (Figure 6). This irradiation time was sufficient to achieve photostationary conditions (*6*), thus transforming a fraction of the non fluorescent Z- and E,Z-isomers to fluorescent E- and E,E-isomers. The irradiated eluate was then monitored with a Hewlett Packard model 1046A fluorescence detector at an excitation wavelength of 350 nm and an emission wavelength of 430 nm. This detection method was adopted from (6,19) and will be published elsewhere. A typical result is shown in Figure 5.

Standard solutions of FWAs were prepared in DMF/water (1:1). Concentration series for external calibration curves were prepared by dilution of the standard solution in 0.27 M TBA in DMF/water (1:1), producing linear correlation from the limits of quantitation (DAS 1: 10 μg/kg dry sediment, DSBP and BLS: 2 μg/kg dry sediment) up to signals corresponding to concentrations of 4 mg/kg dry sediment (DAS 1 and DSBP) and 0.4 mg/kg dry sediment (BLS). Solvent blanks never exceeded a value corresponding to 2 μg FWA / kg dry sediment and were subtracted from the measured signals. Confidence intervals (95 %, n=8) of individual measurements were ± 8 % ((*E*)- and (*Z*)-DAS 1), ± 6 % ((*E*,*E*)-DSBP), ± 15 % ((*E*,*Z*)-DSBP), and ± 10 % (BLS). Recovery rates were 93 - 99 % (DAS 1), 95 - 97 % (DSBP), and 97 - 100 % (BLS), and were taken into consideration for reported results.

Results and Discussion

FWAs as Molecular Markers for Domestic Waste Water. Gravity cores were collected at 16 different locations of Greifensee. Every core was analyzed from the sediment water interface down to a layer where no more FWAs could be found (normally below 20 cm). The concentrations were integrated over the whole core in order to obtain total FWA inventories (in mg FWA/m²) for every core. As illustrated in Figure 6, the sample sites were divided into 3 groups, being under the influence of tributary Aabach (group A), and being under the influence of tributary Aa (group B towards south, group C towards north). Table II shows the FWA inventories, depending on the distance of the sample site from the mouth of the corresponding tributary.

FWA inventories were generally found to decrease with increasing distance from wastewater inputs. To visualize this effect, contours to different DSBP concentrations were manually drawn on Figure 6. A second effect, that can be seen in Table II, is a change in isomer ratios with increasing distance from tributaries. Since (*E*)-DAS 1 and (*E*,*E*)-DSBP have higher sorption coefficients, they sediment faster to the lake bottom than the corresponding (*Z*)-DAS 1 and (*E*,*Z*)-DSBP. Hence (*E*)-DAS 1 and (*E*,*E*)-DSBP are enriched compared to (*Z*)-DAS 1 and (*E*,*Z*)-DSBP in areas close to the tributaries.

Sedimentary Archive and Emission History of FWAs. One freeze core was taken in the middle of the lake. Layers corresponding to periods of two years were analyzed. The so obtained vertical concentration and flux profiles of FWAs in the sediments of Greifensee document the emission history of the FWAs (Figure 7): DAS 1 is first found in sediment layers that were deposited in 1965, DSBP in layers from 1973. BLS is only found in layers dating from the 1980s. These findings match the very limited information which was available from the FWA manufacturing industry. Concerning the general development of FWA concentrations in the sediment, the two expected trend changes in 1967/71 and in 1981/84 can be observed. Laundry detergent use increased continuously during this century, but FWA discharge to the environment, as recorded in the sediment, only increased until 1971 and then more or less stagnated until 1983, after which a decrease in FWA concentrations can be observed. The first change in 1971 corresponds to the beginning of sewage treatment in 1967/71. The second one in 1983 reflects the addition of direct filtration to sewage treatment in 1981/84.

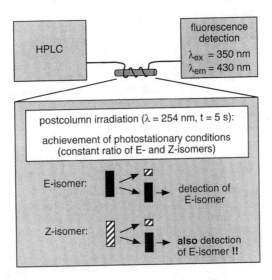

Figure 4. Post-column irradiation and detection of FWAs.

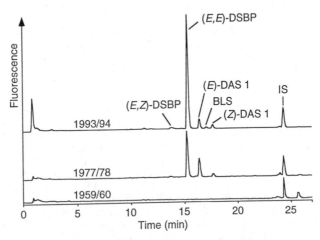

Figure 5. High performance liquid chromatograms of extracts from Greifensee sediments. IS: internal standard. Years given represent time of deposition. Fluorescence: λ(ex) = 350 nm, λ(em) = 430 nm. For chromatographic conditions, see text.

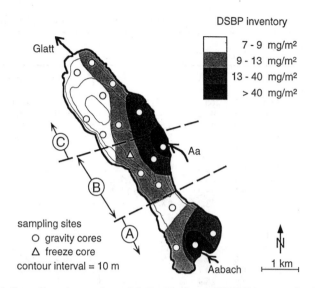

Figure 6. Sampling sites and spatial distribution of DSBP inventories in the sediments of Greifensee. A: Sampling sites under the influence of tributary Aabach, B and C: Sampling sites under the influence of tributary Aa towards south and north, respectively.

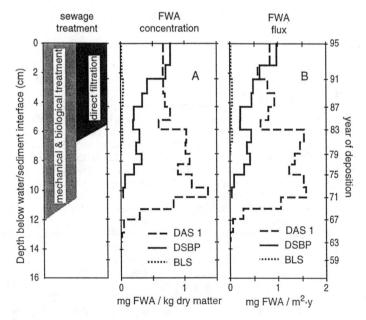

Figure 7. Historic record of FWAs in the sediments of Greifensee measured as concentrations (A) and fluxes (B). The years plotted on the right-side scales mark the borders of analyzed layers.

Table II. FWA Inventories[a] in Sediments of Greifensee, Switzerland

distance from tributary	DAS 1				DSBP				BLS
	(E)	(Z)	total	(E)/(Z)	(E,E)	(E,Z)	total	(E,E)/(E,Z)	total
group A (Aabach)									
0.3 km	243.4	21.9	265.2	11.1	77.0	1.2	78.2	63.1	11.5
0.6 km	35.6	6.6	42.2	5.4	21.7	0.8	22.5	28.2	1.8
0.7 km	14.9	3.3	18.2	4.5	9.9	0.5	10.4	20.1	0.5
1.4 km	19.7	5.6	25.4	3.5	8.0	0.4	8.4	18.4	0.3
group B (Aa towards south)									
0.2 km	157.9[b]	15.0[b]	172.9[b]	10.5	43.7[b]	2.6[b]	46.3[b]	16.7	3.8
0.5 km	40.3	8.1	48.4	5.0	12.2	0.8	13.0	15.1	0.7
1.0 km	28.5	6.7	35.2	4.2	10.3	0.7	11.0	14.8	0.5
1.0 km	27.5	5.7	33.2	4.8	9.6	0.8	10.4	12.8	0.4
group C (Aa towards north)									
0.2 km	157.9[b]	15.0[b]	172.9[b]	10.5	43.7[b]	2.6[b]	46.3[b]	16.7	3.8
1.1 km	49.8	7.6	57.4	6.6	16.0	0.9	16.9	17.3	1.0
1.4 km	27.7	5.6	33.3	4.9	9.6	0.7	10.3	13.8	0.5
1.8 km	24.3	6.4	30.7	3.8	7.5	0.5	8.0	15.2	0.3
2.2 km	21.4	5.5	26.9	3.9	8.6	0.4	9.1	19.2	0.4
2.2 km	26.5	6.6	33.1	4.0	7.6	0.5	8.1	15.8	0.5
2.3 km	20.9	5.9	26.8	3.6	6.9	0.6	7.5	12.1	0.3
2.9 km	18.9	4.8	23.7	4.0	9.0	0.5	9.5	19.0	0.4
3.1 km	21.1	6.1	27.2	3.5	8.3	0.5	8.8	16.3	0.4

[a] Inventories of cores from the sediment/water interface down to a depth where no more FWAs could be detected, unit: mg/m^2
[b] Minimum values, FWAs could still be detected in the lowest layer analyzed (40 cm)

Conclusions

Detergent-derived FWAs could be determined in sediment cores of Greifensee. The known properties of FWAs indicate a conservative behavior in the sediments. Moreover, historic events like the introduction of a new FWA or the improvement of sewage treatment, that can be documented by the measured FWA concentration profiles in the sediment, make processes that would affect the FWA concentrations in the sediment core unlikely. Because FWAs are strongly sorbed to particles, sedimentation is a rapid process compared to the horizontal mixing in Greifensee, and, hence, concentrations in the sediments can provide information about the transport of particles in the lake. Another property that could provide information, is the ratio of E- and Z-isomers of the FWAs, since exposure to sunlight causes isomerization and different isomers show different sorption behavior. For all these reasons we propose the use of FWAs as molecular markers for domestic wastewater.

The investigations and results reported here are part of a comprehensive study on the fate and behavior of FWAs in Greifensee (*15*). Concentrations of FWAs have also been measured in samples from seasonal depth profiles of the water column, from the inflows to the lake, and in particles which were collected in sediment traps deployed in the lake. These data, together with solid/aqueous phase distribution coefficients and photochemical reaction constants determined in the laboratory, will be fed into a mathematical simulation model in order to evaluate processes determining the fate of FWAs in the aquatic environment.

Acknowledgment

We thank the Chemical Industry in Basel, Switzerland for their financial support, Hewlett Packard for the donation of the HPLC equipment within the Rhine Basin Program and Ciba-Geigy for providing the FWA reference chemicals. In addition, we thank R. Berger, A. Lück, J. Maag, R. Reiser, R. Stöckli, and A. Zwyssig for their help during sampling. A. Albrecht, J. Kaschig, and H. Kramer are acknowledged for their assistance during the preparation of this manuscript.

Literature Cited

1) Jakobi, G.; Löhr, A. *Detergents and Textile Washing*; VCH Verlagsgesellschaft mbH: Weinheim, D, 1987.
2) Kramer, J. B. In: *The Handbook of Environmental Chemistry;* Hutzinger, O., Ed.; Anthropogenic Compounds, Detergents; Springer: Berlin, D, 1992; Vol. 3, F; pp 351-366.
3) Anliker, R. In: *Fluorescent Whitening Agents*; Anliker R. and Müller G., Ed.; Environmental Quality and Safety; Georg Thieme Publishers: Stuttgart, D, 1975, Sup. Vol. IV; pp 12-18.
4) Personal communication; Kaschig, J.; Ciba Geigy AG, Switzerland **1996**.
5) Siegrist, A. E.; Eckhardt, C.; Kaschig, J.; Schmidt, E. In: *Ullmann's Encyclopedia of Industrial Chemistry*; VCH Verlagsgesellschaft mbH: Weinheim, D, 1991, Vol. A18, pp 153-176.
6) Poiger, T.; Field, J. A.; Field, T. M.; Giger, W. *Anal. Methods Instrum.* **1993**, *1*, 104-113.
7) Kramer, J. B.; Canonica, S.; Hoigné, J.; Kaschig, J. *Environ. Sci. Technol.* **1996**, *30*, 2227-2234.
8) Gold, H. In: *Fluorescent Whitening Agents*; Anliker R. and Müller G., Ed.; Environmental Quality and Safety; Georg Thieme Publishers: Stuttgart, D, 1975, Sup. Vol. IV; pp 25-46.
9) Poiger, T. *Ph. D. Thesis, ETH Zürich, No. 10832* **1994**.
10) Ganz, C. R.; Liebert, C.; Schulze, J.; Stensby, S. *J. Wat. Pollution Control Federation* **1975**, *47*, 2834-2849.
11) Katayama, M. *Nippon Kasei Gakkaishi* **1989**, *40*, 1025-1028.
12) Dojlido, J. R. *EPA-Report 600/2-79-163* **1979**.
13) Burg, A. W.; Rohovsky, M. W.; Kensler, C. J. *Critical Reviews in Environmental Control* **1977**, *7*, 91-120.
14) Komaki, M.; Kawasaki, M.; Yabe, A. *Sen-I Gakkaishi* **1981**, *37*, 489-495.
15) Stoll, J. M. A. *Ph. D. Thesis, ETH Zürich, in preparation* .
16) Reiser, R.; Toljander, H. O.; Albrecht, A.; Giger, W. In: *Molecular Markers in Environmental Geochemistry*; Eganhouse, R. P., Ed.; ACS Symposium Series, 1997.
17) Liechti, P.; *Der Zustand der Seen in der Schweiz*; Schriftenreihe Umwelt; BUWAL: Bern, CH, 1994; Vol. 237, pp 131-138.
18) Lotter, A. F.; Renberg, I.; Stöckli, R.; Sturm, M. *submitted to Aquatic Sciences* **1997**.
19) Poiger, T.; Field, J. A.; Field, T. M.; Siegrist, H.; Giger, W. *submitted to Water Res.* **1997**.

Chapter 16

Interpretations of Contaminant Sources to San Pedro Shelf Sediments Using Molecular Markers and Principal Component Analysis

Charles R. Phillips[1], M. Indira Venkatesan[2], and Robert Bowen[3]

[1]Science Applications International Corporation, 10260 Campus Point Drive, San Diego, CA 92121
[2]Institute of Geophysics and Planetary Physics, University of California, Los Angeles, CA 90024
[3]Science Applications International Corporation, 165 Dean Knauss Drive, Narragansett, RI 02882

The relative contributions of wastewater discharges and river runoff to measured contaminant concentrations in marine sediments on the San Pedro Shelf, California, were evaluated using molecular markers, principal component analysis (PCA) and soft independent modeling of class analogies (SIMCA). Distributions of sewage markers, linear alkylbenzenes (LABs) and coprostanol+epicoprostanol, were clearly influenced by the wastewater discharges. However, no corresponding patterns in concentrations of chlorinated pesticides, polychlorinated biphenyls (PCBs), or polycyclic aromatic hydrocarbons (PAHs) were evident, with the exception of correlations between concentrations of parent plus alkyl-substituted naphthalenes and both summed LABs and coprostanol+ epicoprostanol. PCA results indicated three station groupings: (1) a wastewater discharge footprint that was associated with distributions of LABs, coprostanol, and naphthalenes; (2) nearshore stations potentially influenced by riverine inputs of petroleum hydrocarbons; and (3) a deep slope and canyon region possibly affected by regional contaminant and natural seep inputs and historical pesticide and hydrocarbon residues. SIMCA analyses did not indicate strong correspondences between chemical signatures for the sewage and river runoff end-members and those for the shelf and slope sediments. The absence of a better fit may be attributable to changes in abundances of sediment contaminants due to weathering or diagenesis, variability in hydrocarbon composition among end-member samples, and the greater relative importance of other, as yet unknown, contaminant sources.

The southern San Pedro Shelf is adjacent to a highly urbanized area of Orange County in southern California that receives contaminant inputs from multiple sources, including approximately 10^9 L/day of treated wastewaters from the County Sanitation Districts of Orange County (CSDOC) ocean outfall. CSDOC routinely monitors the effects of their wastewater discharge. However, evaluations of source contributions have been difficult because the contaminant characteristics for many potential sources are similar, and the analytical methods used previously to analyze wastewaters and sediments are not particularly sensitive. Nevertheless, an understanding of the relative contributions of these sources to sediment contamination is important for wastewater management decisions.

Molecular markers have previously been used to distinguish sewage-derived materials in marine sediments (*1-6*). Applications of molecular markers as tags are based on their enrichment in sewage compared to other source inputs, their strong affinity for particles (analogous to other anthropogenic contaminants), and their resistance to biological degradation (*7*). Multiple suites of markers have also been used to distinguish different contaminant input sources (*8-10*). Similarly, multivariate techniques for data analyses, including PCA, have been applied to evaluate spatial patterns in sediment metals (*11, 12*), petroleum hydrocarbons (*e.g., 13 and references therein*), and organic pollutants (*e.g., 14, 15*). PCA is a particularly effective tool for assessing spatial relationships in compositional patterns from large and complex data sets.

The objective of this study was to evaluate the relative contributions of wastewater discharges to concentrations of anthropogenic contaminants in sediments on the southern San Pedro Shelf. Organochlorine, saturated, and aromatic hydrocarbon concentrations, along with several molecular markers, including LABs, fecal sterols, and selected triterpanes and steranes, were quantified in wastewaters, grease particles (globules) from the outfall, and sediments from the Santa Ana River and Newport Bay (within the immediate drainage basin), as well as offshore sediments, using appropriately sensitive analytical methods. The molecular marker and contaminant data were analyzed using PCA and SIMCA as a descriptive and quantitative approach to improve our ability to resolve the multiple contaminant input sources to the southern San Pedro shelf.

Study Area

The study area is the southern portion of the San Pedro shelf, south of Los Angeles and Long Beach, and bounded by the Newport and San Gabriel submarine canyons that cross the shelf to the east and west of the CSDOC outfall (Figure 1). The narrow shelf consists primarily of soft substrate comprising relict sediments with sedimentary materials from storm runoff, coastal erosion, and aeolian sources (*16, 17*). Sediment grain size generally decreases with increasing water depth, grading from medium sands inshore to fine sands and clays at the upper slope (*18*). Sediment transport on the shelf is influenced by wave-generated currents, primarily during winter storms (*19*). Nontidal currents in the bottom layers are northwesterly during most of the year.

The CSDOC outfall terminates 8 km from shore with a multi-port diffuser at a depth of 60 m. Continuous discharge of treated wastewaters results in a suspended solids mass emission of 42,300 kg/day. Discharges from the Santa Ana River are seasonally variable, but average 72 x 10^6 L/day, and represent average emissions of 200,000 kg/day of suspended solids and 5 g/day each of total DDTs and PCBs (*20*). Newport Bay is a small coastal embayment that receives urban, agricultural, and industrial wastes. Sediments

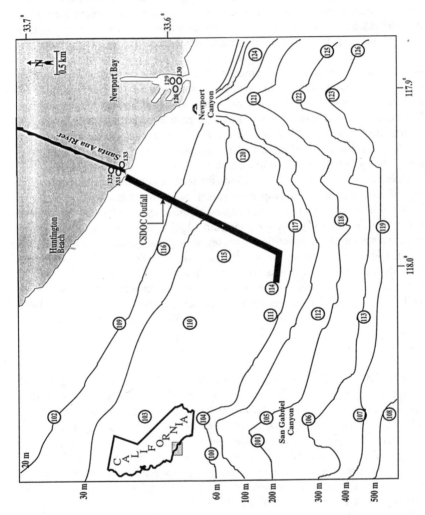

Figure 1. Sediment sampling locations on the San Pedro Shelf and slope, Santa Ana River, and Newport Bay.

within the bay historically have contained elevated concentrations of chlorinated pesticides, organotins, and metals (21). We are not aware of any information regarding export rates to the ocean of bay sediments or associated contaminants. Additional potential sources of contaminants to the region include discharges from commercial vessels, oil and gas production operations at offshore platforms, dredged material disposal operations at a site in Newport Canyon, historical dumping of DDT at sites within San Pedro Channel, and atmospheric deposition (*21, 22*).

Methods and Materials

Sample Collection. Surficial sediments were collected from 27 locations on the San Pedro Shelf, including the San Gabriel and Newport submarine canyons, in depths from 20-500 m (Figure 1). Sediments were collected using a 0.1 m^2, Kynar-coated, Van-Veen grab sampler that was rinsed with filtered seawater, methanol, and hexane between successive grabs. Samples were removed from the surface 2 cm of undisturbed grabs using a Teflon-coated scoop with 2 cm-high sides that provided a guide for achieving a consistent sample thickness. Recent sediments were also collected at three locations within Newport Bay, using the Van-Veen grab, and at three locations near the mouth of the Santa Ana River using a hand-held scoop. Samples for determinations of organic contaminants and markers (approximately 500 g) were placed in pre-cleaned glass jars with Teflon-lined lids and frozen. Separate samples (approximately 10-20 g) taken for CHN (carbon, hydrogen, nitrogen) determinations were placed in precleaned 40 mL glass vials. Globules were obtained from sediment grab (Van Veen) samples collected near the terminus of the outfall diffuser in January 1995 following periods of high flow associated with stormwater runoff. Samples of the final wastewater effluent represented three 24-hour composites collected over a nine-day period (October 1/2, 4/5, 8/9, 1994).

Sample Analyses. Sediment, effluent, and globule samples were extracted, extracts were fractionated, and final fractions were purified according to procedures described previously by Venkatesan (*23-25*). Instrumentation used for identification and quantitation of target compounds are identified below. Detailed descriptions are provided elsewhere (*ref. 24, 25: sterols; ref. 23, 24: n-alkanes, terpanes, steranes, PAHs, PCBs, and chlorinated pesticides; ref. 26: LABs*).

Saturated Fraction. Normal (C_{12}-C_{36}) and branched alkanes were quantitated by gas chromatography (GC) with a flame ionization detector. Triterpanes (13 compounds) and steranes (16 compounds) were quantitated by gas chromatography/mass spectrometry (GC/MS) using extracted ion current profiles (m/z 191 and 217/218/245, respectively). Linear alkylbenzenes (C_{11}-C_{14}; 20 compounds) were quantitated by GC/MS from extracted ion current profiles (*26*). Compound concentrations were corrected for losses during workup based on recoveries of a spiked LAB calibration standard solution.

Aromatic Hydrocarbon Fraction. Aromatic hydrocarbon fractions were analyzed by GC/MS in full scan mode, and the extracted ion current profiles were used for identification and quantitation of PAHs (43 compounds). Sample fractions were initially screened by comparison with a standard PAH mixture containing all of the target compounds. Ratios of response factors for the parent and alkylated homologs of naphthalene in the standard were used to extrapolate response factors for alkylated homologs of other PAHs. Final concentrations were corrected for matrix spike compound recoveries. Chlorinated pesticides (16 compounds including DDTs, hexachlorobenzene, aldrin, dieldrin, endrin, mirex, heptachlor, heptachlor epoxide, alpha-chlordane, trans-nonachlor, and lindane) and PCBs (19 congeners) were quantitated on a dual column (DB-5 and DB-1701 connected with an effluent Y splitter) GC with an electron capture detector.

Polar Fraction. Sterols (10 compounds including coprostanol, epicoprostanol, coprostanone, cholesterol, cholestanol, brassicasterol, campesterol, stigmasterol, β-sitosterol, and dinosterol) were derivatized to silylethers, quantitated by GC, and confirmed by GC/MS.

Carbon and Nitrogen Analyses. Separate sediment samples were analyzed for total carbon and nitrogen using a CHN analyzer. Organic carbon was determined by acidifying sample aliquots, to remove inorganic carbon, and reanalyzing the sample.

Quality Assurance Summary. Procedure blanks, matrix spikes, and reference materials (where available) were analyzed with each batch of samples. Target analytes were not detected in procedure blanks, with the exceptions of concentrations near the respective limits of quantitation of a single PCB congener and 2,4'-DDE in one or two blanks. Measured concentrations of 15 PCB congeners in a reference material (National Institute of Standards and Technology, SRM # 1941) were within 86-161% of certified values, and coefficients of variation (n=4) were from 6-30%. Concentrations of four of six chlorinated pesticides in SRM # 1941 were 72-119% of the certified values. Higher values for two other pesticides, dieldrin and heptachlor epoxide, were likely due to matrix interferences. Coefficients of variation for replicate measurements (n=4) of individual pesticides in the SRM ranged from 6-34%. Concentrations of 18 PAHs in SRM #1941 were within 58-129% of certified values, and coefficients of variation (n=4) were 5-56%, although concentrations of all but four compounds were within 80-110% of certified values and coefficients of variation for all but five compounds were ≤ 20%. Measured carbon concentrations in SRM #1572 were 99-100% of the certified values. No certified reference materials are available for sterols, LABs, terpanes, or steranes. Analytical accuracy of sterol determinations was evaluated from matrix spike recoveries. Mean recoveries of individual sterols ranged from 72-99%, with coefficients of variation for replicate measurements (n=4) ≤ 10%. Matrix spike recovery was not assessed for terpanes and sterenes because no highly concentrated standard spike solution was available.

Data Analysis. Principal component and SIMCA analysis of the compound concentration data for source and sediment samples was performed using SIRIUS software (Pattern Recognition Systems, A/P Breton, Norway). Multiple PCA runs were performed using different subsets (classes) of compounds. Variables used for the two PCA classes described here are defined in Table I. The original data set contained a total of 181 analytes. Initial analyses of these data showed that a large number of compounds (e.g., steranes and triterpanes) had little variability among sediment stations and/or covaried within their compound class. Because these would provide little additional information above that found in the class, they were removed from the working data set. Other selected compounds were summed into composite variables as shown in Table I. While the preselection of compounds used for this PCA could have a significant effect, care was taken in the selection of compounds and compound classes to avoid biasing the results.

Concentrations of compounds, and compound classes, within the group of sediment samples, effluents and globule samples varied over a large range. Because the PCA technique is a least squares method, samples with higher concentrations will significantly influence the results. To prevent this, each sample was normalized using a technique called mid-range normalization. This method has the advantage of eliminating spurious correlations due to the effects of closure (27). Analyte concentration data were log-transformed and then standardized to unit variance by subtracting the mean and dividing by the standard deviation (28, 29). This allows each of the variables to contribute evenly in the PCA (30).

Additional composite variables for petroleum hydrocarbons, sterols, and LABs were evaluated as source and degradation indicators. These included summed C_{10} - C_{36} n-alkanes ($\sum$Alk), ratios of odd to even n-alkanes within the C_{15} - C_{34} range (odd:even), and summed C_{12} - C_{19} alkanes (lower molecular weight alkanes), as described by Steinhauer and Boehm (31) and Tissot and Welte (32). Also, ratios of external (6-C_{12} + 5-C_{12}) to internal (4-C_{12} + 3-C_{12} + 2-C_{12}) phenyldodecane isomers (I/E) were evaluated as an indicator of LAB degradation (33).

Table I. Variables Used for PCA Analyses, Classes 0 and SS

Alkanes: n-C_{12}, n-C_{17}, n-C_{18}, n-C_{31}, n-C_{32}; pristane (pris); phytane (phyt).

PAHs: summed (C_0-C_4) naphthalenes (Naph); summed (C_0-C_3) fluorenes (Fluo); summed (C_0-C_4) phenanthrene/anthracene (Phen); summed (C_0-C_3) dibenzothiophenes (DBT); sum of fluoranthene, pyrene, benzo(a)anthracene, chrysene, benzo(b+k) fluoranthenes, benzo(a)pyrene, benzo(e)pyrene, and perylene (PAH45); sum of 43 PAHs, including alkylated homologs (Tpah).

LABs: total C_{11}-LABs (LAB11); total C_{12}-LABs (LAB12); total C_{13}-LABs (LAB13); total C_{14}-LABs (LAB14); total LABs (TotLAB).

PCBs: sum of 19 PCB congeners (totPCB); sum of PCB congeners # 77, 105, 118, and 126 (coplPCB).

Pesticides: sum of 2,4'- and 4,4'- isomers of DDT, DDD, and DDE (totDDT); sum of 16 pesticides, including DDTs (totpest); hexachlorobenzene (HCB);α-chlordane (alpha-ch).

Sterols: coprostanol (Coprnol); epicoprostanol (Epicopr); coprostanone (Coprone); dinosterol (Dinoster).

Following PCA, a SIMCA analysis was performed to partition the relative contributions from source classes to the chemical signatures of individual stations. SIMCA is a technique of classification by means of disjoint cross-validated principal component models (34). Sediment samples were assigned to one of three classes based upon their geographical associations and results of the preliminary PCA: class SR consisted of river/bay stations 128-133 and all other stations were grouped as class S0. Effluent samples were assigned to class EF. These classes were standardized, and a PCA was performed on each. The method of cross validation was employed to indicate a stopping point in the accumulation of components. The class of sediment samples, S0, was fit to each of these 'source classes', SR and EF, and a measure of the class membership was obtained for the fit of each sample in class S0 to each of the two source classes.

Results and Discussion

Chemical Composition of Source Materials: Effluent and Globules. Concentrations of LABs and fecal sterols in CSDOC effluent and globules (Table II) were comparable to

Table II. Concentrations of Organic Contaminants and Sewage Markers in Source Materials: Effluent, Globules, and Santa Ana River and Newport Bay Sediments. Effluent concentrations normalized to average suspended solids concentrations are shown in parentheses.

	$\sum$DDT	$\sum$PCB	$\sum$Alk	$\sum$PAH	$\sum$LAB	Cop+ Ecop
Effluent (μg/L; n = 3)	0.017 ± 0.004	0.079 ± 0.009	25 ± 4.6	2.0 ± 0.1	8.2 ± 1.8	270 ± 28
(Effluent, μg/g)	(0.25 - 0.48)	(1.4 - 1.8)	(400 - 610)	(40 - 45)	(150 - 230)	(4800 - 6200)
Globules (μg/g, n = 3)	0.53 ± 0.26	10 ± 3.7	860 ± 240	870 ± 230	29 ± 9.6	990 ± 330
Newport Bay (μg/g; n = 3)	0.040 ± 0.036	0.022 ± 0.012	2.4 ± 1.4	1.8 ± 0.89	0.018 ± 0.012	0.56 ± 0.43
Santa Ana R. (μg/g; n=3)	0.006 ± 0.002	0.010 ± 0.002	5.0 ± 5.8	1.7 ± 1.8	0.010 ± 0.007	0.17 ± 0.15

those reported previously in other Southern California wastewater and sludge matrices (*1, 6, 7, 35*). $\sum$LABs in effluent, normalized to average effluent solids concentrations (47 mg/L) and expressed on a dry weight basis (mean: 175 μg/g; range:150-230 μg/g), were severalfold higher than those in globules (mean: 28.9 μg/g; range: 15.5-37.6 μg/g). Coprostanol concentrations (227-295 μg/L) represented 50-53% of the total sterols (i.e., summed concentration of 10 compounds), whereas epicoprostanol was not detected in the effluent. Coprostanol+epicoprostanol (638-1435 $\mu g/g$) were 59-63% of the total sterol concentration in globules.

The relative abundances of individual saturated and aromatic hydrocarbons in the wastewater effluent and globules were generally similar and characteristic of petroleum-derived materials, as indicated by: (1) homologous series of normal alkanes with low odd:even-carbon ratios (~1.2) and high proportions (34-37%) of the n-C_{12} - n-C_{19} to $\sum$Alk; (2) parent plus C_1- through C_4-substituted naphthalenes comprised a large proportion (68-74%) of the $\sum$PAH concentration; and (3) abundances of alkylated homologs typically exceeded those of the parent PAH compounds (Figure 2). Similar hydrocarbon compositions were reported for effluents from the Los Angeles County Sanitation Districts (*35, 36*) and San Diego (Point Loma) Treatment Plant (*37*).

Concentrations of individual and summed chlorinated hydrocarbons in the effluent were low (<0.1 µg/L; Table II). However, $\sum$DDT and $\sum$PCB concentrations, expressed on a dry weight/particle basis, were one to two orders of magnitude higher than average concentrations in Newport Bay, Santa Ana River, and San Pedro Shelf sediments.

Newport Bay and Santa Ana River Sediments. Comparatively low concentrations of sewage markers in Newport Bay and Santa Ana River sediments indicated only minor inputs of domestic waste (Table II). Total LAB concentrations ranged from 0.002-0.031 $\mu g/g$ and from 0.004-0.021 $\mu g/g$, respectively. Coprostanol concentrations in the Santa Ana River sediments ranged from 0.067-0.38 $\mu g/g$, with only trace amounts of epico-prostanol. Coprostanol+epicoprostanol concentrations were nondetectable at one site (station 130), and ranged from 0.13-1.0 $\mu g/g$ at the other sites in Newport Bay. Sources of detectable but low LAB concentrations are unknown.

The saturated and aromatic hydrocarbon compositions of Newport Bay and Santa Ana River sediments were generally similar to those described from other urbanized watersheds (*e.g., 38*). At two of the three sites within the Santa Ana River/flood channel (stations 132 and 133), $\sum$Alk concentrations were 0.36-1.0 $\mu g/g$, with major contributions from the plant wax-derived n-C_{29} and n-C_{31}. Total alkane concentrations were slightly higher (3.3-3.5 $\mu g/g$) at the three sites (stations 128-130) within Newport Bay. Total PAH concentrations at these sites ranged from 0.14-4.2 $\mu g/g$, and unsubstituted fluoranthene and pyrene were the dominant PAHs, suggesting primarily combustion sources (*e.g., ref 38*). At station 131 (Talbert Marsh), adjacent to the Santa Ana River, the $\sum$Alk concentration (13.3 $\mu g/g$) was an order of magnitude higher than those at the other two sites, with only a slight odd-carbon preference within the n-C_{22} to n-C_{32} range (odd/even: 1.2) and a low pristane/phytane ratio (0.59). PAH concentrations also were one order of magnitude higher than those at the other two Santa Ana River sites, and characterized by relatively high proportions of alkylated phenanthrenes (C_1- through C_4-), dibenzothiophenes, and fluoranthenes/pyrenes (C_2- through C_4-), indicative of the presence of petroleum. Average $\sum$DDT and $\sum$PCB concentrations in all Newport Bay and Santa Ana River sediments were <0.1 $\mu g/g$ (Table II).

None of the contaminant or marker compounds or compound classes were unique to a particular matrix or conspicuously absent, with the exception that epicoprostanol and dinosterol were not present in the effluent.

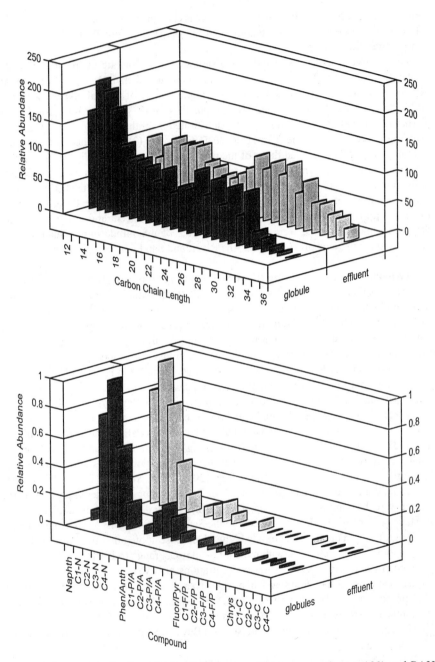

Figure 2. Abundances of n -alkanes (relative to n-C$_{27}$ concentrations x 100) and PAH alkyl homog distributions for CSDOC effluent and globules.

Shelf, Slope, and Canyon Sediments. Organic carbon (OC) concentrations in the shelf, slope, and canyon sediments ranged from 0.2-2.5%, and concentrations increased with water depth between 20 and 400 m (Table III). Sediment OC concentrations in the submarine canyons were elevated, and concentrations near the CSDOC outfall (station 114) were approximately twofold higher than those in adjacent areas of the shelf.

Sediments near the CSDOC outfall were obviously influenced by the wastewater discharge, as indicated by elevated $\sum$LABs and coprostanol+epicoprostanol (Table III). Concentrations of $\sum$LABs exhibited order of magnitude differences within the area, with highest concentrations near the outfall and, to a lesser extent, within the submarine canyons (Figure 3). Spatial patterns in coprostanol+epicoprostanol concentrations generally were similar to those for $\sum$LABs. The highest concentrations occurred near the outfall and decreased by one to two orders of magnitude at the 400-500 m sites. No sewage signature was evident at the 20 m sites; $\sum$LAB and coprostanol+epicoprostanol concentrations were <20 ng/g and 82-155 ng/g, respectively.

Other than patterns in the sewage markers, spatial distributions of organic contaminants did not exhibit large gradients indicative of significant point and non-point source inputs. Concentrations of $\sum$DDT and $\sum$PCB covaried (r^2=0.70), and were well correlated (r^2=0.6-0.7) with sediment OC concentrations (Figure 4). Thus, $\sum$DDT and $\sum$PCB concentrations generally increased between 20-400 m, whereas OC-normalized concentrations did not exhibit significant spatial gradients or trends. Absolute concentrations for summed DDTs and PCBs generally were consistent with values reported by Thompson et al. (*39*) for slope and deep canyon sediments offshore from the CSDOC outfall.

Total PAH and alkane concentrations increased between 20-400 m, similar to depth trends in sediment OC concentrations. Naphthalenes were high near the outfall and in deeper portions of San Gabriel Canyon, as well as portions of the shelf and into Newport Canyon, downcoast from the outfall. Total naphthalene concentrations were moderately correlated with $\sum$LABs (r^2=0.78; Figure 4) and with coprostanol+epicoprostanol (r^2=0.78). The elevated naphthalene and fluorene concentrations, as well as alkyl homolog distributions, suggested relatively larger contributions of petroleum-derived PAHs near the outfall and in Newport Canyon in comparison to abundances of combustion PAHs in sediments from other parts of the Shelf. Slightly higher proportions of low molecular weight alkanes and PAHs also were apparent at nearshore 20-m sites (e.g., station 116), which potentially were attributable to exports of petroleum-contaminated sediments from Talbert Marsh. However, operational discharges from an offshore oil and gas platform upcoast from these sites (*e.g., ref. 40*) also could have contributed. In contrast, concentrations of individual steranes and terpanes covaried, and no spatial patterns in component ratios were evident. The pentacyclic triterpane $17\alpha(H),18\alpha(H),21\beta(H)$-28,30-bisnorhopane (BNH), considered a diagnostic marker for California oils and shales (41), was present at all sites, and concentrations were highly correlated with those of other triterpanes (Table III).

The background hydrocarbon signal represents a probable mixture of combustion- and stormwater-derived high molecular weight PAHs, organochlorines from historical sediment reservoirs, and petroleum hydrocarbons from natural seeps and chronic, low-level discharges. These hydrocarbons, which generally have strong affinities for fine-grained particles, appear to have been redistributed and deposited by prevailing sediment transport

Table III. Organic Carbon (%) and Selected Markers and Contaminant Concentrations (ng/g) in San Pedro Shelf and Slope Sediments

Station	Depth (m)	OC (%)	ΣLAB	I/E	Cop + ecop	Cop+ ecop /Dino	OI/C30 -Hop	BNH /C30- Hop	ΣALK	LMW	odd/ even	ΣPAH	ΣPAH /peryl	4,5 ring	ΣN	ΣF	ΣP/A	ΣPCB	ΣDDT	HCB
100	60	0.3	35.7	1.8	801	2.4	0.25	0.50	356	14	2.7	119	7.4	80	0.0	6.5	24.0	5.67	6.46	0.37
101	200	0.6	49.8	1.7	500	0.71	0.27	0.65	643	17	3.8	251	12	126	0.0	10.4	57.5	10.3	14.2	0.57
102	20	0.2	4.9	8.6	155	0.22	0.27	0.52	268	19	4.8	80	18	33	0.0	10.4	25.5	5.21	3.37	0.21
103	40	0.2	16.6	2.2	296	0.90	0.22	0.40	309	19	2.2	234	30	23	0.0	8.3	23.0	3.42	3.11	0.04
104	60	0.5	60.4	1.8	640	2.1	0.25	0.55	364	11	3.4	147	14	72	0.0	11.4	39.5	6.81	8.57	0.30
105	200	1.6	457	1.6	1846	1.2	0.23	0.57	3687	125	2.4	1565	29	406	70.7	92.6	298	40.2	47.3	0.86
106	300	2.1	157	1.6	892	0.86	0.28	0.57	2672	89	5.4	1590	21	533	16.9	52.0	596	35.0	60.8	1.00
107	400	2.4	83.0	1.6	471	0.36	0.22	0.57	4811	88	2.7	1162	21	526	44.0	20.4	227	34.5	29.8	0.96
108	500	1.7	48.1	1.4	107	0.18	0.24	0.55	1549	53	4.6	325	16	162	12.6	2.8	83.8	14.3	28.2	1.17
109	20	0.2	6.0	1.8	108	0.18	0.26	0.50	311	20	3.7	78	19	51	0.0	2.7	21.9	7.18	3.92	0.24
110	40	0.7	98.4	1.8	1559	3.6	0.23	0.52	497	27	3.8	200	19	114	0.0	3.4	61.9	11.0	9.57	0.36
111	60	0.3	156	1.8	503	2.6	0.25	0.51	410	44	7.1	307	23	97	13.2	5.3	86.2	10.2	4.25	0.33
112	200	0.9	132	1.8	797	0.93	0.25	0.52	1002	50	9.3	329	13	152	1.4	1.2	43.1	29.7	21.9	0.35
113	300	1.4	119	1.7	366	1.1	0.26	0.48	1665	46	4.2	348	15	180	4.0	5.0	69.0	21.6	24.8	0.55
114	60	1.1	584	1.4	8729	19	0.29	0.58	1552	179	1.6	851	72	211	177	82.3	164	17.6	8.58	1.03
115	40	0.4	80.3	1.8	1896	5.1	0.21	0.56	457	26	17	180	22	117	7.2	0.0	40.0	15.2	9.52	0.43
116	20	0.3	19.8	1.3	82	0.18	0.24	0.65	857	20	1.9	229	25	107	19.6	6.5	50.8	9.07	4.74	0.21
117	60	0.4	84.8	1.4	1440	3.3	0.21	0.56	450	37	3.2	108	17	34	0.0	6.6	49.5	5.84	3.13	0.44
118	200	0.7	87.6	1.7	867	1.1	0.24	0.65	1060	38	4.6	333	18	116	9.3	10.5	76.2	11.0	6.70	0.30
119	400	1.3	55.7	1.1	108	0.20	0.25	0.66	1583	43	3.6	259	12	148	0.5	1.5	57.7	14.2	11.5	0.44
120	40	0.3	25.0	1.4	392	1.4	0.27	0.54	652	18	4.0	55	22	76	5.0	0.4	33.5	10.0	4.09	0.25
121	200	1.6	168	1.6	2531	0.84	0.27	0.66	3117	61	5.5	974	18	325	34.5	19.6	190	31.6	18.5	0.42
122	300	1.9	286	1.7	862	0.55	0.24	0.53	2721	120	6.9	1069	21	348	33.4	32.5	265	27.3	21.4	6.28
123	400	1.5	207	2.1	492	0.43	0.23	0.64	3315	70	3.7	1029	22	376	24.3	57.5	216	28.1	21.2	1.45
124	60	0.9	32.8	1.8	567	0.38	0.37	0.94	1534	64	3.6	727	14	245	10.3	4.4	121	15.8	14.3	0.55
125	300	1.4	110	2.3	847	0.65	0.18	0.53	3025	54	5.9	668	22	234	5.8	16.3	178	16.3	13.5	0.64
126	400	1.8	166	2.2	495	0.34	0.25	0.63	3387	79	4.2	724	21	247	17.1	30.0	155	21.5	22.8	0.62

OC = total organic carbon; ΣLAB = sum of 20 C_{11} - C_{14} LABs; I/E = [6 - C_{12} + 5 - C_{12}]/[4 - C_{12} + 3 - C_{12} + 2 - C_{12}] LABs;

Cop + ecop = coprostanol + epicoprostanol; Cop + ecop/Dino = coprostanol + epicoprostanol/dinosterol;

OI/C30-Hop = Oleanane/17α(H),21β(H) - Hopane; BNH/C30-Hop = 17α(H),18α(H),21β(H) - 28,30 - bisnorhopane/17α(H),21β(H) - Hopane;

ΣALK = Sum of C_{10} - C_{36} n-alkanes; odd/even = $C_{15,17,19,21,23,25,27,29,31,33}$/$C_{16,18,20,22,24,26,28,30,32,34}$ n-alkanes;

ΣPAH = Sum of 43 compounds; ΣPAH/peryl = ΣPAH/perylene; 4,5 ring = sum of fluoranthene, pyrene, benz(a)anthracene, chrysene, benzo(b+k)fluoranthenes, benzo(a)pyrene, benzo(e)pyrene, and perylene;

ΣN = C_0 - C_4 naphthalenes; ΣF = C_0 - C_3 fluorenes; ΣP/A = phenanthrene, anthracene, and C_1 - C_4 phenanthrene/anthracene;

ΣPCB = sum of 19 congeners; ΣDDT = sum of 2,4- and 4,4- DDT,DDD,DDE;

ΣChlor = sum of α-chlordane and trans-nonachlor; HCB = hexachlorobenzene.

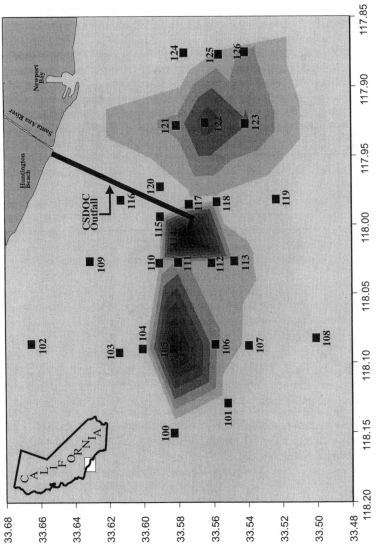

Figure 3. Total LAB concentrations (ng/g) in San Pedro Shelf and slope sediments.

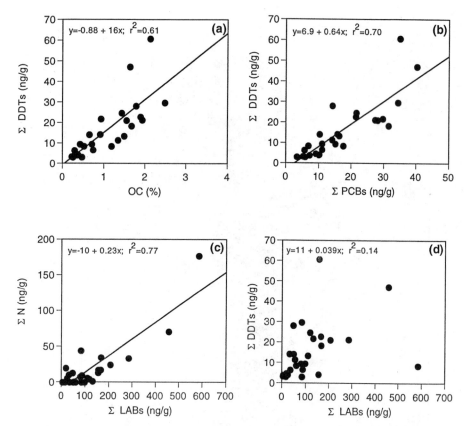

Figure 4. Linear regressions for (a) ΣDDT vs Organic Carbon (OC); (b) ΣDDT vs ΣPCB; (c) Total Naphthalenes (ΣN) vs ΣLAB; and (d) ΣDDT vs ΣLAB concentrations in shelf, slope, and canyon sediments (n=27).

processes. The absence of strong point-source gradients is consistent with sediment metal patterns (*42*), and this spatial uniformity makes interpretation and quantitative evaluations of relative contributions from multiple contaminant inputs difficult. For this reason, PCA analyses of the data are especially helpful for distinguishing and quantifying these contributions.

PCA Results. Multiple PCA runs were performed using different subsets of variables to partition chemical signatures at individual stations. Because the results were largely consistent, only selected PCA classes are discussed below.

Class 0 consisted of 28 variables (Table I) for all samples, including effluents and globules. The first two principal components combined explained 66% of the total variance. Globules and effluent separated from sediment samples along PC1 and PC2, respectively. Sediment sample scores formed a cluster aligned along PC2, with the near-outfall stations (114, 111, and 117) and Santa Ana River/Newport Bay stations (128-133) projected at opposite ends (Figure 5a). Variable loadings on PC1 and PC2 (Figure 5b) indicated that LABs and, to a lesser extent, hexachlorobenzene (HCB), naphthalenes, and fluorenes, were responsible for separation of the effluent samples, as well as distinctions between the near-outfall station group from all other stations. None of the sediments, including those near the outfall, showed close affinities with the globules. The remaining shelf and canyon stations were not clearly differentiated, which reflects the general absence of large differences in relative abundances of contaminants. A second PCA (Class SS) evaluated the same variables used for Class 0, but excluded the effluent and globule data to emphasize differences between the shelf sediment samples. The first two principal components explained 49% of the total variance. Similar to the Class 0 PCA, sample scores projected on PC1 and PC2 showed a station group corresponding to the wastewater footprint (stations 114 and 117), in addition to a river/bay discharge footprint (102, 103, and 109) and a deep canyon and slope group (106-108, 112, 113, 119, 122, 123, and 126). Summed LABs, coprostanol, and epicoprostanol, corresponding to the wastewater footprint, were negatively loaded on PC1 (25% of the variance). The river/bay discharge footprint was associated with positive loadings from petroleum hydrocarbons, including dibenzothiophenes, pristane and phytane, and some medium molecular weight alkanes, as well as dinosterol. The correspondence between positive scores for the nearshore and Santa Ana River sites with loadings for petroleum hydrocarbons suggested apparent influences by river/bay discharges, consistent with the strong petrogenic signature in Talbert Marsh sediments. The deep slope and canyon station group was associated with positive loadings on PC2 for chlorinated pesticides, PCBs, and higher molecular weight PAHs. The remaining stations did not appear to contribute strongly to the principal components.

SIMCA Results. Memberships of S0 samples/sites to the sewage (EF) and river runoff (SR) classes ranged from approximately 2-5% and from 13-28%, respectively. While memberships of the shelf sediment samples in class EF were low and did not indicate appreciable spatial gradients, the strongest fits (3-5%) corresponded to stations between 40-200 m up and downcoast from the outfall. Class EF membership values were relatively lower (<3%) at the shallow, nearshore sites and at depths greater than 300 m, with the

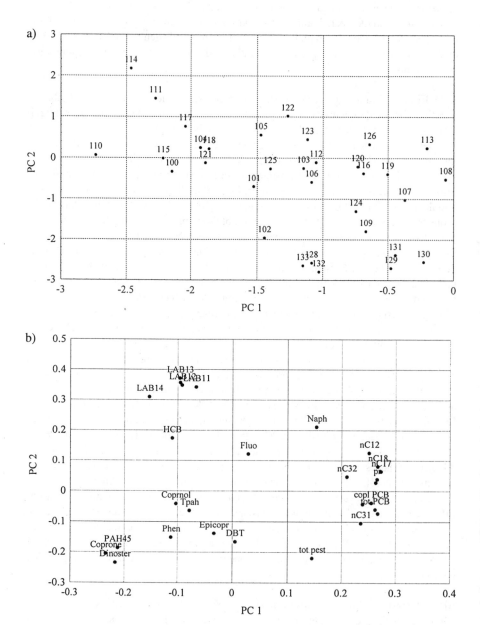

Figure 5. PCA sample scores (5a) and variable loadings (5b) for PC1 vs PC2 for the class 0 sample set.

exception of the higher value for station 122 in Newport Canyon. Significant differences in membership strengths were not apparent for sites east and west of the outfall, suggesting comparable influences from wastewater discharges. The class SR memberships were highest at station 124, east of Newport Canyon and, to a lesser extent, at the other 20-40 m stations, but then decreased with depth and distance off shore.

Cumulative memberships of the class S0 sites to classes EF and SR described by SIMCA typically were <30%. Class membership values do not represent the proportional contribution from specific source materials; rather, the values are interpreted as the fit or signature in the sediments matching that of the source class. Low EF and SR membership values for the class S0 sites were not surprising given the absence of strong gradients related to wastewater and river/bay discharges. Further, the large, unexplained variance associated with PCA and SIMCA is attributable to several factors, including (1) possible variability in the temporal period represented by the surficial sediment layers analyzed during this study, (2) variability in compositions of effluents and river/bay sediment end members, (3) biodegradative changes in chemical signatures, and (4) significant contributions from sources other than those analyzed (e.g., atmospheric and/or historical), Changes over the past three decades in the composition of Southern California wastewaters have been well-documented (*e.g., 43*). The extent to which the surface sediments have integrated these changes, and their relationship to effluents characterized for this study can not be determined. However, the relatively small range in most LAB-I/E values (typically 1.3 - 1.8, although five values were from 2.0 - 8.6), as well as in ratios of selected terpanes, in shelf and slope sediments (Table III) suggests that the degradative state of effluent particles incorporated in bottom sediments were generally uniform. For comparison, Takada and Ishiwatari (*33*) estimated that anaerobic sediments from Tokyo Bay, with mean I/E values of 2.0, had been deposited within a period exceeding one year. Because San Pedro Shelf sediments are aerobic, and LAB degradation rates would be higher than those in anaerobic sediments, it is likely that the effluent particles had been deposited within a relatively shorter period.

These multivariate approaches also assume that the chemical characteristics of end members will be preserved in the surficial shelf and slope sediments. This assumption is appropriate for the more stable sewage markers and chlorinated and high molecular weight hydrocarbons. However, naphthalenes, which are important constituents of wastewaters, are preferentially lost from sediments due to solubilization and microbial degradation (*36*). Thus, sewage imprints on sediment PAHs should be evident only in areas influenced by recently deposited effluent particles. Elevated naphthalene concentrations within San Gabriel canyon suggest rapid transport and deposition of sewage particles, which was also consistent with the relatively low and spatially uniform LAB-I/E ratios up- and down-coast from the outfall. Similarly, the strength of the river discharge signature in nearshore sediments likely will vary seasonally with respect to river runoff volumes.

The present study did not account for atmospheric deposition, which may represent an important source of combustion-derived PAHs (*44*) as well as chlorinated pesticides and PCBs (*45*), or for historical contaminant reservoirs with chemical signatures that are significantly different from those of present effluents. Although they were not character-ized as part of this study, regional and historical sources remobilized by bioturbation and resuspension likely are of primary importance for inputs of chlorinated pesticides and high molecular weight PAHs. Nevertheless, the use of molecular markers and multivariate techniques helped resolve a weak sewage signature in the shelf and slope sediments, and provides a useful perspective on the relative magnitude of wastewater-derived contaminant inputs to the San Pedro Shelf.

Conclusions

This study provided an initial, semi-quantitative assessment of contributions to sediment contaminants from multiple sources with different chemical compositions. Results indicated that shelf and canyon sites up and down coast from the CSDOC outfall at depths of 40-200 m contained a weak sewage signature, and nearshore sites at depths of 20 m contained a moderate river discharge signature. Large portions of the variance appeared to be unrelated to the wastewater discharge or river runoff, and may instead reflect historical residues, including those from past wastewater discharges, and/or regional input sources. The results probably can be improved by incorporating data on the composition of atmospheric particle deposition and dated sediment cores to characterize historical inputs.

Acknowledgements

The authors wish to acknowledge J. Evans (SAIC) and crew of the R/V CRUSADER (MEC) for assistance with sample collection; E. Ruth (UCLA) for GC/MS analyses; R. Petty (UCSB MSI Analytical Lab.) for CHN analyses; R. Gossett and G. Robertson (CSDOC) for providing the effluent and globule samples; and R. Eganhouse and two anonymous reviewers for helpful comments on an earlier version of this paper. CSDOC provided financial support for this study. This is Publication No. 4863 of IGPP, UCLA, Los Angeles, CA.

Literature Cited

(1)   Eganhouse, R.P., Blumfield, D.L., and Kaplan, I.R. *Environ. Sci. Technol.* **1983**, *17*, 523-537.
(2)   Eganhouse, R.P., Olaguer, D.P., Gould, B.R., and Phinney, C.S. *Mar. Environ. Res.* **1988**, *25*, 1-22.
(3)   Grimalt, J.O., Fernandez, P., Bayona, J.M., and Albaigés, J. *Environ. Sci. Technol.* **1990**, *24*, 357-363.
(4)   Jeng, W.-L. and Han, B.C. *Mar. Poll. Bull.* **1994**, *28*, 494-499.
(5)   Takada, H., Farrington, J.W., Bothner, M.H., Johnson, C.G., and Tripp, B.W. *Environ. Sci. Technol.* **1994**, *28*, 1062-1072.
(6)   Venkatesan, M.I. and Kaplan, I.R. *Environ. Sci. Technol.* **1990,** *24*, 208-214.
(7)   Chalaux, N., Bayona, J.M., Venkatesan, M.I., and Albaigés, J. *Mar. Poll. Bull.* **1992,** *24*, 403-407.
(8)   Van Vleet, E.S., Fossato, V.U., Sherwin, M.R., Lovett, H.B., and Dolci, F. *Adv. Org. Geochem.* **1987**, *13*, 757-763.

(9)   Kennicutt II, M.C. and Comet, P.A. In *Organic Productivity, Accumulation, and Preservation in Recent and Ancient Sediments.* Whelan, J.K. and Farrington, J.W. Eds. Columbia Univ. Press: Columbia, SC, 1993, pp. 308-332

(10)   Chalaux, N., Takada, H., and Bayona, J.M. *Mar. Environ. Res.* **1995,** *40,* 77-92.

(11)   Samhan, O., Zarba, M., and Anderlini, V. *Mar. Environ. Res.* **1987,** *21,* 31-48.

(12)   Montalvan, P., Vitturi, L.M., Pavoni, B., and Rabitti, S. *Cont. Shelf Res.* **1985,** *4,* 321-340.

(13)   Yunker, M.B., Snowdon, L.R. MacDonald, R.W. Smith, J.N., Fowler, M.G., Skibo, D.N., McLaughlin, F.A., Danyushevskaya, A.I., Petrova, V.I. , and Ivanov, G.I. *Environ. Sci. Technol.* **1996,** *30,* 1310-1320.

(14)   Grimalt, J.O. and Olivé, J. *Anal. Chim. Acta* **1993,** *278,* 159-176.

(15)   Grimalt, J.O., Canton, L., Olivé, J. *Chemom. Intelligent Lab. Sys.* **1993,** *18,* 93-109.

(16)   Shepard, F.A. *Submarine Geology.* 3rd Ed. Harper and Row: New York, NY. 1973.

(17)   Brownlie, W.R. and Taylor, B.D. *Sediment Management for Southern California Mountains, Coastal Plains, and Shorelines. Part C: Coastal Sediment Delivery by Major Rivers in Southern California.* Environmental Quality Laboratory Report 17C. California Institute of Technology: Pasadena, CA. 1981.

(18)   Emery, K.O. *The Sea off Southern California.* Wiley: New York, NY. 1960.

(19)   Drake, D.E., Cacchione, D.A. and Karl, H.A. *J. Sed. Petrol.* **1985,** *55,* 15-28.

(20)   SCCWRP. *Coastal Water Research Project Biennial Report, 1991-1992.* Southern California Coastal Water Research Project: Long Beach, CA. 1992.

(21)   Mearns, A.J., Matta, M., Shigenaka, G. , MacDonald, D., Buchman, M., Harris, H., Golas, J., and Lauenstein, G. *Contaminant Trends in the Southern California Bight: Inventory and Assessment.* NOAA Technical Memorandum NOS ORCA 62. NOAA Ocean Service: Seattle, WA. 1991.

(22)   Eganhouse, R.P. and Venkatesan, M.I. In *Ecology of the Southern California Bight: A Synthesis and Interpretation.* Daily, M.D., Reish, D.J., and Anderson, J.W. Eds. Univ. Of Calif. Press: Berkeley, CA. 1993.

(23)   Venkatesan, M.I. *Historical Trends in the Deposition of Organic Pollutants in the Southern California Bight.* Final Report to National Oceanic and Atmospheric Administration. 1994.

(24)   Venkatesan, M.I., Ruth, E.C., Steinberg, S., and Kaplan, I.R. *Mar. Chem.* **1987,** *21,* 267-299.

(25)   Venkatesan, M.I., Ruth, E.C., and Kaplan, I.R. *Mar. Poll. Bull.* **1986,** *12,* 554-557.

(26)   Eganhouse, R.P., Ruth, E.C., and Kaplan, I.R. *Anal. Chem.* **1983,** *55,* 2120-2126.

(27)   Yunker, M.B., MacDonald, R.W., Veltkamp, D.J., and Cretney, W.J. *Mar. Chem.* **1995,** *49,* 1-50.

(28)   Kvalheim, O.M. *Anal. Chim. Acta* **1985,** *177,* 71-79.

(29)   Johansson, E. and Wold, S. *Anal. Chem.* **1984,** *56,* 1685-1688.

(30)   Zitko, V. *Mar. Poll. Bull.* **1994,** *28,* 718-722.

(31)   Steinhauer, M.S. and Boehm, P.D. *Mar. Environ. Res.* **1992,** *33,* 223-253.

(32)   Tissot, B.P. and Welte, D.H. *Petroleum Formation and Occurrence*, Springer-Verlag: Berlin. 1984, p. 699.

(33)   Takada, H. and Ishiwatari, R. *Environ. Sci. Technol.* **1990**, *24*, 86-91.

(34)   Wold, S. *Pattern Recognition* **1976**, *8*, 127-139.

(35)   Eganhouse, R.P. and Kaplan, I.R. *Environ. Sci. Technol.* **1982**, *16*, 541-551.

(36)   Eganhouse, R.P. and Gossett, R. In *Organic Substances and Sediments in Water. Vol. 2. Processes and Analytical.* Baker, R.A., Ed. Lewis Publ.: Boca Raton, FL, 1991, pp 191-220.

(37)   Zeng, E.Y. and Vista, C.L. *Environ. Toxicol. Chem.* (in press).

(38)   Hostettler, F.D., Rapp, J.B., Kvenvolden, K.A., and Luoma, S.N. *Geochim. Cosmochim. Acta* **1989**, *53*, 1563-1576.

(39)   Thompson, B.E., Hershelman, G.P., and Gossett, R.W. *Mar. Poll. Bull.* **1986**, *17*, 404-409.

(40)   O'Reilly, J.E., Sauer, T.C., Ayers, R.C. Jr., Brandsma, M.G., and Meek, R. In *Drilling Wastes*, Engelhardt, F.R. , Ray, J.P., and Gillam , A.H. Eds. Elsevier Applied Science: London, 1988, pp 647-665.

(41)   Simoneit, B.R.T. and Kaplan, I.R. *Mar. Environ. Res.* **1982**, *3*, 113-128.

(42)   Phillips, C.R. and Hershelman, G.P. *Water Environ. Res.* **1996**, *68*, 105-114.

(43)   Schafer, H. *J. Water Poll. Control Fed.* **1989**, *61*, 1395-1401.

(44)   Cross, J.N., Hardy, J.T., Hose, J.E., Hershelman, G.P., Antrim, L.D., Gossett, R.W., and Crecelius, E.A. *Mar. Environ. Res.* **1987**, *23*, 307-323.

(45)   Hom, W., Risebrough, R.W., Soutar, A., and Young, D.R. *Science* **1974**, *184*, 1197-1199.

Chapter 17

Use of Trialkylamines as a Marker of Sewage Addition into the Marine Environment

J. M. Bayona[1], N. Chalaux[1], J. Dachs[1], C. Maldonado[1], M. Indira Venkatesan[2], and J. Albaigés[1]

[1]Department of Environmental Chemistry, Centro de Investigacion y Desarrollo, Consejo Superior de Investigaciones Cientificas, Jordi Girona 18, E–08034 Barcelona, Spain
[2]Institute of Geophysics and Planetary Physics, University of California, Los Angeles, CA 90095–1567

Molecular markers of sewage in the ocean, namely, coprostanol, linear alkylbenzenes (LABs) and trialkylamines (TAMs) have been determined in pollution sources (i.e. wastewaters, sewage sludges, river water and sediments), seawater and sediments collected in bays (Tokyo and Guanabara) and in the open sea (western Mediterranean and southern California, Pacific Ocean). The TAM concentrations found in wastewater and sewage sludge are of the same order of magnitude as coprostanol and one order of magnitude higher than LABs. The correlation between coprostanol and TAM concentrations in wastewater ($r^2 = 0.89$, n = 7) demonstrates a similar urban source. In enclosed bays, TAMs exhibited the highest concentrations presumably attributable to their higher stability than other molecular markers investigated. In open sea transects, the distribution of LABs in surficial sediments is consistent with their enrichment in the colloidal fraction of wastewaters causing a concentration maxima in sediments farther from the source than the other molecular markers. Furthermore, TAMs and LABs enabled us to trace long-range transport of pollution from land-based sources to the open ocean due to their stability in a variety of depositional environments.

Source-specific and conservative molecular markers are needed for the assessment of anthropogenic pollution in the marine environment. Coprostanols (5β(H)-cholestan-3β-ol and 5β(H)-cholestan-3α-ol, Figure 1) have traditionally been used for this purpose [1-10]. However, the stability of these compounds under oxic conditions is believed to be limited [9], and *in situ* formation can occur under strongly anoxic conditions [11]. In order to improve the source specificity, we have defined several indices based on the epimeric 5α(H)/5β(H) ratio of stanols and stanones which are useful for slightly to moderately polluted areas [12]. On the other hand, Venkatesan and Kaplan have used the ratio of coprostanol to planktonic sterols (i.e. dinosterol) to trace the addition of sewage into the open ocean [8].

The large amounts of surfactants and related products used in domestic applications makes the hydrophobic intermediates found in surfactant formulations attractive as tracers of sewage pollution in the marine environment. Linear alkylbenzenes (LABs, Figure 1) and trialkylamines (TAMs, Figure 1), which are trace constituents of anionic and cationic surfactants, respectively, have been used to trace sewage pollution in the marine environment [13-25].

The rationale for using hydrophobic intermediates of surfactants to trace sewage pollution is their high affinity for particles. This could result in the enhanced accumulation of the intermediates in sediments. For instance, a greater persistence of LABs vs. linear alkylbenzenesulfonates (LAS) has been elegantly illustrated by Takada and coworkers in the Tokyo Bay sediments [17]. Whereas LAS exhibited higher concentrations than LABs in river water, LAS concentrations declined more rapidly than the LABs in open bay waters. Another advantage of surfactant related products as tracers of sewage pollution is the possibility to date a sedimentary record from the concentration profile with depth since surfactant formulations are often changing at specific times [14].

In this paper, we discuss the occurrence of TAMs in pollution sources (i.e. wastewater and sewage sludges) and their fate in receiving aquatic systems (i.e. estuaries, bays and in the open sea) of different climatic regions (i.e. tropical and temperate). Furthermore, we assess their source specificity and stability in the marine environment as compared to other molecular sewage pollution markers, namely, coprostanol and LABs.

Environmental Setting

Tokyo Bay (Japan). The surface area of Tokyo Bay is ca. 1200 km^2 with an average water column depth of 15 m [15-17]. The residence time of water in the bay is estimated to be 0.13 years. The population of the metropolitan area of Tokyo is ca. 2.9×10^7 inhabitants. Rivers receiving partially treated wastes are the major point sources of pollution to the bay. Wastewater samples were collected at the influent from a Tokyo sewage treatment plant (24 h composite) in August 1984 and May 1986. Suspended particulate samples were collected in the Tamagawa River in February, 1985 and August, 1985. Eleven surficial sediment samples were collected with a grab sampler along a transect from the inner to the outer bay.

Guanabara Bay (Brazil). The surface area of Guanabara Bay is ca. 400 km^2 and the average depth is 8 m. The bay is eutrophic, and the surficial sediments often become anoxic. The bay is flushed by strong tidal currents (0.6-0.7 m^3 s^{-1}). The exchange volume during tides is ca. 10^8 m^3 which represents 10% of the bay volume. Population density surrounding the bay is high (2×10^7 inhabitants), Rio de Janeiro and Niteroi being the largest riparian cities. Further, a large industrial area is located near the western part of the bay. Untreated urban and industrial wastes (20 m^3 s^{-1}) are mainly disposed into the bay. A total of six surface sediment samples covering the whole bay were collected in 1988 using a grab sampler.

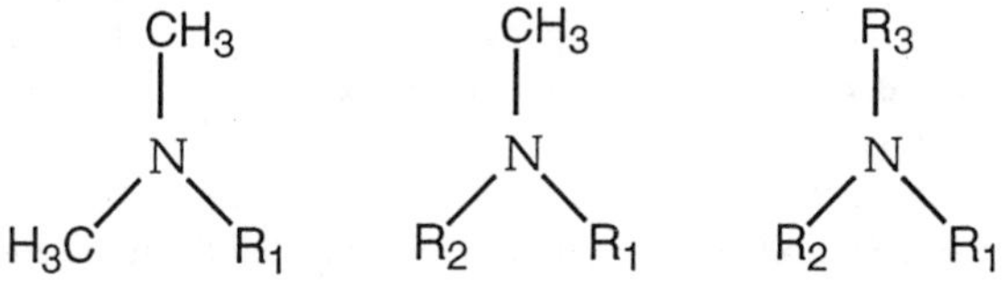

TAMs $R_1, R_2, R_3 = C_{14}H_{29}, C_{16}H_{33}$ or $C_{18}H_{37}$

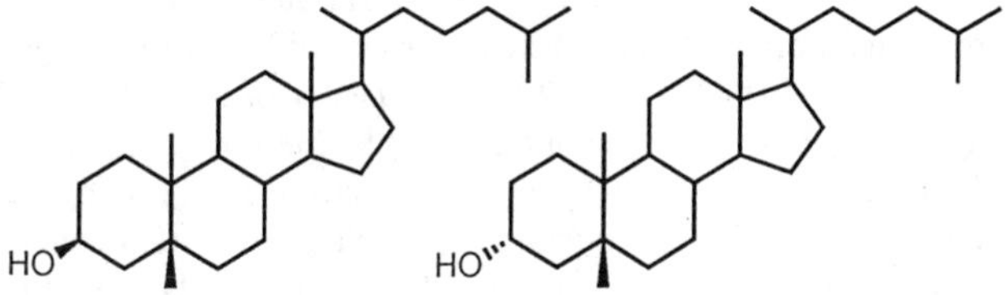

COPROSTANOL ($3\beta, 5\beta$)   **EPICOPROSTANOL** ($3\alpha, 5\beta$)

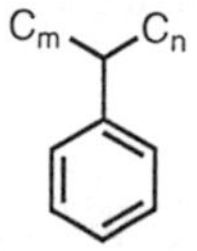

LABs $m + n = 9 - 13$

Figure 1. Structure of trialkylamines (TAMs), coprostanol, epicoprostanol, and linear alkylbenzenes (LABs).

Western Mediterranean. The surface area of the western Mediterranean is ca. 2.8 x 10^5 km^2 [21-22]. The predominant surface current along the continental shelf and slope is from the northeast to the southwest sweeping the coasts of northern Italy, France and Spain. Wastewaters of the largest cities are subjected to primary treatment processes, and the resultant treated sewage is dumped into the continental shelf by submarine outfalls (ca. 3-4 km off-shore). Surficial sediments are oxic except in the proximities of the sewage disposal sites. Surficial seawater samples (14 particulate and 7 dissolved) and nine sediment samples were collected along cross-shelf transects originating at large cities (i.e. Barcelona and Valencia) and extending to the open sea and the Gulf of Lions (Figure 2A). Sampling cruises took place from 1988 to 1992 in the framework of the European River Ocean Systems (EROS-2000) project of the European Union (EU).

Southern California (USA). Sludge sample was collected from the advanced primary partial-secondary treatment plant of the Joint Water Pollution Control Plant (JWPCP) of Los Angeles County Sanitation District in November of 1987 [14, 24]. The composition of the sewage influent is ca. 18% industrial and 82% domestic. The predominant subsurface current at the discharge depth (60 m) is to the northwest, parallel to the coast [26]. Box cores of sediment samples were collected from Santa Monica Basin along a northeast to southwest transect offshore Los Angeles (southern California) in October of 1985 (Figure 2 B). The center of the basin is about 30 km from land, and the maximum water column depth is 910 m. The bottom water is anoxic below the sill depth of approx. 730 m. Data from surface sections (0-4 cm) are discussed in this report. Box core sediments were collected (water depth of 458 to 906 m) from slightly different sites in Santa Monica Basin and also in San Pedro Basin (southern California) in September, 1991 as part of the National Oceanic and Atmospheric Administration (NOAA) / National Status and Trends program. Here, only the data from surface horizons (0-1 cm) are reported for comparison. A detailed discussion of the contaminants from these cores will be published elsewhere.

Analytical Procedures

Sample Handling. Aliquots (500 mg) of fabric softeners (Quanto and Flor from Camp-Benckisser; Mimosin from Lever; Perlan and Vernel from Henkel) were homogenized and freeze-dried. Composite wastewaters (6 h sampling period) and primary-partial secondary treated sewage sludges were collected with a home-made sampling device at the influent and effluent of treatment plants, respectively. The sampling devices were rinsed with distilled water between samples. Wastewaters were homogenized and filtered through a precombusted (450 °C) glass fiber filter rated at 0.7 μm nominal pore size (Whatman GF/F). Hydrophobic organic substances from the filtrate (200 mL) were extracted with a 90 mm Empore C18 disk which was previously rinsed with 15 mL of methanol. Recoveries of TAMs in the filtrate were higher than 80% according to the standard addition procedure in fortified 500 mL samples. Sewage sludges were centrifuged and freeze-dried. Particulate and dissolved phases in surficial seawater (5 m depth) were collected with stand-alone submersible pumps (Axys Environmental Systems, BC, Canada) equipped with glass fiber filters and a PTFE column (20 mm i.d. x 300 mm length) packed with 50 g XAD-2. Technical grade

XAD-2 (Rohm and Haas, Barcelona, Spain) was Soxhlet extracted with CH_2Cl_2:MeOH (1:1) until successful blanks were achieved (i.e. 1-5 days). Precleaned XAD-2 columns were end capped with PTFE stoppers and stored in the refrigerator. TAM recoveries obtained by the Empore disk (<10 L) were similar to those obtained by the XAD-2 column (<50 L). The latter extraction system was preferred for seawater analysis due to its faster extraction rate (400 mL min^{-1}) and large breakthrough volumes (50-100 L). After sampling, the XAD-2 columns were stored at 4 °C until analysis (e.g. < 2 weeks). A sequential counter-flow column elution was performed (2 mL min^{-1}) with 200 mL of MeOH followed by 200 mL CH_2Cl_2 supplied by a reciprocating pump. The methanolic fraction was concentrated by rotary evaporation to half the original volume, and it was extracted three times with 25 mL of n-hexane. Hexane extracts were dried over anhydrous Na_2SO_4, combined with the dichloromethane extract and reduced by rotary vacuum evaporation to a volume of 1-2 mL for column chromatographic fractionation (see analytical procedures section).

All sediments were sampled with a box-corer except those collected in Tokyo and Guanabara Bays which were collected with a grab. Box-core samples were subsampled with a stainless steel or aluminum tubing endcapped with PTFE stoppers and kept frozen at -20°C until analysis.

Molecular Marker Analysis. Freeze-dried fabric softeners were extracted with 20 mL hexane by sonication. Freeze-dried sediments and suspended particulate matter were Soxhlet extracted for 24 h with dichloromethane-methanol (2:1). Pacific Ocean sediments were wet-extracted with methanol and then dichloromethane. A hexane back extract of the methanol and water phase was added to the dichloromethane extract and processed as described below. Tridecylamine, 1-dodecylbenzene and 5α(H)-androstan-3β-ol were used as analyte surrogates and were added to the sample prior to the extraction. Detailed analytical procedures and mass spectral data of TAMs are reported elsewhere [25,27]. Briefly, organic extracts was fractionated by column chromatography using neutral alumina activated at 120 °C (5 g). LABs were eluted with 10 mL of hexane, and the fraction containing TAMs and coprostanol was eluted with 12 mL of methanol-dichloromethane (1:1 v/v). Sterols were analyzed as trimethylsilyl ethers. The analytical determination of LABs, TAMs and coprostanol was performed by capillary gas chromatography coupled to electron impact mass spectrometry (cGC-EI-MS), nitrogen-phosphorus detector (NPD) and flame ionization detector (FID), respectively. Surrogate recoveries ranged from the 70-90%. Analytical results were corrected for recovery. Procedural blanks were obtained from all sample sets. The relative standard deviation (RSD) of three independent replicates was <10%. A control sediment sample was spiked with methanolic solution of surrogates at two concentration levels (50 and 800 ng g^{-1}) and were analyzed to assess the accuracy of the analytical procedure. Recoveries were higher than 80%. Typical TAM detection limits are 0.5 ng L^{-1} and 2 ng g^{-1} for seawater (50 L) XAD-2 column preconcentrated, and sediments (5 g dry wt.), respectively.

Results and Discussion

Sources of Trialkylamines(TAMs). The TAMs considered in this study contain one methyl group and two longer alkyl substituents having 14, 16 or 18 carbon atoms.

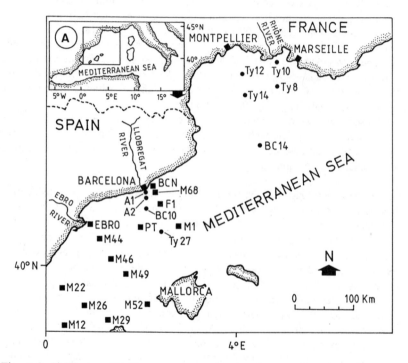

Figure 2. A) Western Mediterranean map showing the sampling sites of seawater (BCN, M68, F1, M1, PT, EBRO, M44, M46, M49, M52, M29, M26, M22, M12) and sediments (A1, A2, BC10, Ty27, Ty8, Ty10, Ty12, BC14 and Ty14) collected during several sampling cruises carried out from 1988 to 1992. B) southern California map showing the sediment sampling sites (filled squares: CABs and filled circles: NOAA samples). Sanitation district of the City of Los Angeles, County of Los Angeles and Orange County are indicated as 1, 2 and 3 respectively.

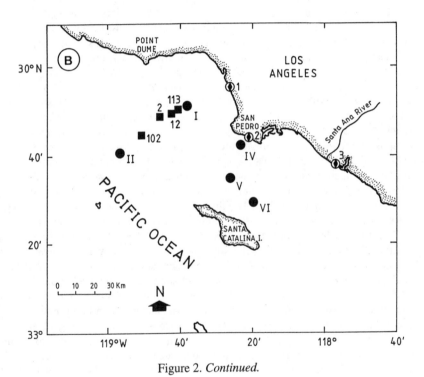

Figure 2. *Continued.*

The other two TAM subclasses containing 2 or 0 methyl groups in the molecule are not considered because either they are only moderately hydrophobic or they are present at very low concentrations in sediments [22,28]. Hereafter, TAMs denote only the sum of tertiary amines containing a single methyl group (Figure 1).

TAMs were identified at variable concentrations (1.2 to 2700 mg L^{-1}) at average value of 1630 mg L^{-1} in five fabric softeners widely used throughout the European Union (EU). The concentrations found in urban-industrial sewage sludges ranged from 0.11 to 1.01 mg g^{-1} (Table 1). In untreated wastewater, the TAM concentration ranged from 28 to 160 μg L^{-1}. These concentrations are comparable to other trace organic constituents of sewage (e.g. coprostanol). The LABs were also detected in these samples but usually at one or two orders of magnitude lower concentration than TAMs and coprostanol (Table 1). Only in the sewage sludges from the JWPCP, the LAB concentrations exceed that of TAMs. This is probably attributable to differences in the constituents of fabric softener formulations and to the contribution of industrial effluents of LAB production [R. Eganhouse, personal comm.]. The higher concentrations of TAMs than LABs in wastewaters and sewage sludges contrast with the industrial output of the related surfactants. Anionic surfactants make up 64% of the industrial output, whereas cationic surfactants only account for 9% in US during 1994 [29] which represents 1.3 x 10^5 t [30]. However, it is impossible to accurately estimate the relative inputs of surfactant precursors (e.g. TAMs and LABs) into the marine environment due to limited data base and the variability of their concentrations with time. Surfactant formulations may also differ on geographical basis [31]. Therefore, the higher concentrations of TAMs in sewage and wastewaters may be attributable either i) to higher inputs of TAMs or ii) to higher persistence in wastewater systems. Arguments supporting the latter statement are i) the higher molecular weight (MW = 423-535) of the TAMs limits diffusion through bacterial membranes [32] with only the dimethyl tertiary alkylamines being readily biodegradable (kinetics are controlled by the length and number of alkyl substituents [33,34]) and ii) the almost constant concentration ratios between homologs (i.e. $C_1NC_{16}C_{16}$, $C_1NC_{18}C_{16}$ and $C_1NC_{18}C_{18}$) that are found in environmental samples. Photooxidation experiments (λ>290 nm) of TAMs in aqueous solution have shown a remarkable stability. Only in presence of a sensitizer (i.e. H_2O_2, acetone) is photooxidation observed [37]. At present, a realistic environmental fate assessment of TAMs in the aquatic environment is precluded due to the lack of knowledge of key physico-chemical properties. Although similar behavior to TAMs has been reported for the photooxidation of LABs [38], they still show a preferential depletion of the internal substituted isomers in either in vitro cultures [35] or in environmental samples [16,18,23,25]. The ratio of internal to external isomers has indeed proposed as a degradation index [36]. Gledhill et al [38] have applied a multimedia fugacity model to evaluate the fate of LABs in the environment and found that the atmospheric compartment is by far the most favored at equilibrium.

The identification of TAMs as well as coprostanol in the suspended particulate phase of polluted river water (i.e. Tamagawa River) and in estuarine sediments (e.g. Nile, Rhône and Ebro Rivers, Figure 2 and Table 1) suggests the input of sewage (Table 1). Industrial sources of TAMs cannot be overlooked because cationic surfactants (i.e. dimethylditallowammonium chloride) are also used in other applications [31]. However, Chalaux [39] has found a significantly higher correlation

($r^2 = 0.89$, n = 7) between coprostanol-TAMs than coprostanol-LABs ($r^2 = 0.51$) in untreated wastewaters which strongly supports the usage of TAMs as sewage marker since the remaining sources of TAMs are negligible *vs* domestic sources [30].

Fate of Trialkylamines (TAMs) in the Open Sea. TAMs were identified in the suspended particulate and dissolved phases of coastal Mediterranean seawater at average concentrations of 30 and 2 ng L^{-1}, respectively (Table 1). The higher concentrations of TAMs observed in the particulate phase is consistent with their hydrophobic nature (log $K_{ow} > 8$, according to the Hansch and Leo prediction method [40]).

Table 1. Occurrence of sewage markers in the aquatic environment*

Sample type	Characteristics	Area of study	TAMs	LABs	Coprostanol
untreated wastewater (μg L^{-1})	dissolved particulate	Barcelona (n=7)**	28±15 159±80	46±14 10±4	40±24 125±71
	particulate	Tokyo (n=1)	88	0.94	89
sewage sludge (mg g^{-1})	primary	Barcelona (n=2)	1.01	0.05	1.4
	secondary	Munich (n=1)	0.38	NA	NA
	primary-secondary	JWPCP (n=1)	0.11	0.20	0.26
river water (μg L^{-1})	particulate	Tamagawa (Japan) (n=2)	13.3	0.12	1.24
seawater (ng L^{-1})	dissolved particulate	Mediterranean (n= 14)	2.3±1.1 31±10	2±0.5 10±6	NA
sediments (ng g^{-1})	grab (0-5 cm)	Nile estuary (n=2) (Egypt)	588	876	1587
	grab (0-5 cm)	Guanabara Bay (n=6) (Brasil)	19080± 8350	199± 130	1236±723
	grab (0-5 cm)	Tokyo Bay (n=11)	4430± 1450	479± 325	122±54
	box-core (0-2.5 cm)	Mediterranean (n=9)	980± 245	587± 354	177±71
	box-core (0-1 cm)	S. California S. Pedro Basin (n=3)	5440± 850	NA	4220±2430
	box-core (0-1, 0-4 cm)	S. California S. Monica Basin (n=5)	585± 245	236± 124	2218±1242

NA: not available.
*Arithmetic mean of total concentrations ± standard deviation.
**number of samples analyzed.

Furthermore, their distribution in the surficial waters is consistent with a land based source because a negative gradient of concentrations are usually found with off-shore distance (Figure 3). Association with particles and subsequent settling through the water column may be the pathway for their incorporation into sediments. The increase in the TAM concentration at the station located at 20 km offshore (Figure 3) has also been found for other persistent organic pollutants (i.e. PCBs and PAHs, 41, 42) and is attributable to a persistent thermohaline front leading to a relatively higher concentration of particles at this location than at other open ocean stations.

The occurrence and fate of urban molecular markers in an open system was investigated in the Pacific Ocean (southern California) and in the western Mediterranean. In the latter case, sedimentary concentrations of sewage markers, coprostanol and TAMs sharply decreased with off-shore distance, while LABs exhibited a maximum concentration at deeper stations (10 km off-shore, Figure 4A). LABs are the sewage markers with the lowest apparent distribution coefficient (K'd) in untreated wastewater (Table 2). However, all sewage markers were identified at low concentrations even at remote stations located in the deep sea pointing to their transport from continental shelf to the open sea. This long-range transport of land-derived contaminants is also evident in the southern California basins. However, in contrast to the western Mediterranean, a concentration increase, particularly in TAMs and LABs, is apparent in the deeper waters (900 m) of Santa Monica Basin (Figure 4B). The smaller particle size content of the deep sea sediments and the prevailing anoxic conditions below the water sill depth of $\approx$ 730 m might aid in better preservation of organic matter [8]. Because the southern California countercurrent flows upcoast [26], particles containing effluent-derived contaminants from the JWPCP can be transported into Santa Monica Basin. The enrichment of molecular markers at farther stations in Santa Monica Basin is, thus consistent with a dual point emission of sewage (e.g. Hyperion and JWPCP) in the farther stations compared to the nearshore stations which are mostly impacted by Hyperion (Figure 2B). The positive concentration gradient seaward is more apparent in the case of the TAMs than LABs in Santa Monica Bay sediments. In fact, the ratio of the contents of LABs and TAMs from remote to nearshore stations for LABs and TAMs are of 1.4 and 55.8, respectively [24]. This could be explained by a stronger affinity of TAMs for fine clay particles as has been found for the tetraalkylammonium salts [43]. These fine particles can be easily advected farther offshore than coarse sand materials. In the case of the western Mediterranean, a steeper decline in TAM concentrations than in LABs with increasing off-shore distance was found (Figure 4A). This could be attributed to a higher enrichment of LABs in the dissolved-colloidal phases of wastewaters which allow transport to greater distance than TAMs associated with the particulate phase of seawater. Takada et al [17] have also found a similar distribution of LABs at the northern Tokyo Bay where LABs associated with low-density particles are preferentially being transported to off-shore regions.

Table 2. Apparent partition coefficients (K'd, L g^{-1}) in urban raw sewage*

	n-alkanes	PAHs	Coprostanol	LABs	TAMs
K'd	4.2±0.4	5.9±0.6	19.3±2.3	1.2±.2	30.1±3.4

*Samples collected at the influent of Barcelona treatment plant (n=7)

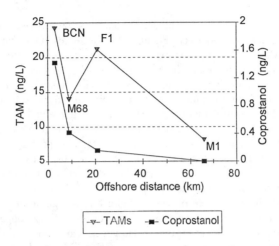

Figure 3. Concentrations of TAMs and coprostanol in the particulate phase of surficial seawater collected in a transect off Barcelona. Sampling sites are indicated in Figure 2A.

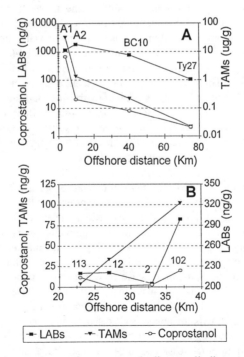

Figure 4. Concentrations of coprostanol, linear alkylbenzenes (LABs) and trialkylamines (TAMs) in sediments according with their offshore distance. A) transect offshore Barcelona (western Mediterranean) and B) Santa Monica Basin, southern California (Pacific Ocean). The sampling site location is shown in Figures 2A and B.

Coprostanol exhibited only a moderate concentration increase with water depth in the Santa Monica Bay transect (Figure 4B). These results emphasize a higher stability of TAMs and LABs in the marine environment allowing the recognition of sewage inputs in the deep ocean. The concentration of coprostanol in Santa Monica Basin is similar to that previously reported from the same area [8]. The strong decline in concentrations of coprostanol in remote sediments from southern California contrasts with the enhanced incorporation of coprostanol and LABs in sediments collected at Deepwater Dumpsite 106 located 185 km offshore New Jersey in water depths of 2400 and 2900 m [18]. The different fate of coprostanol between both marine systems could be accounted for by the relatively fast sinking of coprostanol associated with large sewage particles into the latter region in comparison to slow advective transport in the former basin [44].

In general, the prevalence of sewage indicators is significant both in the San Pedro and Santa Monica Basins, reflecting the importance of anthropogenic activities associated with waste disposal in the region. However, San Pedro Basin sediments are uniformly enriched with urban molecular markers (i.e. TAMs, LABs and coprostanol) relative to Santa Monica Basin. This is also consistent with the rates of petroleum deposition found to be larger in San Pedro than in the Santa Monica Basin as reflected by particle trap studies which is strongly correlated with human activities [45]. The LAB concentrations in Santa Monica Basin sediments are comparable to those reported for a core (SPB1) from San Pedro Basin at a water depth of 869 m [14] and that from station E3, from Santa Monica Basin at 150 m water depth, about 8 km northwest from the end of the Hyperion 7 mile sludge outfall [46]. That the LAB contents in the sediments from the centre of the Santa Monica Basin are still comparable or greater than that from the inner shelf demonstrates that LABs are relatively freely advected offshore, associated with fine particles and colloids as discussed earlier. However, LAB concentration in the JWPCP effluent from the current study is one order of magnitude lower than the values reported about a decade earlier [14, R. Eganhouse, personal comm.]. This decline is most likely correlated with recent wastewater treatment plant improvements.

Fate of Trialkylamines (TAMs) in Semienclosed Systems. The distribution of TAMs in sediments was investigated in two semi-enclosed systems: Tokyo and Guanabara Bays (Table 1). LAS is the most widely used surfactant in Brazil. In contrast cationic surfactants with quaternary ammonium salts are not used as commonly [A. Wagener, personal comm.]. In Japan, the market share of cationic surfactants during 1985 accounts for 0.6×10^5 t y^{-1} while anionic surfactants represent 5.2×10^5 t y^{-1} [47]. In the particulate phase of wastewaters, coprostanol and TAMs predominate over LABs. In the suspended particulate phase of a river water outflowing into the Tokyo Bay, TAMs also predominate over the remaining molecular markers. Accordingly, TAMs exhibited higher sedimentary concentrations than the other sewage markers investigated in both bays. In the Tokyo Bay sediments, TAMs occurred at concentrations al least one order of magnitude higher than the other markers investigated. In order to investigate the processes affecting to the fate of LABs, the ratio between the internal and external (I/E) substituted isomers are used as a degradation index [36]. Whereas in waste water and river water the I/E ratio is close to unity, the average value obtained in eleven sediments collected in the inner bay was as high as four. According to the biodegradation index defined by Takada and Ishiwatari [36], the obtained I/E value amounts for the 60% LAB removal from estuarine sediments. This is consistent with a removal rate of 40% obtained in sediments collected from nearby estuaries [16]. Similarly, the LAB average I/E ratio found in the Guanabara Bay sediments is 3.2 which accounts for the 45% removal. Conversely, the homolog distribution found in the case of TAMs is almost constant among the different environmental compartments investigated with a predominance of $CH_3NC_{16}H_{33}C_{18}H_{37}$. In contrast, the biodegradation kinetics of TAMs in aerobic conditions are dependent on alkyl substitution [33,34]. The low volatility and preferred particulate phase association of TAMs with particles (Tables 1 and 2) in which biotic or abiotic degradation are less favored could explain the apparently refractory behavior of these markers. Other hydrophobic markers which are more associated with the dissolved and colloidal phases are prone to a variety of biotic or abiotic transformations in the aquatic environment. Among the molecular markers investigated, coprostanol is the least resistant to degradation in both bays since its concentration strongly dereased compared to its sources (Table 1). As a consequence, in Tokyo Bay sediments collected farther from the source coprostanol is not detectable, whereas TAMs and LABs were still present at measurable concentrations, enabling one to trace the lateral transport of organic contaminants from the inner bay to the open ocean [25].

Conclusions

Trialkylamines have been found ubiquitously distributed in the aquatic environment being detected both in the dissolved as well as in the particulate phases of the open seawater and in the underlying sediments. Their identification in fabric softener formulations and in untreated wastewaters and in primary-secondary treated sewage allow us to propose TAMs as molecular marker to trace urban pollution in the aquatic environment. In this regard, the significant correlation between coprostanol and TAMs supports a predominant sewage origin of TAMs rather than from non domestic sources.

Concentrations of TAMs in suspended particles of surficial seawater transect collected in the western Mediterranean maximized in the thermohaline frontal area enriched in particles due to resuspension processes. This fact is consistent with the high apparent partition coefficient of TAMs found in seawater.

Sediments collected in the open ocean revealed different fates for the molecular markers investigated here namely, coprostanol, TAMS and LABs supporting an integrated usage in marine pollution studies. Further studies are needed to evaluate the fate of urban pollution markers considering a three compartment model (e.g. dissolved, particulate and colloidal) either in wasterwaters or in the receiving systems.

Acknowledgments

Authors are indebted to Dr. H. Takada from Tokyo University of Agriculture and Technology and Prof. A. Wagener from Pontificia Universidade Catolica do Rio de Janeiro for providing sediment samples. Comments and suggestions of the anonymous reviewers and editor are kindly acknowledged. Partial funding was provided by CICYT (AMB96-0926), NOAA (No 170A0479-01), SEA (R/CZ-126-PD) and NATO (No 920437). This is publication # 4861 from the Inst. Geophysiscs and Planetary Physics, UCLA, California.

Literature Cited

1. Goodfellow, R.M.; Cardoso, J.; Eglinton, G.; Dawson, J.P.; Best, G.A. *Mar Pollut. Bull.* **1977**, *8*, 272-276.
2. Hatcher, P.G.; McGillivary, P.A. *Environ. Sci. Technol.* **1979**, *13*, 1225-1229.
3. McCalley, D.V.; Cooke, M.; Nickless, G. *Water Res.* **1981**, *15*, 1019-1025.
4. Brown, R.C.; Wade, T.L. *Water Res.* **1984**, *18*, 621-632.
5. Readman, J.W.; Preston, M.R.; Mantoura, R.F.C. *Mar. Pollut. Bull.* **1986**, *17*, 298-308.
6. Eganhouse, R.P.; Olaguer, D.P.; Gould, B.R.; Phinney, C.S. *Mar. Environ. Res.* **1988**, *25*, 1-22.
7. LeBlanc, L.A.; Latimer, J.S.; Ellis, J.T; Quinn, J.G. *Estuarine Coastal Shelf Sci.* **1992**, *34*, 439-458.
8. Venkatesan, M.I.; Kaplan, I. *Environ. Sci. Technol.* **1990**, *24*, 208-214.
9. Walker, R.W.; Wun, C.K.; Litsky, W. *CRC Crit. Rev. Environ. Control* **1982**, *12*, 91-112.
10. Vivian, C.M.G. *Sci. Total Environ.* **1986**, *53*, 5-40.
11. Nishimura, M.; Koyama, T. *Geochim. Cosmochim. Acta* **1977**, *41*, 379-385.
12. Grimalt, J.; Fernández, P.; Bayona, J.M.; Albaigés, J. *Environ. Sci. Technol.* **1990**, *24*, 357-363.
13. Ishiwatari, R.; Takada, H.; Yun, S.; Matsumoto, E. *Nature* **1983**, *301*, 599-600.
14. Eganhouse, R.P.; Blumfield, D.L.; Kaplan, I.R. *Environ. Sci. Technol.* **1983**, *17*, 523-530.
15. Takada, H.; Ishiwatari, R. *Environ. Sci. Technol.* **1987**, *21*, 875-883.
16. Takada, H.; Ishiwatari, R. *Water Sci. Technol.* **1991**, *23*, 437-446.
17. Takada, H.; Ishiwatari, R. *Estuarine Coastal Shelf Sci.* **1992**, *35*, 141-156.
18. Takada, H.; Farrington, J.W.; Bothner, M.H.; Johnson, C.G.; Tripp, B.W. *Environ. Sci. Technol.* **1994**, *28*, 1062-1076.
19. Raymundo, C.C.; Preston, M.R. *Mar. Pollut. Bull.* **1992**, 24, 138-146.
20. Murray, A.P.; Gibbs, C.F.; Kavanagh, P.E. *Mar. Environ. Res.* **1987**, *23*, 65-76.
21. Valls, M.; Bayona J.M.; Albaigés, J. *Nature* **1989,** 337, 722-724.
22. Fernández, P.; Valls, M.; Bayona J.M.; Albaigés, J. *Environ. Sci. Technol.* **1991,** *25*, 547-550.
23. Takada, H.; Ogura, N.; Ishiwatari, R. *Environ. Sci. Technol.* **1992**, *26*, 2517-2523..
24. Chalaux, N.; Bayona, J.M.; Venkatesan M.I.; Albaigés J. *Mar. Pollut. Bull.* **1992,** *24*, 403-407.
25. Chalaux N.; Takada H.; Bayona J.M. *Mar. Environ. Res.* **1995**, *40*, 77-92.
26. Hickey, B.M. In *Ecology of Southern California Bight*; Ed M.D. Dailey, D.J. Reish; J.W. Anderson; Univ. California Press. Berkeley, CA. 1993, pp 19-70.
27. Valls, M; Bayona, J.M. *Fresenius J Anal. Chem.* **1991**, 339, 212-217.
28. Valls, M; Fernández, P.; Bayona, J.M. *Chemosphere* **1989**, *19*, 1819-1827.
29. Ainsworth, S.J. *Chem. Eng. News* **1996**, *74* (Jan 22), 32-54.

30. Cross, J. In *Cationic Surfactants*. Ed. J. Cross, E.J. Singer; Surfactant Science Series, vol. 53; Marcel Dekker, New York. 1994, pp 4-28.

31. Greek, B.F.; Layman, P.L. *Chem. Eng. News* **1989**, *67* (Jan 23), 29, 49.

32. Scow, K.M. In *Chemical Property Estimation Methods*, Lyman, W.J., Reehl, W.F., Rosenblatt, D.H. ACS, Washington, DC, 1990, p 9-1.

33. van Ginkel, C.G.; Pomper, M.A.; Stroo, C.A.; Kroon, A.G.M. *Tenside Surf. Det.* **1995**, *32*, 355-359.

34. Yoshimura, K.; Machida, S.; Masuda, F. *J. Am. Oil Chem. Soc.* **1980**, *57*, 238-241.

35. Bayona, J.M.; Albaigés, J.; Solanas, A.M.; Grifoll, M. *Chemosphere* **1986**, *15*, 595-598.

36. Takada, H.; Ishiwatari, R. *Environ. Sci. Technol.* **1989**, *24*, 86-91.

37. Valls, M.; Bayona, J.M.; Albaigés, J.; Mansour, M. *Chemosphere* **1990**, *20*, 599-607.

38. Gledhill, W.E.; Saeger, V.W.; Trehy, M.L. *Environ. Toxicol. Chem.* **1991**, *10*, 169-178.

39. Chalaux, N. Ph.D. Dissertation, University of Barcelona at Barcelona, 1996.

40. Hans, C.; Leo A.J., *Substituent Constants for Correlation Analysis in Chemistry and Biology*, John Wiley, New York, 1979.

41. Dachs, J.; Bayona, J.M.; Albaigés, J. *Mar. Chem.* In press.

42. Dachs, J.; Bayona, J.M.; Raoux, C.; Albaigés, J. *Environ. Sci. Technol.* **1997**, March issue.

43. Zhong Zhang, Z.; Sparks, D.L.; Scrivner, N.C. *Environ. Sci. Technol.* **1993**, *27*, 1625-1631.

44. Lavelle, J.W.; Ozturgut, E.; Baker, E.T.; Tennant, D.A.; Walker, S.L. *Environ. Sci. Technol.* **1988**, *22*, 1201-1207.

45. Crisp, P.T.; Brenner, S.; Venkatesan, M.I.; Ruth, E.; Kaplan, I.R. *Geochim. Cosmochim. Acta* **1979**, *43*, 1791-1801.

46. Zeng, E.; Yu, C.; Vista,C. Annual Report of the Southern California Coastal Water Research Project, **1993-1994**, 43-54, Westminister, CA, USA.

47. Ministry of Trade and Industry of Japan. Year Book of Chemical Industries Statistics Department. **1986**, pp. 141-155.

Chapter 18

Aliphatic Hydrocarbons, Linear Alkylbenzenes, and Highly Branched Isoprenoids in Molluscs of the Adriatic Sea, Italy

G. P. Serrazanetti, R. Artusi, and C. Pagnucco

Section of Veterinary Biochemistry, Department of Biochemistry, University of Bologna, Via Tolara di Sopra 50, 40064 Ozzano Emilia, Bologna, Italy

Aliphatic hydrocarbons and linear alkylbenzenes (LABs) were determined in four species of molluscs of the Central Adriatic Sea: two bivalves, *Mytilus galloprovincialis, Solen vagina* and two gastropods, *Patella vulgata, Cassidaria echinophora*. The concentrations of total aliphatic hydrocarbons determined in the mollusc samples were between 35.0 and 68.1 μg g^{-1} dry wt, and they may be considered within normal limits for areas reported to be mildly polluted. In *P. vulgata* and in *C. echinophora* over 70% and 50%, respectively, of the total aliphatic hydrocarbons are represented by highly branched isoprenoids (HBIs). In *M. galloprovincialis* 30% is made up of squalene. These hydrocarbons are usually considered of recent biogenic origin, and in particular HBIs found in *P. vulgata* support previous suggestions that these alkenes might be considered molecular markers of the presence of some diatom species. LAB concentrations are between 3.2 and 15.9 μg g^{-1} dry wt and show that all four mollusc species are contaminated by these molecular tracers of domestic wastes. Based on their composition, the LABs appear to have been recently discharged into the marine environment. The high concentration of external isomers indicates that LABs have not been biodegraded. This is probably due to the fact that the wastes were not treated or were only partially treated prior to reaching the sampling area.

Oil pollution in some coastal areas of the Mediterranean Sea is considered to be serious requiring specific studies on the sources, effects, and fate of fossil fuel compounds discharged in this environment (*1*). Research has been carried out to understand the origins of hydrocarbons. In particular, it is of interest to know whether the fossil hydrocarbons indicative of petroleum contamination are present along with recent biogenic hydrocarbons in order to understand what role these compounds

276

might play as biochemical markers in the marine environment. Moreover, it is known that the determination of linear alkylbenzenes (LABs) allows one to verify the presence of municipal waste contamination   because these compounds are manufactured for production of linear alkylbenzenesulfonate (LAS) surfactants used in commercial detergent formulations (*2, 3*).

Previous research has shown that bivalves concentrate many pollutants to a marked degree above sea water concentrations. This ability provides an alternative to collecting large volumes of sea water and transporting them to the laboratory for extraction and analysis (*4*). The determination of hydrocarbons in molluscs is consequently of great interest both because organisms belonging to this phylum play an important role as monitors of contamination in the coastal zone (*4, 5*) and because several species are consumed by humans.

Therefore, we decided to analyze aliphatic hydrocarbons with gas chromatographic (GC) retention times from n-C_{15} to n-C_{30} and LABs in two species of filter-feeding bivalves: *Mytilus galloprovincialis*, widely distributed and frequently utilized in environmental studies, and *Solen vagina*, a species living in the sandy sea floor. Both species are widespread along the Italian coasts, and they are also interesting to analyze as they constitute human seafood (*6*). Also, we deemed it useful to determine the hydrocarbons in two species of gastropod molluscs that occupy different positions in the marine trophic chain (*7*). Specifically, we analyzed specimens of the herbivorous gastropod *Patella vulgata*, which lives on intertidal rocks, and the carnivorous gastropod, *Cassidaria echinophora*, which lives on the sea floor. All specimens of the four species were collected in the Adriatic Sea. This is a semi-enclosed basin with poor exchange, heavy traffic from oil tankers and a lot of human activity associated with urban centers and harbors located along its coasts.

In this work we determined saturated and unsaturated aliphatic hydrocarbons with branched or straight chains and LABs in different species of molluscs because information on concentrations of these compounds in organisms from the Adriatic Sea is limited. This research aims to establish the presence in these organisms of biogenic hydrocarbons, in order to obtain information on the transfer of the organic matter to the marine trophic chain, of fossil hydrocarbons to detect the presence of oil-contamination, and finally of LABs because they are to be considered very important as molecular tracers of domestic wastes.

Materials and Methods

Study Site. All four mollusc species were collected in summer by divers, well away from motor boats, in the central Adriatic Sea about 100 m off the coast opposite the town of Pescara. In this area the water depth is about 5 m, a north/south current is present and in this season the water temperature is always over 20°C (Figure 1). The specimens of *P. vulgata* were sampled on the artificial protective reef made up of natural rocks, those of *M. galloprovincialis* on the same reef at a water depth of about 3 m (subtidal rocks) and those of *S. vagina* and *C. echinophora* on the sandy sea floor at a water depth of about 5 m.

Sample Treatment. All samples were manually collected, washed with distilled water, enclosed in aluminium foil and kept at about 4°C. At the end of the sampling,

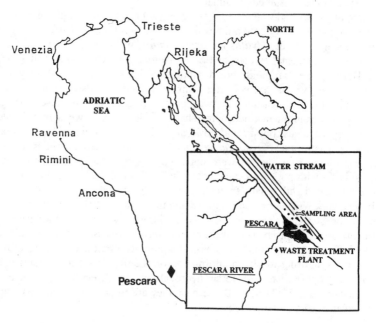

Figure 1 - The sampling area in the central Adriatic Sea coast.

the specimens were frozen and kept at -20°C until dissections were performed. After thawing, the soft tissues were excised, and surplus fluid from the molluscs was discarded. The total soft tissues were homogenized with a Waring blender so that for each of the four species we could prepare 3 aliquots of whole soft parts from which approximately 5g were used for dry weight determination and 100g were used for the lipid extraction. The dry weight of the tissues was determined gravimetrically after freeze drying.

Lipid Extraction. All glassware and steel tools were carefully washed, rinsed with distilled water, acetone, ethyl ether/*n*-hexane then dried at 100°C for an hour at least, (all solvents were obtained from Carlo Erba, Milano, Italy).

The lipids were extracted from wet tissues with chloroform-methanol (2:1) (*8*) with propyl gallate as antioxidant, measured gravimetrically and saponified (40°C) by refluxing for 2 h in 150 ml round-bottom flask with methanolic ($CH_3OH:H_2O = 95:5$) KOH (2N). The flask was cooled at room temperature, and then 50 ml of *n*-hexane-ethyl ether (1:1) were poured into the flask. The contents were transferred to a separatory funnel, and the flask was rinsed with 50 ml of *n*-hexane-ethyl ether (1:1) which was then combined with the digestates in the funnel. Distilled water (100 ml saturated with NaCl) was subsequently added to the funnel. The digestates were shaken for 2 min and allowed to settle. The clearly separated digestates were extracted again twice with 60 ml of *n*-hexane-ethyl ether (1:1). The combined extracts were backwashed three times with 100 ml distilled water. After dehydrating with anydrous sodium sulphate, the combined extracts were transferred to a round-bottom flask and concentrated in a rotary evaporator (40°C) under vacuum.

Hydrocarbon Separation. The hydrocarbons were separated from the other lipid classes of unsaponifiable matter (about 40 mg per plate) by thin layer chromatography (*9*) using *n*-hexane-ethyl ether (1:1) as the mobile phase, and n-C_{19}, squalene, 1-phenyldodecane and chrysene as external standards to check the positions of the hydrocarbons on the plates. After elution, the plates were sprayed with 2,7 dichlorofluoroscein (0.02% in EtOH) and examined under an ultra-violet lamp. The band with Rf higher than 0.90 containing aliphatic hydrocarbons, squalene and LABs was scraped off and extracted with *n*-hexane-ethyl ether (1:1). The aromatic hydrocarbons with two or three rings, which could have been collected along with the unsaturated aliphatic fraction and the polycyclic aromatic hydrocarbons that show Rf values lower than 0.90, were present in amounts not detectable by GC-MS analysis.

Quantitative and qualitative analyses were carried out by means of capillary gas chromatography/mass spectrometry (GC/MS).

Gas Chromatography. The hydrocarbons were analyzed by gas chomatography using a Carlo Erba HRGC 5160 MEGA, equipped with a flame ionization detector (FID) and a fused silica capillary column coated with OV1 (25m x 0.32mm; film thickness = 0.4μm). The oven temperature was programmed from 120°C to 295°C at the rate of 3.5°C min^{-1} and final isothermal hold of 4 min. The injector (splitter) and detector temperature was 310°C, the carrier gas flow was 1.8 ml min^{-1} and the split ratio was 1:80.

Mass Spectrometry (MS). Mass spectra were obtained by GC-MS using either a Carlo Erba Mega QMD 1000, 70 eV coupled to a Carlo Erba MEGA 5300 gas chromatograph with a fused silica capillary column coated with SE52 (25m x 0.25mm; film thickness = 0.20μm) or a Finnigan MAT ITS40 coupled to a Varian 3300/3400 GC with a fused silica capillary column coated with SPB5 (30m x 0.25mm; film thickness = 0.25μm).

The aliphatic hydrocarbon concentrations were determined by GC using n-C_{24} as an internal standard, and the LAB concentrations were determined using two daily injected solutions of 1-phenyldodecane, at known concentration, close to those of the analytes, as an external standard. All samples were injected at least three times and linearity of detector response was periodically checked.

Procedural blanks were determined at the start of the work and for every six samples. Blanks and samples with n-C_{24} and 1-phenyldodecane were processed as real samples. No contaminations were evident as well as no LABs were determined in several samples of macroalgae analyzed with the same method in our laboratory. Recoveries for n-C_{24} were 87 ± 9% and for 1-phenyldodecane were 85 ± 11% (n-3). As we calculated the recoveries only for n-C_{24} among the aliphatic hydrocarbons, and for 1-phenyldodecane among the LABs, none of the data presented here were adjusted for recovery. The detection limits are 200 ng of total aliphatic hydrocarbons and 100 ng of total LABs.

Aliphatic hydrocarbon identification was performed by the use of retention indices and coinjection with a standard mixture containing all n-alkanes from n-C_{12} to n-C_{30} (minus 25, 27, 29 n-alkanes), pristane, phytane and n-$C_{20:1}$. By GC-MS, previous identifications were confirmed, and the identification of the different olefins and of LABs was carried out. The use of columns with slightly different polarity has permitted us to avoid interferences with coeluting peaks. Hereafter, LAB isomers will be symbolized as nφ-C_m, where n = position of phenyl substitution along the linear alkyl chain and m = number of carbon atoms in the alkyl chain.

Total Aliphatic Hydrocarbons

The total aliphatic hydrocarbon concentrations determined in the four species of molluscs are reported in Table I. The concentrations of these compounds appear rather uniform and range between 35.0 and 68.1 μg g $^{-1}$ dry wt.

Our results are difficult to compare with those obtained by other researchers because the only available data concern *Mytilus spp.* and not the species analyzed in this study. However, it is known that the hydrocarbon concentrations reported in the literature for molluscs may vary by as much as a factor of 10^4 (*10*) because the data on these compounds include different kinds of hydrocarbons (saturated, unsaturated, only aliphatic, aliphatic plus aromatic, etc.).

Comparing the concentrations of total aliphatic hydrocarbons in the whole soft tissues of molluscs from the central Adriatic Sea with those in specimens of *M. galloprovincialis* from other areas affected to various degrees by oil pollution, we consider those given in Table I (ranging from 8 to 20 μg g^{-1} wet wt) to be within normal limits for areas regarded as mildly polluted (1-46 μg g^{-1} wet wt) (*11*). However, Ameijeiras et al. (*11*) only examined saturated aliphatic hydrocarbons, whereas the present work also includes unsaturated hydrocarbons of recent biological origin (for instance squalene and highly branched isoprenoids). For this reason, the apparent level of contamination by petroleum of some species examined here is considered to be low. The concentration of n-alkanes found in *M. galloprovincialis* from the Adriatic Sea (17.2 µg g^{-1} dry wt) is similar to that of *M. californianus* (14.8 µg g^{-1} dry wt) exposed to a moderate level of pollution (*12*), and is equivalent to that of *M. edulis* (17.3 µg g^{-1} dry wt) exposed to low levels of chronic oil contamination (12) and lower than those found in certain polluted areas of Australia: Cockburn Sound (50 µg g^{-1}) (13) and Westernport Bay (over 100 µg g^{-1}) (14).

Table I. Concentrations of total lipids, aliphatic hydrocarbons, and linear alkylbenzenes in Molluscs of Adriatic Sea

	Solen vagina	*Mytilus galloprovincialis*	*Patella vulgata*	*Cassidaria echinophora*
wet wt / dry wt	5.05	5.35	3.55	3.35
Total lipids (mg g^{-1} dry wt)	44 ± 6*	68 ± 10*	129 ± 5*	165 ± 16*
Aliphatic hydrocarbons (µg g^{-1} dry wt)	38.7 ± 2.6	35.0 ± 15.8	68.1 ± 19.5	49.3 ± 18.3
(ng mg^{-1} lipid)	879 ± 59	515 ± 232	528 ± 151	299 ± 111
Σ n-alkanes (µg g^{-1} dry wt)	29.8 ± 2.1	17.2 ± 5.9	9.3 ± 2.6	15.2 ± 3.4
ΣLABs (µg g^{-1} dry wt)	15.9 ± 2.9°	6.3 ± 2.6	3.2 ± 1,4	4.7 ± 2.7
(ng mg^{-1} lipid)	361 ± 85	93 ± 38	25 ± 11	29 ± 17
I/E ratio	1.12	0.98	0.91	1.18

Values are means ± SD (n = 3)
* All the values are significatively different (P < 0.05)
° Value significatively higher than others (P < 0.05)
I/E = [6ϕ-C$_{12}$ + 5ϕ-C$_{12}$] / [4ϕ-C$_{12}$ + 3ϕ-C$_{12}$ + 2ϕ-C$_{12}$]

In *M. galloprovincialis* from the western Mediterranean others (*15*) have reported petrogenic aliphatic hydrocarbon concentrations of 25-40 µg g^{-1} dry wt and have suggested that these may represent background concentrations of aliphatic hydrocarbons occurring near oil platforms. The concentration of total aliphatic hydrocarbons (including biogenic hydrocarbons) in *M. galloprovincialis* from the Adriatic Sea is similar to those reported for the western Mediterranean (*15*) and generally lower than the concentration of complex hydrocarbon mixtures, principally of petroleum (130 ± 109 μg g^{-1} dry wt), determined in other samples of the same species from the vicinity of Ebro Delta (Spain) (*16*).

Hydrocarbon Profiles. The examination of individual hydrocarbons in aquatic organisms is critical for establishing whether there is possible petroleum contamination of the samples or the presence of recent biogenic hydrocarbons. The distinction is possible by means of a careful examination of the characteristic differences between hydrocarbons from living organisms and from oil pollution (*1*).

Of the four species tested, only *S. vagina* exhibited a hydrocarbon distribution dominated by petroleum contamination. In fact, in this mollusc, only small quantities of biogenic hydrocarbons were found, while phytane was identified and an unresolved complex mixture (UCM) was evident (Table II; Figure 2B). The presence of the isoprenoid hydrocarbon, phytane, a significant UCM and both even and odd n-alkanes with CPI (from n-C$_{19}$ to n-C$_{30}$) near unity is considered to be indicative of petroleum contamination (*1, 17-19*). It is known than the isoprenoid hydrocarbons like pristane and phytane are degraded more slowly than the straight hydrocarbons. The presence of these paired compounds at relatively high proportions along with the UCM (Table II; Figure 2B) in this detritivorous benthonic mollusc is comparable with patterns observed for sediments contaminated by biodegraded petroleum hydrocarbons (*19*).

In the other three species, quite low quantities of phytane were present, the UCM was less abundant and the CPI was generally greater than in *S. vagina*, with the highest value (3.0) in *P. vulgata*. Based on these observations, we believe that there is limited petroleum contamination of these species (Table II; Figure 2).

Aside from the high quantities of branched alkenes in *P. vulgata* and in *C. echinophora* and of squalene in *M. galloprovincialis*, the amounts of biogenic hydrocarbons in all four species appear minimal. For example, pristane, biosynthesized in the marine environment by zooplankton (*20*) and isolated in high quantities in the northern Adriatic Sea in the planktivorous fish *Sardina pilchardus* (*21*) and in zooplankton (*22*), was always found in very low percentages in the mollusc species analyzed here. The highest amount was found in *S. vagina* (5.1%), but in this species, like in all the others, it was present in concentrations lower than or close to that of phytane (*1*).

In the present study we also isolated very small quantities of *cis*-3,6,9,12,15,18-heneicosahexaene (HEH), a hydrocarbon which, in zooplankton and in a teleost from the northern Adriatic Sea, had been found at percentages near 50% of the total hydrocarbons determined (*21, 22*). The almost total absence of this 21:6 olefin is particularly significant in the case of the two filter-feeding molluscs which would be expected to have incorporated it from the several diatom species present in blooms that characterize the Adriatic Sea (*23, 24*). However, algal blooms usually occur in the Adriatic Sea in spring and in fall, while the Pescara samples were collected in summer.

Other hydrocarbons characteristic of phytoplankton (*25*) and macroalgae (*26*), such as n-C$_{17}$ and C$_{17}$ olefins were absent as well. These hydrocarbons have been found to be abundant in macroalgae (*27*), zooplankton (*22*), a sea urchin (*28*) and a

Table II. Aliphatic hydrocarbons in molluscs of Adriatic Sea

	Solen vagina	*Mytilus galloprovincialis*	*Patella vulgata*	*Cassidaria echinophora*
n-C_{15}	1.1 ± 0.7	tr	0.9 ± 0.5	tr
n-C_{16}	4.5 ± 0.0	2.4 ± 1.0	0.7 ± 0.2	2.0 ± 0.1
n-$C_{17:1}$	-	-	1.2 ± 0.3	-
n-$C_{17:1}$	-	-	0.9 ± 0.1	-
n-C_{17}	10.8 ± 4.5	3.0 ± 1.4	1.3 ± 0.2	3.0 ± 1.0
Pristane	5.1 ± 0.8	1.1 ± 0.5	0.5 ± 0.4	1.9 ± 0.7
n-C_{18}	13.3 ± 2.7	3.6 ± 1.2	1.1 ± 0.4	3.1 ± 0.1
Phytane	12.9 ± 2.3	3.1 ± 1.5	1.0 ± 0.6	3.2 ± 0.3
n-C_{19}	8.8 ± 2.1	5.8 ± 2.2	0.5 ± 0.2	2.0 ± 0.1
n-C_{20}	5.0 ± 0.3	tr	-	2.0 ± 0.6
n-$C_{21:6}$	-	0.9 ± 1.3	-	-
n-C_{21}	3.2 ± 0.1	2.5 ± 0.1	3.3 ± 0.1	1.8 ± 0.7
n-C_{22}	4.2 ± 0.4	2.8 ± 1.0	0.5 ± 0.1	2.2 ± 1.0
n-C_{23}	4.0 ± 0.1	3.1 ± 0.8	0.7 ± 0.2	2.4 ± 1.7
n-C_{24}	4.2 ± 0.5	4.5 ± 1.3	1.0 ± 0.4	2.3 ± 1.7
n-C_{25}	4.6 ± 1.4	6.7 ± 1.6	0.9 ± 0.3	3.0 ± 0.4
n-C_{26}	6.8 ± 2.3	4.7 ± 1.1	0.5 ± 0.2	2.4 ± 2.7
n-C_{27}	1.8 ± 1.3	3.7 ± 1.2	1.4 ± 0.2	1.4 ± 1.4
n-C_{28}	2.0 ± 1.5	2.2 ± 0.8	0.4 ± 0.2	1.3 ± 1.2
Squalene	5.4 ± 4.2	28.9 ± 3.7	2.8 ± 0.1	7.9 ± 8.2
n-C_{29}	1.4 ± 0.9	3.0 ± 0.3	0.4 ± 0.1	1.2 ± 1.0
n-C_{30}	1.2 ± 0.6	1.1 ± 0.6	tr	0.8 ± 0.5
$\Sigma C_{25:?}$	-	7.0 ± 0.2	0.7 ± 0.1	3.9 ± 0.0
$\Sigma C_{25:2}$	-	3.6 ± 0.4	42.6 ± 1.6	21.4 ± 4.9
$\Sigma C_{25:3}$	-	3.6 ± 0.2	11.4 ± 0.0	24.8 ± 8.2
$C_{25:4}$	-	-	19.7 ± 1.0	1.9 ± 0.0
$C_{25:5}$	-	-	2.8 ± 0.2	0.9 ± 0.6
$C_{30:x}$	-	2.0 ± 2.9	0.7 ± 0.5	3.1 ± 0.0
CPI	1.0	1.6	3.0	1.1

All values are expressed as % of total hydrocarbons - Values are mean ± SD (n=3)
tr = trace. un = unidentified. CPI = carbon preference index from n-C_{19} to n-C_{30} (*19*)

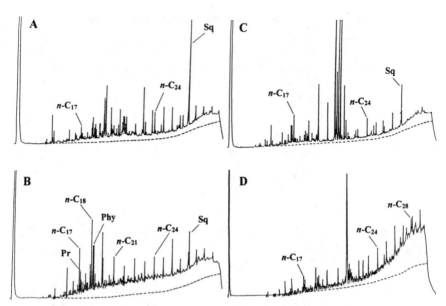

Figure 2 - Gas chromatograms of the aliphatic hydrocarbons isolated from molluscs of the Adriatic Sea - A., *Mytilus galloprovincialis*. B., *Solen vagina*. C., *Patella vulgata*. D., *Cassidaria echinophora*.
Pr., Pristane. Phy., Phytane. Sq., Squalene. ----- baseline

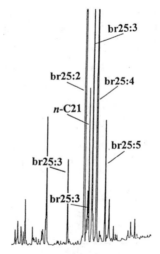

Figure 3 - Gas chromatogram of the highly branched hydrocarbons isolated from *P. vulgata*

planktivorous fish (*21*) of the northern Adriatic. The lack of these hydrocarbons in the molluscs of the central Adriatic Sea and the presence of other biogenic hydrocarbons (see below) confirms the role of the different hydrocarbons as molecular tracers in the marine trophic chain.

Highly Branched Hydrocarbons

The highly branched isoprenoid (HBI) alkenes with 20, 25, 30 carbon atoms and varying degrees of unsaturation (from 1 to 5) are considered biogenic hydrocarbons (*29*). In fact, most authors have assumed cyclic or branched alkenes to be of natural biological origin rather than pollutants, since several researchers had found them in recent sediments, sediment trap particles and animal organisms (*29*) but particularly in different diatom communities (30, 31) and in laboratory cultures of two marine diatoms (*32*).

As mentioned above, the highest concentration of aliphatic hydrocarbons was found in the herbivore, *P. vulgata* (Table I), but in this species more than 70% was represented by C_{25} HBI alkenes. The HBI alkenes were also found in high proportions (over 50%) in *C. echinophora* and in lower percentages in *M. galloprovincialis* (about 10%) (Table II, Figure 2). However, in *M. galloprovincialis* this percentage, added to that of squalene (28.9%) (Table II), exceeded 40% of the total aliphatic hydrocarbons determined. The high percentages of these recent biogenic hydrocarbons in *P. vulgata*, *C. echinophora* and *M. galloprovincialis* suggest that the presence of petroleum residues in these species is very limited.

The large quantities of C_{25} HBIs in *P. vulgata* (Table II, Figures 2C, 3) would seem to confirm their role as biological markers. The samples of this mollusc were collected on rocks where diatoms containing HBIs probably were living. Because only a small number of diatoms are known to synthesize these unusual hydrocarbons (*30-32*), further studies of the microflora present in this environment may allow us to relate the HBIs in *P. vulgata* to its food sources. These sources could also contribute to the particulate matter consumed by *M. galloprovincialis*.

High proportions of HBIs were also found in the carnivorous mollusc, *C. echinophora* (Table II), living on the sea floor. In this species the quantities of br25:4 and br25:5 were lower than those of *P. vulgata*. This may be due to the loss of these more unsaturated alkenes during their transfer to the food chain or to a different source. These results offer insights to the environment where *C. echinophora* is living, and Echinoderma, which are the main food source of this species.

HBIs were not observed in the filter-feeding mollusc, *S. vagina*, which lives in the surface layers of sediments (Table II). The lack of HBIs in this mollusc is difficult to explain but may reflect a difference in diet or metabolism.

Linear Alkylbenzenes

Unlike many toxic organic substances, such as polychlorinated or polynuclear aromatic hydrocarbons, long-chain linear alkylbenzenes (LABs) have been shown to be useful as waste-specific markers for detection of sludge at sea (*33*). The LABs have been synthesized since the mid 1960s in the production of linear alkylbenzenesulfonates (LAS), the most widely used anionic surfactants. LABs are found as unsulfonated residues in detergents along with the LAS (*34*). Although it is believed the LABs do not represent an environmental threat (*34*), they degrade much more slowly than LAS (*3*) and accumulate in marine sediments (*35*). For this reason and owing to the fact that LABs are always present along with the alkylbenzene sulfonates in the common domestic and industrial detergents, they have been suggested as specific molecular markers of anthropogenic waste contamination (*2, 3, 36, 37*).

Several studies have been carried out on LABs. They have been isolated in sediments (*2, 3, 33-41*), sediment trap particles, in Mediterranean fish (*42*) and in mussels (*5, 43*). Although many reports on the presence of these molecular tracers of domestic wastes in sediments are available, data on their occurrence in molluscs (particularly *Mytilus* spp.) are limited. Therefore, we deemed it interesting to measure LABs in the same four mollusc samples collected along the coast opposite the town of Pescara.

LABs were found in all four species of molluscs (Table I). The highest concentration of these compounds was found in *S. vagina*. Among the other species, the lowest concentrations were observed in the herbivore *P. vulgata* (Table I). Concentrations in *M. galloprovincialis* (6.3 μg g^{-1} dry wt, 93 ng mg^{-1} lipid) are lower than those determined in *M. edulis* from Port Philip Bay, Victoria, Australia, collected near an oil refinery with a LAB plant (23-39 μg g^{-1} dry wt) (*5*) and lower than those found in the digestive gland tissues of specimens from Boston Harbor (125 to 275 ng mg^{-1} lipid) (*43*).

The observation of higher concentrations of these hydrocarbons in *S. vagina* compared with those of the other species from the Adriatic Sea is consistent with the feeding behavior of this detritivorous mollusc. It lives in the surface layers of sediments where LABs are likely to be concentrated. The relatively low concentrations in *P. vulgata* may be explained by the essentially plant diet of this mollusc living on the intertidal rocks. With regard to this point it is important to emphasize that LABs were not isolated from samples of *Ulva lactuca* collected on the same reef (Serrazanetti *et al.*, *in preparation*) and in other macroalgae from the Adriatic Sea (*27, unpublished results*).

Although the differences are not statistically significant and the data are limited, the higher quantities of LABs in the carnivore *C. echinophora* compared with those found in the herbivore *P. vulgata* could mean the possibility of accumulation of these compounds in species belonging to the highest levels of the trophic chain.

Among the various surfactant residues reaching the sea, some are mainly composed of linear alkylbenzenes with 13 carbon chains (C_{13}LABs) (*40*) and others by C_{12}LABs and/or C_{11}LABs (*2, 40*). In all four species from the Adriatic Sea, C_{13}LABs are dominant although relatively large amounts of C_{14} LABs are also observed (Figure 4).

In Italy most LABs are synthesized by alkylation of benzene using chloroparaffins and AlCl$_3$. The LABs derived from this process, are dominated by homologs having 11 and 12 carbon aliphatic chains (31% and 32% respectively). Those with 13 carbon chains show a lower percentage (about 22%) while those with 10 carbon chains are 15% (data based on the analysis of 35 common commercial Italian detergents) (*44*). Much smaller quantities of LABs are produced in Italy by alkylation of benzene with olefins and HF as a catalyst. In this case, LABs with 11 or 12 carbon chains are still dominant, while homologs with 14 carbon chains are almost absent.

The I/E ratio (I= isomers with n= 5 and 6; E= isomers with n= 2, 3 and 4) calculated for C_{12}LABs in 14 Italian commercial detergents is 0.73±0.15 (*44*). The I/E ratios reported here range from 0.91 to 1.18 show that LABs identified in the four species of molluscs are almost undegraded (*38*). In fact, aerobic microorganisms degrade LAB isomers with the phenyl group closer to the end of the alkyl chain (E) most rapidly (*38, 45*) The highest values of I/E ratio were found in the two mollusc species which live on the sea floor and where LABs may be more degraded.

A plant for treatment of sludge is located about 4 km from Pescara. After being treated, the wastes flow into Pescara river and finally to the Adriatic Sea (Figure 1). The low I/E ratios, an indication of a limited LAB degradation, in the molluscs from Pescara coast, may be explained by the presence of southward flowing currents along the Adriatic coast that would take the municipal treated wastes fr from the sampling area located to the north of the river mouth; wastes produced to the north of Pescara town, where there are several hotel settlements and where there are no plants for treatment of wastes probably reach the protective reef. These wastes, untreated or only partially treated, because of the strong currents and the limited water volume (a few meters of water depth) would rapidly reach the sampling area and would be responsible for the contamination of the samples. Under these conditions extensive LAB degradation would not occur as it does when particles settle through the water column, over days or even weeks (*2*). In addition to the rapid transfer of LABs from the sources to the sampling area, in summer in the Adriatic Sea, and mainly in the coastal areas, anoxia (*24*) may inhibit the activity of aerobic bacteria responsible for degradation of these compounds. This process also helps explain the particular distribution of LAB isomers in the molluscs analyzed.

In all four mollusc species from the Adriatic Sea C_{13} LAB homologs dominated along with high percentages of C_{14} LABs, with the exception of *P. vulgata* (Figure 4). These distributions do not correspond to the Italian surfactant composition where C_{11-12} LABs and LAS are dominant (*44*). The molluscs were collected in an environment characterized by shallow water (about 5m) and high water temperature (over 20°C). In this condition the high vapor-pressure of alkylbenzenes with the short alkyl chain (*46*) could involve loss of $C_{10-11-12}$ LABs and an enrichment of C_{13} LAB homologs. This is consistent with the distribution of LABs in *P. vulgata* (Figure 4), where C_{13} LABs are dominant and where this composition reflects only that present in water because these hydrocarbons were not found in the macroalgae living on the rocks where *P. vulgata* feeds. While C_{14} LABs were limited in *P. vulgata*, these and remarkably C_{13} LABs were isolated in very high amounts in the two filter-feeding *S. vagina* and *M. galloprovincialis* (Figure 4) whereas the more volatile C_{10} LABs were practically absent in these organisms.

C_{13} and C_{14} LABs are less volatile and less soluble in water than $C_{10-11-12}$ LABs (*46*) and, owing to their higher hydrophobicity, they sorb more readily to the particulate matter that reaches filter-feeding organisms. This may explain the higher C_{13} and C_{14} LAB levels determined in the two filter-feeding molluscs (e.g. *S. vagina*) living on the sea floor. In *C. echinophora*, the C_{10} LABs are absent and C_{13} LABs are abundant. This distribution is difficult to explain because it is probably affected by the composition of LABs in the water and in the diet of this carnivorous species.

The accumulation of the hydrophobic LAB homologs in molluscs is related to the

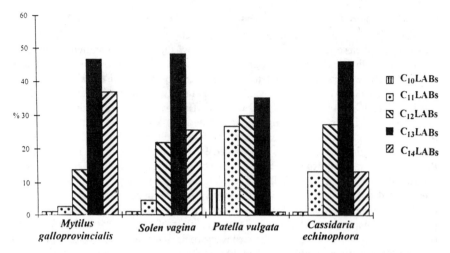

Figure 4 - Distribution of LAB homologous groups (as % of total LABs)

dietary lipids, to sorption processes, that in the filter-feeding molluscs occurs by the gills, and to the physiological fluctuations of these constituents. The lower LAB concentrations in the lipids of the two gastropods than in those of the two bivalves (Table I) may be explained with the different metabolic and physiological lipid pathway in the species belonging to two different classes of molluscs.

Conclusions

The aliphatic hydrocarbon concentrations determined in four species of molluscs sampled in the central Adriatic Sea are equivalent to those of areas considered mildly polluted with oil. Only the benthonic mollusc, *S. vagina,* was heavily contaminated by hydrocarbons of petroleum origin.

Large quantities of HBIs of autochthonous origin were determined in the herbivore *P. vulgata,* and in the carnivore, *C. echinophora,* while squalene was the dominant hydrocarbon in *M. galloprovincialis.* None of the four species showed high concentrations of other hydrocarbons of recent biogenic origin .

In all the four species analyzed in this study we could identify LABs which are considered markers of domestic wastes with the highest concentrations in the two filter-feeding species. The isomeric distribution of these compounds suggests that they have not undergone significant biodegradation. The composition of different homologs, with higher concentrations of longer chain LABs, appears to reflect physical processes such as volatilization and sorption.

Our results confirm the utility of molluscs as sentinel organisms in environmental research and of hydrocarbons as molecular tracers in aquatic environmental studies.

Acknowledgments - The authors thank Dr Robert Eganhouse for his criticisms, suggestions and encouragements and Ms Gladys Putaturo for sample collection and preparation. This work was supported by a Grant from the M.U.R.S.T. (60%), Roma.

Literature cited

1. Farrington, J. W.; Meyer, P. A. In: *Environmental Chemistry*, Eglinton G. Ed.; The Chemical Society, London, En, **1975**; vol. 1; Chapter 5; pp. 109-136.
2. Eganhouse, R. P.; Blumfield, D. L.; Kaplan, I. R. *Environ. Sci. Technol.* **1983**, *17*, 523-530.
3. Ishiwatari, R.; Takada, H.; Yun, S. J. *Nature* **1983**, *301*, 599-600.
4. Goldberg, E. D.; Bowen, V. T.; Farrington, J. W.; Harvey, G.; Martin, J. H.; Parker, P. L.; Risebrough, R. W.; Robertson, W.; Schneider, E.; Gamble, E. *Environ. Conservation* **1978**, *5*, 101-125.
5. Murray, A. P.; Richardson, B. J.; Gibbs, C. F. *Mar. Poll. Bull.* **1991**, *22*, 595-603.
6. Grzimek, B. In *Vita degli Animali*; Bramante Ed.; Varese, It, **1972**, vol.3, pp 149-195.
7. Fischer, W.; Schneider, M.; Bauchot, M.-L. *Mediterranee et Mer Noire;* FAO, Roma, It, **1987**; vol.1.
8. Bligh, E. G.; Dyer, W. J. *Can. J. Biochem. Physiol.* **1959**, *37*, 911-917 .
9. Paoletti, C.; Pushparaj, B.; Florenzano, G.; Capella, P.; Lercker, G. *Lipids* **1975**, *10*, 258-265.
10. Nevenzel, J. C. In *Marine Biogenic Lipids, Fats and Oils;* Ackman, R. G. Ed.; CRC Press, Boca Raton, Fl, **1989**, Vol. II; pp. 3-71.
11. Ameijeiras, A. H.; Gandara, J. S.; Hernandez, J. L.; Lozano, J. S. *Mar. Poll. Bull* **1994**, *28*, 396-398.
12. Clark, R. C.; Finley, J. S. *Mar. Poll. Bull.* **1973**, *4*, 172-176.
13. Alexander, R.; Cumbers, M.; Kagi, R.; Offer, M.; Taylor, R. In *Chemistry and analysis of hydrocarbons in the environment* Albaigés, J.; Frei, R. W.; Merian, E. Eds.; Gordon and Breach Science Publishers, New York, U.S., **1981**, Vol. 5; pp 273-297.
14. Burns, K. A.; Smith, J. L. *Estuar. Coast. Shelf Sci.* **1981**, *13*, 433-443.
15. Soler, M.; Grimalt, J. O.; Albaigés J. *Chemosphere* **1989**, *18*, 1809-1819.
16. Risebrough, R. W.; de Lappe, B. W.; Walker II, W.; Simoneit, B. R. T.; Grimalt, J.; Albaigés, J.; Regueiro, J. A. G.; Nolla, A. B.; Fernandez, M. M. *Mar. Poll. Bull.* **1983**, *14*, 181-187.
17. Farrington, J. W.; Davis, A. C.; Frew, N. M.; Rabin, K. S. *Mar. Biol.* **1982**, *66*, 15-26.
18. Saliot, A. In *Marine Organic Chemistry*; Duursma, E. K.; Dawson, R. Eds.; Elsevier, Amsterdam, NL, **1981**, pp. 327-374.
19. Aboul-Kassim, T. A. T.; Simoneit, B. R. T. *Mar. Poll. Bull.* **1995**, *30*, 63-73.
20. Corner, E. D. S. In: *Adv. Mar. Biol.*, Russell, F. S.; Yonge, M. Eds.; Academic Press, London, En, **1978**; 15, 289-380.
21. Serrazanetti, G. P.; Conte, L. S.; Cortesi, P.; Totti, C.; Viviani, R. *Mar. Chem.* **1991**, *32*, 9-18.
22. Serrazanetti, G. P.; Pagnucco, C.; Conte, L. S.; Artusi, R. *Chemosphere* **1994**, *28*, 1119-1126.
23. Blumer, M.; Guillard, R R L.; Chase, T. *Mar. Biol.* **1971**, *8*, 183-189.
24. Marine Coastal Eutrophication, Vollenweider, R. A.; Marchetti, R.; Viviani, R. Eds.; Science of the Total Environment, Supplement **1992**; Elsevier, Amsterdan, Nl.

25. Shaw, D. G.; Wiggs, J. N. *Phytochemistry* **1979**, *18*, 2025-2027.
26. Yougblood, W. W.; Blumer, R. *Mar. Biol.* **1973**, *21*, 163-172.
27. Serrazanetti, G. P.; Conte, L. S.; Pagnucco, C.; Bergami, C. *It. J. Biochem.* **1992**,*41*, 264A-265A.
28. Serrazanetti, G. P.; Pagnucco, C.; Conte, L. S.; Cattani, O. *Chemosphere*, **1995**, *30*, 1453-1461.
29. Rowland, S. J.; Robson, J. N. *Mar. Environ. Res.* **1990**, *30*, 191-216.
30. Nichols, P. D.; Volkman, J. K.; Palmisano, A. C.; Smith, G. A.; White, D. C. *J. Phycol.* **1988**, *24*, 90-96.
31. Hird, S. J.; Rowland, S. J. *Mar. Environ. Res.* **1995**, *40*, 423-437.
32. Volkman, J. K.; Barrett, S. M.; Dunstan, G. A. *Org. Geochem.* **1994**, *21*, 407-413.
33. Bothner, M. H.; Takada, H.; Knight, I. T.; Hill, R. T.; Butman, B.; Farrington, J. W.; Colwell, R. R.; Grassle, J. F. *Mar. Environ. Res.* **1994**, *38*, 43-59.
34. Eganhouse, R. P.; Ruth, E. C.; Kaplan, I. R. *Analyt. Chem.* **1983**, *55*, 2120-2126.
35. Chalaux, N.; Bayona, M. I.; Venkatesan, M. I.; Albaigés, J. *Mar. Poll. Bull.* **1992**, *24*, 403-407.
36. Eganhouse, R. P.; Olaguer, D. P.; Gould, B. R.; Phinney, C. S. *Mar. Environ.Res.* **1988**, *25*, 1-22.
37. Eganhouse, R. P. *Intern. J. Environ. Anal. Chem.* **1986**, *26*, 241-263.
38. Takada, H.; Ishiwatari R. *Environ. Sci. Technol.* **1990**, *24*, 86-91.
39. Eganhouse, R. P.; Kaplan, I. R. *Mar. Chem.* **1988**, *24*, 163-191.
40. Takada, H.; Ishiwatari R. *Environ. Sci. Technol.* **1987**, *21*, 875-883.
41. Raymundo. C. C.; Preston, M. R. *Mar. Poll. Bull.* **1992**, *24*, 138-146.
42. Albaigés, J.; Farran, A.; Soler, M.; Gallifa, A. *Mar. Environ. Res.* **1987**, *22*, 1-18.
43. Sherblom, P. M.; Eganhouse, R. P. In *Organic Substances and Sediments in Water*; Baker, R. A. Ed.; Lewis Publishers, 1991; Vol. 3.
44. Ballarin, B.; Stelluto, S.; Marcomini, A. *Rivista Ital. Sostanze Grasse* **1989**, *66*, 349-353.
45. Bayona, J. M.; Albaigés, J.; Solanas, A. M.; Grifoll, M. *Chemosphere* **1986**,*15*, 595-598.
46. Sherblom, P. M.; Gschwend, P. M.; Eganhouse, R. P. *J. Chem. Eng. Data* **1992**, *37*, 394-399.

Chapter 19

2-(4-Morpholinyl)benzothiazole as an Indicator of Tire-Wear Particles and Road Dust in the Urban Environment

Hidetoshi Kumata, Hideshige Takada, and Norio Ogura

Laboratory of Aquatic Environment Conservation, Faculty of Agriculture, Tokyo University of Agriculture and Technology, 3–5–8 Saiwai, Fuchu, Tokyo 183, Japan

As an indicator of tire wear debris and road dust, 2-(4-morpholinyl)-benzothiazole (24MoBT) was determined in environmental samples collected from urban Tokyo in 1989 and 1993/94. 24MoBT existed in tire tread rubber, road dust, runoff particles, river water particles, river sediment and aerosols in the widely varying concentrations (~ng g^{-1} to ~μg g^{-1}). Comparison of 24MoBT concentrations in tire tread with those in environmental samples showed that tire debris were a significant component of ambient particulate matter. The contribution of tire debris to runoff was greatest during the early and middle stages of storms. Comparison of 24MoBT concentrations in road dust samples and river samples collected in 1993/94 showed that road dust may comprise as much as 28-65% of suspended river particles during storm-flow and 5.5-41% of river sediments.

Various pollutants (e.g., heavy metals and polycyclic aromatic hydrocarbons) are present in road dust on urban road surfaces. These pollutants are derived from weathered materials of the road surface, various types of vehicle emissions (e.g., exhaust, crank case oil, uncombusted fuel, abraded tire tread debris, brake linings, etc.) and atmospheric deposition (*1-4*). Road dust is washed off road surfaces during heavy rain storms and transported to rivers, wastewater treatment plants, and estuaries (*5-11*). Consequently, road dust and road runoff are considered a major source of various pollutants in aquatic environments (*6, 12-16*).

It is often difficult to differentiate and to quantify the magnitude and the contributions from runoff and/or road dust to aquatic environments. A molecular marker would greatly aid in the assessment of the relative importance of this source. We focused on 2-(4-morpholinyl)benzothiazole (24MoBT) as a molecular marker of road dust and/or runoff (Figure 1a). 24MoBT is a minor component of a commercially used vulcanization accelerator, OBS. OBS consists mainly of 2-morpholinothiobenzo-thiazole (Figure 1b), and was one of the most widely used vulcanization accelerators, used predominantly in tire tread rubber. Because 24MoBT is contained in tire tread rubber, it is deposited on road surfaces as vehicle tire abrasion debris (referred to as tire wear particles). Hence, it is primarily a marker of tire particles, which finds its way into various media (road dust, runoff particles, river sediments, aerosols, etc.) as part of the transport process. Measuring 24MoBT in such media would allow us to trace not only the process of tire wear particle transportation, but also of road dust transportation as tire wear particles are primarily accumulated on road surfaces. Tire particles have attracted attention as a possible source of atmospheric aerosols (17-25). Recently, Rogge et al. (4) investigated high molecular weight n-alkanes as molecular markers of tire wear particles in aerosols. However, thus far no studies have demonstrated the utility of tire wear particles to differentiate sources of pollutants in aquatic environments.

Spies et al. (26) demonstrated 24MoBT's existence in a vulcanization accelerator, roadside soils, and estuarine sediments. They also showed the persistent nature of 24MoBT relative to the parent compound (i.e., 2-morpholinothiobenzothiazole) and proposed that 24MoBT might be used as a molecular marker of road dust and/or road runoff.

Since that work, however, there have been few papers on the environmental distribution and fate of 24MoBT, and its utility as an indicator has not been evaluated (26, 27). This is due in part to the lack of a quantitative analytical method for determination of 24MoBT concentration in environmental samples. In a previous publication, we presented a quantitative method for analyzing 24MoBT concentrations (28). In the present study, we present 24MoBT concentrations in a wider variety of environmental samples as well as in tire tread rubber, and then relate the abundance of the marker in the source material to that in environmental samples.

The purpose of this study is (1) to examine the distribution of 24MoBT in the urban environment of Tokyo, and (2) to apply 24MoBT as an indicator of tire wear particles and of road dust to estimate their contributions to ambient particulate matter (i.e., road dust, runoff particles, river water particles, river sediment and aerosols).

Experimental.

Sample Collection. The study was conducted in the Tokyo metropolitan area. Sampling locations are shown in Figure 2. Tokyo, with a population of nearly twelve million, is the largest city in Japan and it has high traffic density. Consequently, the adjoining aquatic environments are expected to receive a significant amount of pollution related to motor vehicles and/or road dust delivered via urban runoff.

Road dust was collected from heavily traveled roads in Tokyo carrying >10,000 vehicles day[-1]. To examine the contributions from atmospheric fallout and soil, road dust samples were collected from both inside and outside tunnels. Samples from outside tunnels were taken in 1989 and in 1994, and from inside tunnels in 1989 (Yaesu Tunnel) and 1994 (Shinjukugyoen-shita Tunnel and Yaesu Tunnel). All road dust samples were collected on prebaked (400 °C, 3 hours) glass fiber filters (ADVANTEC GB100R or Whatman GF/C: >1 μm) by vacuuming the road surface. The filters and particles were stored at -20 °C until analysis. Details of the sampling locations for the 1989 samples are given elsewhere (2).

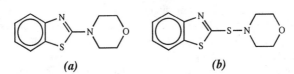

Figure 1. Molecular structures of (**a**) 2-(4-Morpholinyl)-benzothiazole (24MoBT) (MW = 220.30; CAS: 4225-26-7), an impurity of the vulcanization accelerator, OBS, and (**b**) 2-Morpholinothiobenzothiazole (MW = 252.36; CAS: 102-77-2), a major component of the vulcanization accelerator, OBS. OBS are name of vulcanization accelerators defined in the Japan Industrial Standard (JIS K 6202-1979).

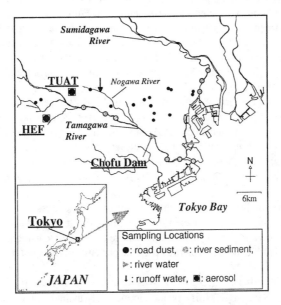

Figure 2. Sampling locations in the Tokyo metropolitan area.

Street runoff was collected in 1989 from a storm drain discharging into the Nogawa River, a tributary of the Tamagawa River. The hydrograph of this drain is illustrated in Figure 3, and it shows the variation of rain intensity and water flow as functions of elapsed time for the runoff event, as well as the variation of suspended solid concentration. The event lasted more than 9 hours, and 10 water samples were collected during the storm at 0.5- to 2-hour intervals. There were two distinctive peaks in flow rate: 3.5 and 6.5 hours from the beginning of this storm event. Suspended solid concentration showed its largest peak at the beginning of the storm in spite of the weak drain flow, and two subsequent distinctive peaks associated with flow rate. Some other aspects of these samples have been given elsewhere (9). The first, fifth and ninth samples (referred to as the initial, middle and final stages, respectively) were analyzed for 24MoBT.

During the period June 1993 to September 1994, river water samples were collected at the Chofu dam located just above the upper limit of the tidal zone of the Tamagawa River. They are categorized into two groups: normal-flow samples, collected at water flows less than 20 m^3 s^{-1}, and storm-flow samples, collected at water flows greater than 20 m^3 s^{-1}. Samples were transported to the laboratory and immediately filtered through prebaked glass fiber filters (GF/C). The filters containing particles were wrapped in aluminum foil and then stored at -20 °C until analysis.

River sediments were taken from the Tamagawa and Sumidagawa Rivers in 1993 using an Ekman-Birge grab sampler (15 X 15 X 15 cm) and the thickness of sediment samples were <10 cm approximately. These are two of the largest rivers discharging into Tokyo Bay. The sediments were transferred into polypropylene containers and stored at -20 °C until analysis.

The atmospheric aerosol samples were collected in January 1994, from the roof-top of our university (referred to as TUAT) and in an experimental forest (Hakyu-chi Experimental Forest, referred to as HEF) in suburban Tokyo. These sites are located about 20 km and 30 km, respectively, from downtown Tokyo. Aerosol samples were collected on prebaked quartz fiber filters (Pallflex 2500 QAT-UP: >0.1 μm) using a high volume air sampler (Shibata HVC-1000) at a flow rate of 1.28 m^3 min^{-1} for 48 hours.

Used automobile tires of four different tire manufacturers, representative of the market share in Japan, were collected in 1989 to provide tire tread samples. Tire tread rubber samples were collected by abrading the tire tread surface using a metal rasp, and stored at -20°C until analysis. The rasp was rinsed with solvents before use.

Analytical Procedure.   24MoBT was analyzed using the previously reported procedure (28). Because 24MoBT is not commercially available, we extracted it from a vulcanization accelerator by liquid-liquid extraction with dilute sulfuric acid. The vulcanization accelerator, OBS (trade name; MSA), was kindly offered by Ohuchi Shinko Kagaku Inc. (Fukushima, Japan). Its purity was confirmed by GC/MS (gas chromatography/mass spectrometry) analysis. The 24MoBT in this solution was quantified by comparing its peak area on an FID chromatogram to that of benzothiazole (BT). Therefore, the concentrations of 24MoBT are presented as BT concentration in this study, and this stock solution was used as the 24MoBT standard.

A dried sample was Soxhlet-extracted with a 6:4 (v/v) mixture of benzene and

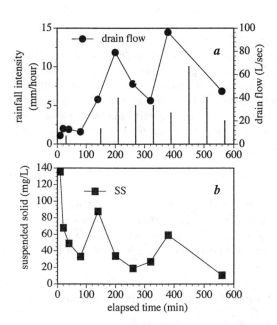

Figure 3. Variation of urban runoff parameters as a function of elapsed time for the storm of Oct 19, 1989. (**a**) Rain intensity and drain flow, (**b**) suspended solids.

methanol for 24 h. The cycling rate for the extraction was 8 min/cycle. (*Caution: Because benzene is a known carcinogen, all handling procedures should be done in a fume hood.*) The extract was concentrated and subjected to liquid-liquid extraction with sulfuric acid (4 X 10 mL of 0.18 M H_2SO_4: pH<2). The combined aqueous solution was made basic (pH>12), back-extracted with dichloromethane (4 X 5 mL), and then purified through a 5% H_2O deactivated silica gel column (1 cm i.d. X 9 cm). The 24MoBT fraction was concentrated to just dryness and redissolved in an appropriate volume of benzene for injection into the gas chromatograph. Concentrations of 24MoBT were determined using a Hewlett Packard 5890 series II GC with a flame photometric detector (FPD). The detection limit of the method was 0.08 ng of injected 24MoBT. Reproducibility and recovery were 1.5% and 85.4 ± 5.2%, respectively (n=4).

Identifications of 24MoBT in the stock solution and ambient samples (i.e., road dust, runoff particles, river particles, river sediments, aerosols) were carried out with a Hewlett Packard 5972A mass spectrometer interfaced to a Hewlett Packard 5890 series II plus GC. The obtained mass spectra, which are presented in Figure 4, are essentially identical. The four major fragments, m/z 108, 135, 163 and 220, are the same as reported in the database (NIST/EPA Library) and in the literature (*26, 29*). However, the mass spectrum for the aerosol sample (Figure 4f) has many additional peaks probably because of the low concentration of 24MoBT.

Results and Discussion.

Environmental Distribution. As shown in Tables I, II and III, we found 24MoBT in tire tread rubber and in a variety of environmental samples. Concentrations varied widely (~ng g[-1] to ~μg g[-1]). However, our data reveal large differences in the 24MoBT concentrations found in samples collected during 1989 and 1994.

Table I. 24MoBT Concentrations in Tire Tread Rubber

brand	no. of samples	μg g[-1]	% market share in 1989[a]
Tire#1	1	2.77	50
Tire#2	1	2.27	18
Tire#3	1	2.23	12
Tire#4	1	1.00	15
weighted mean:		2.3[b]	

[a] market share in Japan (1989) from ref. 8. [b] calculated by weighting individual concentration by fractional market share.

Table I indicates the 24MoBT concentrations in tire tread rubber as well as the market share of each brand (Tire#1 - #4) in Japan (*30*). Because the tire samples were

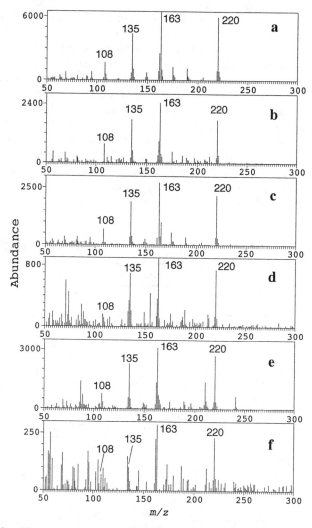

Figure 4. Electron impact ionization mass spectra of 24MoBT in the stock solution prepared from a vulcanization accelerator (**a**) and extracts of environmental samples; (**b**) road dust, (**c**) runoff particles, (**d**) river water particles, (**e**) river sediment, (**f**) aerosol.

obtained in 1989, the 24MoBT concentrations in the tire tread (Table I) are only representative of tires produced and sold in 1989. The weighted mean concentration of 24MoBT in tire tread rubber was calculated as 2.3 μg g^{-1} by weighting individual concentrations by fractional market share. In the calculation, 24MoBT concentrations in tires which occupy the remaining 5% of the market share were assumed to be in the range shown in Table I.

The concentrations of 24MoBT in road dust collected in 1989 ranged from 12 to 490 ng g^{-1} (Table II) and were at approximately the same level as that reported for roadside soil of a highway in San Francisco (273 ng g^{-1}; ref. 26). The concentrations are an order of magnitude lower than that of tire tread rubber mainly because of the dilution of tire wear particles by other particles (e.g., sand, soil, finely ground plant tissue, road surface material, and atmospheric fallout), and partly because of environmental alteration (i.e., degradation, desorption, and leaching). 24MoBT existed at higher concentrations (293 ± 155 ng g^{-1}) in dust from inside tunnels than those from outside tunnels (84 ± 108 ng g^{-1}). The difference between the inside- and outside-tunnel samples may be due to increased "dilution" of tire wear particles by soils and sand in outside tunnels, and/or to environmental alteration.

Table II. 24MoBT Concentrations in Road Dusts and Runoff Particles Collected in 1989

	no. of samples	concentration range	average ± s.d.
road dust[a]			
inside tunnel	5	60-491	293 ± 155 ng g^{-1}
outside tunnel	6	12-289	84 ± 108 ng g^{-1}
runoff particles[b]			
initial stage	1	-	215 ng g^{-1} (29 ng L^{-1})[c]
middle stage	1	-	150 ng g^{-1} (13 ng L^{-1})[c]
final stage	1	-	1.2 ng g^{-1} (0.73 ng L^{-1})[c]

[a] collected by Takada et al. (ref. 2) [b] collected by Ito et al. (ref. 9)
[c] numbers in parentheses are concentrations on a volume basis.

24MoBT was also found in urban runoff (Table II). The concentration of 24MoBT in runoff varied considerably with time during the runoff event (29 ng L^{-1} for the initial stage to 0.73 ng L^{-1} for the final stage). The presence of 24MoBT in runoff demonstrates that urban runoff can transport tire wear particles and road dust into aquatic environments.

The road dust samples collected in 1994 (Table III) contained less 24MoBT than those collected in 1989. The concentrations for the 1994 samples were about a 0.1 (inside tunnels) to 0.2 (outside tunnels) times those observed in the 1989 samples. This decrease in 24MoBT concentrations in 1994 road dust may be attributed predominantly to the decreased usage of the vulcanization accelerator OBS. Production of OBS in 1994 was less than 10% of that in 1989 (Figure 5). This change probably occurred because of toxicological concerns. In 1987, the West German authorities regulated the concentration of certain N-nitrosamines in work area atmospheres (*31*). Because 2-morpholinothiobenzothiazole, a major constituent of OBS, was suspected to be a source of a regulated nitrosamine (*32*), most tire manufacturers in Japan shifted from OBS to other accelerators. Some of the differences may be due to variability of road dust compositions because of the differences in sampling locations.

24MoBT was also found at significant concentrations in other samples collected in 1993 and 1994 (Table III), even though the vulcanization accelerator, OBS, is no longer used in tire manufacturing as described above. It is assumed, however, that significant numbers of vehicles in Tokyo were still equipped with tires containing 24MoBT which were probably manufactured in or before 1989.

In river water, 24MoBT was detected only in the storm-flow samples (0.55-1.0 ng L^{-1}) and not in the normal-flow samples (Table III), suggesting that tire wear particles and road dust are transported to the river primarily during and after heavy rain storms. 24MoBT was measured in river sediment samples in the ng g^{-1} range for both the Tamagawa and Sumidagawa Rivers. These concentrations are one to two orders of magnitude lower than those reported for San Francisco Bay sediments (*26*). This difference may be due to the decrease in overall usage of the vulcanization accelerator OBS and the subsequent decrease in 24MoBT contents in road dust. The 24MoBT concentrations are higher in the Sumidagawa River (5.1 ± 1.7 ng g^{-1}dry) than in the Tamagawa River (2.5 ± 1.5 ng g^{-1}dry). This may be due to the proximity of the Tokyo Expressway to the Sumidagawa River.

24MoBT was also detected in aerosol samples. It has been reported that most tire wear particles are deposited near roads and that only a small fraction is available for a long range transport (*18, 33*). The TUAT station is in an area with roads of high traffic density, while the station HEF is in a forest of 22,000 m^2 which is more than 500 m from the nearest road (which has little traffic). Thus, it is reasonable to expect higher 24MoBT concentrations at the TUAT station (15-18 pg m^{-3}) than at the HEF station (5.6-6.2 pg m^{-3}). These results demonstrate that tire particles and/or road dust contribute significantly to atmospheric aerosols as previously reported by Hildemann *et al.* (*34*).

On a unit weight basis (ng g^{-1}), 24MoBT concentrations in river water particles (5.2-12 ng g^{-1}) are between those in road dust (12-82 ng g^{-1}) and river sediments (1.0-7.6 ng g^{-1}). These results are reasonable considering that both road dust and river sediment contributes to river water particles. At the same time, the difference between sediment and suspended-particle concentrations may also reflect differences in the distribution of tire wear particles among different particle sizes, and differences between suspended-particle and sediment-particle size distributions. Furthermore, the

Table III. 24MoBT Concentrations in Various Environmental Samples Collected in 1993 and 1994

	no. of samples	concentration range	average ± s.d.
road dust			
inside tunnel	8	18-82	30 ± 22 ng g^{-1}
outside tunnel	10	12-30	19 ± 7.0 ng g^{-1}
river particles			
normal-flow[a]	2	nd[b]	nd[b]
storm-flow[c]	4	5.2-12 (0.55-1.0)[d]	7.0 ± 3.4 ng g^{-1} (0.76 ± 0.24 ng L^{-1})[d]
river sediments			
Sumidagawa River	7	2.0-7.6	5.1 ± 1.7 ng g^{-1}
Tamagawa River	7	1.0-5.4	2.5 ± 1.5 ng g^{-1}
aerosols			
HEF	2	91-114 (5.6-6.2)[d]	103 ± 16 ng g^{-1} (5.9 ± 0.4 pg m^{-3})[d]
TUAT	2	180-210 (15-18)[d]	195 ± 21 ng g^{-1} (17 ± 2 pg m^{-3})[d]

River sediment samples and some river particle samples were taken in 1993. All other samples were collected in 1994. [a] At water flow rates less than 20 m^3 s^{-1}. [b] 'nd' not detected (below detection limit). [c] At water flow rates greater than 20 m^3 s^{-1}. [d] numbers in parentheses are concentrations on a volume basis.

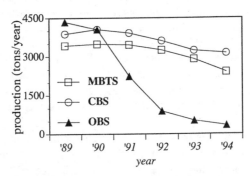

Figure 5. Annual production (tons year^{-1}) of vulcanization accelerators used in tire manufacuture in Japan. MBTS, CBS, and OBS are name of vulcanization accelerators defined in the Japan Industrial Standard (JIS K 6202-1979). Common trade names are DM (MBTS); CM, CZ (CBS); and MSA, NS, NBS, NOB (OBS) respectively.

sediments have a longer residence time and may have been subjected to more leaching. In contrast, 24MoBT concentrations in aerosols on a dry weight basis are much higher (91-210 ng g^{-1}) than in road dust samples (12-82 ng g^{-1}). This may be explained by the fact that aerosols likely consist of fine particles while road dust are "diluted" by larger particles with higher mass per unit particle.

Source Contributions to the Ambient Particulate Matter.   In the following sections, contributions from tire particles and road dust to the ambient particulate matter are calculated by comparing 24MoBT concentrations in various media.   The estimates are based on the assumptions that: (1) there are no significant sources of 24MoBT other than automobile tire abrasion, and (2) 24MoBT behaves conservatively in the environment (i.e., there is no significant degradation, particle desorption and leaching).   The first assumption is probably valid.   Although the vulcanization accelerator OBS is commonly used for rubber products (*35*), nothing but tire abrasion could be expected as such a wide and consistent emission source of 24MoBT.   This does not conflict with our data: 24MoBT was detected in runoff water and in the storm-flow samples of river water, but not in the normal-flow samples.   On the other hand, little is known about the physico-chemical properties of 24MoBT, such as Kow, solubility and vapor pressure, which control its behavior in the environment.   In addition, unlike many other molecular markers, 24MoBT is incorporated into a solid material (tire rubber) before release to the environment.   Thus, the properties of the tire rubber particles are also important during the transport.   All of them could affect the calculations we made here and need further investigation.

Tire Particle Contributions.   By dividing the 24MoBT concentrations in the road dust samples by the weighted mean concentration for the tire tread (Table I), the contribution of tire wear particles to road dust is estimated at 13 ± 7% (inside tunnels), and 3.6 ± 4.7% (outside tunnels; Table IV). These results fall within the same range as has been previously reported by other investigators (i.e., 3-4%; ref. *4*, *25*).

Table IV.   Contribution of Tire Wear Particles to Road Dusts and Runoff Particles Collected in 1989

	% min.-max.	% average ± s.d.
road dust		
inside tunnel	2.6-21	13 ± 7
outside tunnel	0.5-12	3.6 ± 4.7
runoff particles		
initial stage	-	9.2
middle stage	-	6.6
final stage	-	0.05

The estimated contributions of the tire wear particles in runoff particles are relatively high for the initial stage (9.2%) and for the middle stage (6.6%), while they are quite small for the final stage (0.05%). These results imply that a significant fraction of 24MoBT is associated with fine and/or light particles in road dust. The specific gravity of tire tread rubber is typically estimated as <1.3 (JIS K 6313-1981), while that of soil and sand is usually in the range of 2 to 3 (36). Tire wear particles have a bimodal size distribution with maxima in the <0.4-0.5 μm and >5-7 μm size ranges (18, 21, 25, 33). Consequently, 24MoBT is easily mobilized by water flow, fractionation taking place depending on the particle's specific gravity and size. That is to say, during runoff events, tire wear particles are enriched in the initial stage because they are readily mobilized and selectively washed out. By contrast, tire wear particles should be largely depleted in the final stage because they are "diluted" with heavier particles such as sand. The contributions from tire wear particles estimated here could be underestimated, since some 24MoBT might have been lost through degradation, desorption, and leaching from tire wear particles after their emission to the environment.

Because general use of the automobile tires which contain 24MoBT has been decreasing since 1989, the 24MoBT concentrations in tire rubber particles we determined are not representative of general tire usage in 1993 and 1994. Hence, the tire rubber data shown in Table I cannot be used reliably for quantitative source apportionment of the environmental samples collected in 1993 and 1994.

Road Dust Contributions.   Comparing the mean concentration of 24MoBT in road dust with those in river samples, the contribution of road dust to the environmental samples (i.e., river sediment, river water particles) can be estimated (Table V). Our calculations show that about 13 ± 8% of the mass in the Tamagawa River sediments and about 25 ± 10% of the mass in the Sumidagawa sediments are comprised of road dust. In river water particles of the storm-flow samples, 28-65% of the mass is estimated to be contributed by road dust. In the calculation, a mean 24MoBT content of road dust outside tunnels was used, because road dust inside tunnels is expected to be largely unavailable during runoff events.

This approach does not consider the fractionation of road dust particles that would take place during runoff events or as a result of traffic- or wind-induced resuspension. Because some of the large particles in road dust (such as sand) are generally not washed out in runoff events (1), 24MoBT probably becomes enriched in particles discharged into receiving water bodies. This could lead to overestimation of the mass contribution of road dust to runoff particles. At the same time, loss of 24MoBT from tire particles through degradation, desorption, and leaching, which could cause underestimation, should be taken into account. Both fractionation and alteration affects the accuracy of our estimation. Since 24MoBT is expected to be strongly incorporated into the tire rubber matrix, however, such alteration of 24MoBT in tire rubber particles would not significantly affect the source apportionment made here. This may result in overestimation of the road dust contribution to aquatic environments.

Table V. Contribution of Road Dust to River Water Particles and Sediments Collected in 1993 - 1994

	% min.-max.	% average ± s.d.
river particles		
normal-flow	NA[a]	NA[a]
storm-flow	28-65	38 ± 18
river sediments		
Sumidagawa River	11-41	25 ± 10
Tamagawa River	5.5-29	13 ± 8

[a] 'NA' not available because 24MoBT was not detected (below detection limit)

Conclusions.

We have demonstrated that 24MoBT exists in a wide range of environmental samples, indicating that it is a potential molecular marker for tire wear particles and/or road dust. The present study provides the first field application to estimate the contribution of tire wear particles and/or road dust particles to ambient samples. Our estimation suggests that automobile-related particles are widely distributed and contaminate urban environments. However, to increase the accuracy of the estimation, physico-chemical properties (Kow, vapor pressure, photo-degradability) and biodegradability should be determined in future studies. Furthermore, the mode and mechanism of association of 24MoBT with the tire rubber matrix, which could largely control the transport of 24MoBT, should be addressed.

Because the vulcanization accelerator, OBS (containing 24MoBT as an impurity) is no longer used in tire manufacture, its ambient concentrations will continue to decrease in the future. This will lead to the use of 24MoBT in dating recent sections of sediment cores or reconstructing particle sources in a historical context for sediment profiles. We are now investigating other potential molecular markers for tire particles and road dust.

Literature Cited.

1. Sartor, J. D.; Boyd, G. B.; Agardy, F. J., *Journal Water Pollut. Control Fed.*, **1974**, *46*, 458-467.
2. Takada, H.; Onda, T.; Harada, M.; Ogura, N., *Sci. Total Environ.*, **1991**, *107*, 45-69.
3. Takada, H.; Onda, T.; Ogura, N., *Environ. Sci. Technol.*, **1990**, *24*, 1179-1186.
4. Rogge, W. F.; Hildemann, L. M.; Mazurek, M. A.; Cass, G. R.; Simoneit, B. R. T., *Environ. Sci. Technol.*, **1993**, *27*, 1892-1904.

5. MacKenzie, M. J.; Hunter, J. V., *Environ. Sci. Technol.*, **1979**, *13*, 179-183.
6. Hoffman, E. J.; Mills, G. L.; Latimer, J. S.; Quinn, J. G., *Environ. Sci. Technol.*, **1984**, *18*, 580-587.
7. Hoffman, E. J.; Latimer, J. S.; Hunt, C. D.; Mills, G. L.; Quinn, J. G., *Water, Air, and Soil Pollut.*, **1985**, *25*, 349-364.
8. Yamane, A.; Nagashima, I.; Okubo, T.; Okada, N.; Murakami, A., *J. Jap. Soc. Wat. Environ.*, **1989**, *16*, 251-260 (in Japanese).
9. Ito, M.; Takada, H.; Ogura, N., *Chikyukagaku (Geochemistry)*, **1990**, *24*, 105-114 (in Japanese).
10. Hahn, H. H.; Pfeifer, R., *Sci. Total Environ.*, **1994**, *146/147*, 525-533.
11. Wüst, W.; Kern, U.; Herrmann, R., *Sci. Total Environ.*, **1994**, *146/147*, 457-463.
12. Wakeham, S. G.; Schaffner, C.; Giger, W., *Geochim. Cosmochim. Acta*, **1980**, *44*, 403-413.
13. Windsor, J. G., Jr.; Hites, R. A., *Geochim. Cosmochim. Acta*, **1979**, *43*, 27-33.
14. Hunter, J. V.; Sabatino, T.; Gomperts, R.; Mackenzie, M. J., *J. Water Pollut. Control Fed.*, **1979**, *51*, 2129-2138.
15. Eganhouse, R. P.; Kaplan, I. R., *Environ. Sci. Technol.*, **1981**, *15*, 310-315.
16. Eganhouse, R. P.; Simoneit, B. R. T.; Kaplan, I. R., *Environ. Sci. Technol.*, **1981**, *15*, 315-326.
17. Cardina, J. A., *Rubber Chem. Technol.*, **1973**, *46*, 232-241.
18. Pierson, W. R.; Braachaczek, W. W., *Rubber Chem. Technol.*, **1974**, *47*, 1275-1299.
19. Thompson, R. N.; Nau, C. A.; Lawrence, C. H., *American Industrial Hygiene Association Journal*, **1966**, *27*, 488-495.
20. Toyosawa, S.; Umezawa, Y.; Kameyama, Y.; Shirai, T.; Yanagisawa, S., *Bunseki Kagaku*, **1977**, *26*, 38-42 (in Japanese).
21. Toyosawa, S.; Takeuchi, Y.; Fujimura, M.; Inoue, H.; Shirai, T.; Yanagisawa, S., *Nippon Kagaku Kaishi*, **1978**, *1*, 124-127 (in Japanese).
22. Kim, M. G.; Yagawa, K.; Inoue, H.; Shirai, T., *Bunseki Kagaku*, **1989**, *38*, 223-238 (in Japanese).
23. Kim, M. G.; Yagawa, K.; Inoue, H.; Lee, Y. K.; Shirai, T., *Atmos. Environ.*, **1990**, *24 A*, 1417-1422.
24. Kurosaki, H.; Kitajima, E.; Maruyama, T.; Yanaka, T., *J. Jap. Soc. Air Pollut.*, **1986**, *21*, 60-66 (in Japanese).
25. Higashi, S.; Inoue, H.; Shirai, T.; Yanagisawa, S., *J. Jap. Soc. Air Pollut.*, **1981**, *16*, 163-167 (in Japanese).
26. Spies, R. B.; Andresen, B. D.; Jr, D. W. R., *Nature*, **1987**, *327*, 697-699.
27. Jop, K. M.; Kendall, T. Z.; Askew, A. M.; Foster, R. B., *Environ. Toxicol. Chem.*, **1991**, *10*, 981-90.
28. Kumata, H.; Takada, H.; Ogura, N., *Anal. Chem.*, **1996**, *68*, 1976-1981.
29. Ogura, H.; Sugimoto, S.; Ito, T., *Org. Mass Spectrom.*, **1970**, *3*, 1341-1348.
30. *Riekinaki Shea Kyoso (Nonpaying competion for the market share);* Monthly Tire 1994, (July); pp 34-35 (in Japanese).

31. Technische Regeln für Gefahrstoffe 552 "N-Nitrosoamine", BArbBl. (September-1988)
32. Davies, K. M.; Lloyd, D. G.; Orband, A., *Kautsch. Gummi, Kunstst.*, **1989**, *42*, 120-3.
33. Cadle, S. H.; williams, R. L., *J. Air Pollut.Control Association*, **1978**, *28*, 502-507.
34. Hildemann, L. M.; Markowski, G. R.; Cass, G. R., *Environ. Sci. Technol.*, **1991**, *25*, 744-759.
35. *Vulcanization accelerators*;   12394 no Kagaku Shohin (12,394 Chemical Products);   Kagaku Kogyo Nipposha (The Chemical Daily Co., Ltd.): Chuo, Tokyo, Japan 103, 1994. pp 999-1013 (in Japanese).
36. *Specific Gravity*;   Chigaku Dantai Kenkyukai (Society for the Study of Geology), Eds.;   Chigaku Jiten (Encyclopedia of Geology); Heibonsha, Inc.: Chiyoda, Tokyo, Japan 102, 1969. pp 894 (in Japanese).

Chapter 20

Detecting and Distinguishing Sources of Sewage Pollution in Australian Inland and Coastal Waters and Sediments

Rhys Leeming[1], Val Latham[1], Mark Rayner[1], and Peter Nichols[1-3]

[1]Division of Marine Research, CSIRO, GPO Box 1538, Hobart,
Tasmania 7000, Australia
[2]Antarctic Co-operative Research Centre, Hobart, Tasmania 7000, Australia

Sewage pollution often causes public concern, and, therefore, assessment of fecal pollution with the traditionally used indicator microorganisms has been a standard monitoring approach. Molecular markers such as fecal sterols, however, may overcome shortcomings of the classical bacterial indicators. The principal human fecal sterol is coprostanol (5β-cholestan-3β-ol), whilst herbivore feces are dominated by C_{29} sterols (24-ethylcoprostanol and 24-ethylcholesterol). In a pilot investigation, combined fecal sterol and bacterial indicator measurements were used to distinguish fecal pollution from human, herbivore, domestic pet and bird inputs in inland waters during a rain event. Native birds were found to be a major source of the fecal pollution. In a further application of the molecular marker technique in marine waters, Sydney's coastal environment was surveyed before (1989) and after (1992 and 1993) commissioning of deep ocean outfalls in 1991. Most sites showed large increases in the amount of coprostanol present after commissioning of the deep ocean outfalls. Compared to the distribution in 1989, areas of highest concentration of coprostanol were observed further off-shore in 1992 and 1993, adjacent to the diffusers. Minimal dispersion was apparent to regions further off-shore. A decrease in the concentration of coprostanol in sediments in 1993 relative to 1992 reflects a decreased discharge of sludge. These findings suggest the molecular marker approach will be useful to determine future temporal changes in Sydney's coastal environment, including effects due to changing effluent disposal practices.

Sewage tracers can be used to examine the distribution and transport of sewage in the environment (1). One organic tracer (molecular marker) is the sterol, coprostanol (5β-cholestan-3β-ol), which is produced in the digestive tract of humans and is an excellent

[3]Corresponding author

marker for sewage pollution. Use of a molecular marker such as coprostanol has been proposed as an alternative measure of fecal pollution by a large number of researchers. The use of the fecal sterol methodology overcomes shortcomings (e.g. die-off, lack of correlation with fecal pathogens) of classical microbiological indicators of sewage pollution.

Recent studies have highlighted the usefulness of coprostanol for examining sewage pollution in the marine environment. For example, the dispersal and accumulation of untreated sewage in the canals and lagoons of Venice has been examined (*2*), as has the transport of sludge-derived organic pollutants to dump sites off the coast of New Jersey (*3*) and the distribution of sewage contamination in Narragansett Bay (*4*). Similarly, the extent of distribution of sewage-derived material has been traced in the vicinity of a small research station in East Antarctica and at a larger base in McMurdo Sound (*5,6*), although care is needed in interpreting sterol compositional data for Antarctic sediments as marine mammals may be significant contributors in these regions (*7*). A range of other environments have been investigated (e.g. *8-11*). In further applications, archaeological soils have been analysed for coprostanol to detect fecal material (*12*). This approach has been extended to archaeological materials in attempts to distinguish fecal inputs from various animals, e.g. humans (omnivores) *versus* ruminants (herbivores), the basis for which are observed differences in relative levels of coprostanol and 24-ethylcoprostanol (*13*).

In this study, we use the fecal sterol, coprostanol, to measure fecal pollution in the marine environment, specifically in waters off Sydney, Australia. Deep ocean outfalls were commissioned in 1991 to dispose of Sydney's sewage. Further details on the design and use of the outfalls are provided in (*14*). Prior to this date, effluent was discharged through near-shore cliff-face outfalls. Measurement of coprostanol levels in inner-shelf sediments collected before and after commissioning of the outfalls, therefore, provides the opportunity to assess spatial and temporal changes in the quality of sediments in this region. In addition, we present new results from a pilot study that demonstrate the use of a wider range of fecal sterols, in combination with conventional bacterial indicators, to distinguish sources of fecal pollution in environmental samples.

Experimental

Collection of human and animal fecal matter was by Australian Water Technologies (AWT) - Science and Environment, Sydney Water. Samples were freeze-dried and transported to Hobart. Marine sediments from the Sydney region were collected using a stainless steel Smith-McIntyre grab during Research Vessel Franklin cruises 13/89 (December 1989), 1/92 (January 1992) and 9/93 (December 1993). Station locations are shown in Figure 1, and other details are in (*14,15*). Twenty six off-shore Sydney sites were sampled in 1989 and were resampled in 1992 and 1993. Sediments were stored frozen until analysis. Samples from inland waters were collected during a rain event in the Wyong region, New South Wales (Figure 1) using 5 L plastic sampling vessels. Bacterial indicators [thermotolerant coliforms (often termed fecal coliforms) and *Clostridium perfringens* spores] were enumerated by AWT using standard procedures (*14*).

Sediments (0-2 cm), particulate matter obtained by filtering water using glass fibre filters and animal feces were analysed for coprostanol and other sterols as described elsewhere (*14*). Briefly, samples were extracted by a modified Bligh and Dyer (chloroform-methanol-water) procedure (*14,16*). Non-saponifiable lipids (including sterols) were extracted with hexane/chloroform after saponification of an aliquot of the total extract.

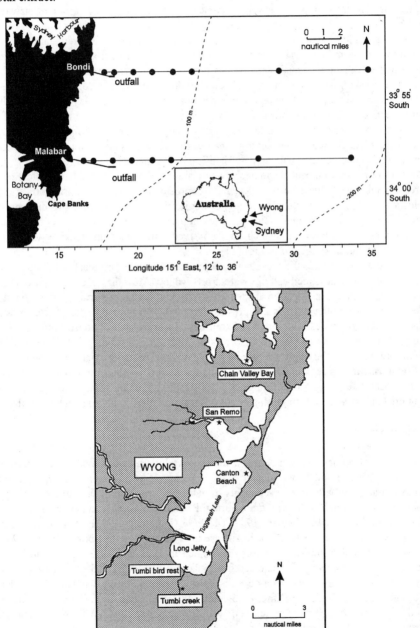

Figure 1. Location of Wyong and Sydney (insert, upper frame) and site locations for inner-shelf sediments (upper frame) and water sampling in Tuggerah Lakes region (lower frame). See also Table I.

Sterols were converted to trimethylsilyl ethers and analysed by capillary gas chromatography (GC) with flame ionisation detection and GC-mass spectrometry as previously described (*14,17*). Coprostanol and epicoprostanol (5β-cholestan-3α-ol) were separated by the GC conditions used, but only data for coprostanol are reported.

Results and Discussion

Distinguishing Human and Other Sources of Sewage Pollution in Inland Waters

Analyses of animal feces. The source specificity of fecal sterols is due to diet, biosynthesis and biotransformations.   These three factors determine "the sterol fingerprint" of fecal matter. Consequently, there can be considerable differences in sterol profiles between warm blooded animals. Analyses were performed of the fecal sterols from humans and from herbivores, domestic pets and birds. These four groups were identified as possible sources of fecal contamination to the Wyong catchment. Results for the concentration of coprostanol, 24-ethylcoprostanol, cholesterol and 24-ethylcholesterol in feces from humans, and representative animals from the other three groups - sheep, dogs and birds are shown in Table I. Data for these four sterols are used in the following sections to estimate the proportion of fecal pollution from these four animal groups. Analyses of fecal sterols from a wider range of animals are reported elsewhere, as are results for other sterols (*18*). Concentrations of other relevant sterols such as coprostanone, 5α-cholestanol and epicoprostanol, catchment data also may be used to assist in the determination of source(s) of fecal pollution.

Table I. Sterol content and bacterial indicator abundance in water samples from Wyong, NSW and in fecal matter. * Site numbers refer to Figures 1 and 2.

	Concentration (µg/g for feces samples, n = 3-6; ng/L for water samples)									
Sterol	Human	Dog	Sheep	Birds	Tumbi Creek (1)*	Tumbi bird rest (2)	Long Jetty (3)	Canton Beach (4)	Chain Valley (5)	San Remo Bay (6)
A	3430	8	170	9	72	106	28	51	9	84
B	746	1570	111	649	1610	3520	1250	3820	868	2560
C	1250	-	246	10	108	61	-	30	2	9
D	121	274	196	967	954	1560	968	2160	313	2240
Bacterial indicators (cfu/100 ml)										
TC	3.6E7	2.0E8	1.7E7	7.6E8	16000	33000	15000	14000	6400	49000
CP	1.9E4	2.5E8	3.2E4	5.5E2	3800	3200	2800	2400	680	11000

Sterols: A, coprostanol; B, cholesterol; C, 24-ethylcoprostanol; D, 24-ethylcholesterol. Bacterial indicators: TC, thermotolerant coliforms; CP, *Clostridium perfringens* spores. SOURCE: Fecal sterol data for selected animals from ref. 18. Birds: magpie, rosella, swan and duck. Seagulls were found to contain 24-ethylcoprostanol (132 ng/l, *18*), however, subsequent analysis of a larger number of samples indicates that this sterol generally is below detection (unpublished data).

The mean sterol content in human feces was 5600 $\mu g/g$ and was around a factor of 5 greater than for the other animals. The concentration ($\mu g/g$, dry weight) of individual sterols also varied between animals (Table I). Major sterols in human feces were: coprostanol, 24-ethylcoprostanol, cholesterol and 24-ethylcholesterol. C_{27} sterols were present in human feces at higher concentrations than C_{29} sterols, whereas the opposite was observed for sheep (herbivore). Although coprostanol was present in feces from other animals, the concentration was considerably lower (1 to 3 orders of magnitude) than in human feces. Major sterols in bird feces were: 24-ethylcholesterol and cholesterol (Table I). Coprostanol generally was absent or present as only a trace component in birds (18). Dogs (carnivores) differed markedly in their sterol composition. Cholesterol, 24-ethylcholesterol and 24-methylcholesterol were abundant. Coprostanol was only a trace component in dog feces. C_{27} sterols were present at higher concentrations than C_{29} sterols in dogs in comparison to sheep. Dogs and birds were distinguished by differences in the relative abundance of thermotolerant coliforms and *Clostridium perfringens* spores (Table I, see below for further discussion).

Bacterial Indicators and Sterols in Australian Inland Waters. When fecal matter is discharged or washed into aquatic systems, the sterol fingerprint of the source animal is diluted and mixed with the sterol profile of autochthonous algae, detritus and other material. Despite this, even low concentrations of fecal sterols can indicate whether fecal pollution from humans or herbivores is present. An important point to note is that experiments on sewerage effluents indicate there is no significant difference in the degradation rate of C_{27} and C_{29} sterols or stanols over a period of 2-3 weeks (Leeming, unpublished), suggesting that the sterol profile of fecal pollution would remain intact in the environment.

In a pilot case study of inland waters from Wyong, NSW, water samples from six sites were collected during a rain event and analysed for fecal sterol composition and bacterial indicators (Table I). The objectives of this pilot investigation were to use combined fecal sterol and bacterial indicator data to elucidate sources of fecal pollution in the Wyong region. Thermotolerant coliform counts for the Wyong water samples were very high, ranging from 6,400-49,000 colony forming units (cfu)/100 ml. *Clostridium perfringens* spores ranged from 680-11,000 cfu/100 ml.

All water samples contained very low concentrations of 5β-stanols (Table I) compared to the high abundance of the bacterial indicators. The concentration of coprostanol was higher than 24-ethylcoprostanol at all sites except Tumbi Creek (Table I). Based on the relationship between fecal sterol concentrations and fecal indicator counts (19), the combined concentration of these two compounds indicates very much lower fecal contamination than that measured by the bacterial indicators. Leeming and Nichols (19) showed that 60 and 400 ng/L of coprostanol corresponded to currently defined primary and secondary contact recreation guidelines (for thermotolerant coliforms, approximately 150 and 1000 cfu/100 mL respectively). Whilst this relationship may vary between regions, the overwhelming difference between the abundance of bacterial indicators and the concentration of fecal sterols present in the Wyong region waters can only be explained by an input of fecal matter from other animals such as birds or dogs.

Human and Herbivore Contributions. It is possible to estimate the relative contributions of fecal pollution of herbivorous animals (e.g. cows and sheep) and humans where fecal matter from both sources are present by comparing the ratios of coprostanol and 24-ethylcoprostanol. Based on the results presented in Table I [also (*18*)], the percentage of coprostanol relative to the sum of coprostanol and 24-ethylcoprostanol is 73% ($\pm 4\%$, n=6) in fecal matter of human origin. Conversely, the percentage of coprostanol relative to the sum of coprostanol and 24-ethylcoprostanol is 38% [$\pm 4\%$, n=19 from 5 species of herbivores (*18*), see also Table I] in fecal matter of herbivorous origin. Where both sources are present the proportions of each source can be calculated between the above percentages relative to each other using the following calculations. It must be stressed that the following calculations provide an empirical basis for determining proportions of fecal pollution and are not used in isolation from other quantitative, qualitative and anecdotal data.

$$\text{If} \left(\frac{5\beta\text{-}C_{27}\Delta^0}{5\beta\text{-}C_{27}\Delta^0 + 5\beta\text{-}C_{29}\Delta^0} \right) \times 100 \ > 73$$

where $5\beta\text{-}C_{27}\Delta^0$ is coprostanol and $5\beta\text{-}C_{29}\Delta^0$ is 24-ethylcoprostanol, then fecal contamination probably originates solely from humans.

$$\text{If} \left(\frac{5\beta\text{-}C_{27}\Delta^0}{5\beta\text{-}C_{27}\Delta^0 + 5\beta\text{-}C_{29}\Delta^0} \right) \times 100 \ < 38$$

then fecal contamination probably originates solely from herbivores.

In the case of a sample containing 45% coprostanol relative to the sum of coprostanol and 24-ethylcoprostanol, a factor of 2.86 [derived from the difference between the mean maximum percentages measured of the two sources divided by 100 (i.e. 73-38)/100] can be added for every one percent below 73% to estimate the proportion of human fecal contamination relative to herbivores. In that instance;

(73 - 45) x 2.86 = 80% herbivore contribution and, therefore, 20% human contribution.

Taking these proportions into account, it is possible to estimate the number of thermotolerant coliforms expected for a given concentration of coprostanol by comparison with experimental data measuring both variables. From the data summarised in Table I, a ratio of ≈ 10 fecal coliforms for every ng of coprostanol was calculated so that contributions of human fecal matter could be quantified. This compares with a ratio of 25 coliforms per ng coprostanol previously calculated from a range of environments (*19*). Similar calculations using the mean concentrations of 24-ethylcoprostanol (Table I) and abundances of bacterial indicators (Table I) from cows and sheep (*18*, the likely possible sources of fecal matter to Tumbi Creek) were performed to estimate the number of fecal coliforms expected for observed concentrations of 24-ethylcoprostanol. The ratio used was 38. The ratios used in this preliminary study are tentative and are presently the focus of further study. Calculating the number of thermotolerant coliforms from human and herbivore sources and dividing by the total number of thermotolerant coliforms measured in receiving waters gives an estimate of the proportion of fecal pollution from these two sources compared to the total.

Bird and Domestic Pet Contributions. Domestic pets and native birds are ubiquitous and numerous in the Tuggerah Lakes region and were also likely alternative sources of fecal matter. The differences in bacterial indicator abundances for birds and domestic pets (Table I) suggest that a water sample which contained large abundances of thermotolerant coliforms and few *Clostridium perfringens* spores (and no fecal sterols) would contain fecal matter predominantly from birds.   This hypothesis is proposed because the faeces of native birds (seagulls, swans, rosellas, magpies and ducks) contained 10^6 - 10^8 cfu / g of fecal coliforms, but generally less than 10^2 cfu / g of *C. perfringens* spores, whereas dog and cat faeces contain roughly equal and comparatively high numbers (10^6 - 10^8 cfu / g) of both fecal coliforms and *C. perfringens* spores. If fecal matter of domestic pets is assumed to contain equivalent numbers of thermotolerant coliforms as *C. perfringens* spores (Table I) and birds, humans and herbivores were assumed to have no *C. perfringens* spores (at most the proportion is < 0.01%), then estimates of the proportions of fecal matter from these two groups also can be calculated.

Proportions of Fecal Matter From Each Source. To quantify the proportions of fecal matter from all four groups (humans, herbivores, birds and domestic pets) in receiving waters, one of the biggest disadvantages of thermotolerant coliforms, their lack of specificity, can be used to advantage. Using thermotolerant coliforms as a common denominator, it is possible to; (i) estimate the number of fecal coliforms expected for given concentrations of fecal sterols, coprostanol and 24-ethylcoprostanol, (ii) estimate the number of thermotolerant coliforms from domestic pets on the basis of *Clostridium perfringens* spore numbers, and (iii) subtract the sum of thermotolerant coliforms from the previous three groups from the total to estimate the number of thermotolerant coliforms derived from birds.   Whilst acknowledging the important assumptions (described above) and the inherent variability observed for environmental samples, the above rationale provides an empirical basis for the estimation of the proportion of fecal pollution from different sources in environmental samples.

Using the ratios of fecal sterols to bacterial indicators, the highest estimate of the possible contribution from human sources at any of the sites was up to approximately 5% (Figure 2). Because the concentration of coprostanol was so low in comparison to the amount of fecal pollution measured by bacterial indicators, it is difficult to confirm the exact source of this fecal sterol. Seepage or overflow from sewers is possible. Some may be derived from cat feces or from trace contributions from bird feces. The low concentrations of the predominantly herbivorous fecal sterol, 24-ethylcoprostanol, (relative to total contamination measured by thermotolerant coliforms) also make it difficult to confirm its exact source.

Only the sample from Tumbi Creek had a significant concentration of 24-ethylcoprostanol compared to the total fecal contamination. At this site herbivores were estimated to contribute 24% of the fecal contamination (Figure 2). This finding agrees with anecdotal evidence on land use patterns in the catchment. The Tumbi Creek sample was the only site with a significant area devoted to agricultural grazing. Overall, the proportion of fecal contamination from herbivore or human sources at most sites was low in comparison to the degree of fecal contamination (based on bacterial indicators) evidenced in this study.

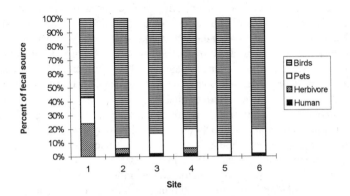

Figure 2. Contribution (% total) of fecal sources to water samples from Wyong, NSW. Sites: 1, Tumbi Creek; 2, Tumbi bird rest; 3, Long Jetty; 4, Canton Beach; 5, Chain Valley Beach; 6, San Remo.   For site locations, see also Figure 1 and Table I. Pets: dogs and cats; Herbivores: sheep and cows.

Fecal matter from domestic pets was estimated to contribute 8-19% of the fecal pollution (Figure 2).   Samples from Chain Valley Bay and Tumbi bird rest had the lowest proportions of fecal matter derived from domestic pets (8 and 9% respectively) which was consistent with lower housing densities (less pets) and larger bird numbers (adjacent to wetlands or bird resting sites). The remaining fecal matter was considered to be from native birds which made up the largest estimated proportions of fecal contamination (79-90%, except for Tumbi Creek which was 57%). The results are supported by the presence of very large bird populations in the area.

Calculations to estimate the proportions of fecal matter from specific sources are based on new, but as yet, limited data. They are naturally subject to considerable variation where errors are multiplied during calculations from the original measurements. The values given for human and herbivore contributions are probably at the high end of possible estimated contributions. The proportions of fecal matter calculated for any source should, therefore, presently be used with appropriate caution. In summary, the results of this pilot case study indicate that human fecal matter was, at worst, only a minor component of the total fecal pollution evident in receiving waters in the Wyong lakes during the rain event. Further research is currently underway to refine the approach presented in this pilot study. A wider range of animals together with increased number of replicates will be analysed in order that the large variation found for bacterial indicators, in particular, can be accounted for.

Sydney's Deep Ocean outfalls

Like many western cities in the twentieth century, Sydney has chosen ocean outfalls as part of its strategy for disposal of sewage effluent. Since the commissioning of Sydney's deep ocean outfalls in mid 1991, it has been recognised that there has been a general improvement in water quality at beaches which were formerly affected by the near-shore cliff-face sewage outfalls located at Malabar, North Head and Bondi (20). Effluent from the former nearshore and new deep ocean outfalls has been subjected to primary treatment only.   A major objective of three oceanographic research cruises conducted in the location of the newly commissioned deep ocean outfalls was to determine, using the fecal sterol molecular marker approach, whether or not there has been an increase in sewage-derived material present in sediments adjacent to these deepwater outfalls since their commissioning in 1991. Further aims were mapping the distribution of sewage-derived material and examining spatial distributions and variations that may have occurred over time.

5β-Stanols such as coprostanol do not to our knowledge occur in uncontaminated fresh or marine waters and generally are detected at only trace concentrations (< 20 ng/g) in aerobic sediments (3,5,8,14,17; see also Table II), with the exception of selected environments where input from marine mammals occurs (7). As part of the broader project objectives, coprostanol concentrations (ng/g) were determined in sediments collected along two cross shelf transects offshore from Sydney in 1989, 1992 and 1993 (Figure 1, Table III, 14).   The latter two samplings took place after commissioning of the deep ocean outfalls.

Coprostanol was present at significant concentrations in sediments collected during 1992 and 1993 relative to that found for reference stations (<20 ng/g, this and related studies, Table II). Based on the coprostanol concentration data, the highest levels of sewage-derived organic matter were detected in inner-shelf (0.5-5 nautical miles offshore) sediments from north of Sydney Harbour to south of Botany Bay. The areas showing the highest levels of coprostanol were generally within 1-3 nautical miles of the deep ocean outfalls.

In selected environments, 5β–stanols may be produced in situ by anaerobic bacteria. However, the ratio of coprostanol to 5α-cholestanol can be used to assess whether coprostanol in surficial sediments is of fecal origin or, from an in situ microbial source. A ratio of <0.3 is generally observed for pristine aerobic sediments, with ratios >0.5 for sites impacted by sewage. For Sydney inner-shelf sediments, the ratio was in most cases >2.   The ratio of coprostanol to epicoprostanol (5β-cholestan-3α-ol) can be similarly used to distinguish fecal pollution derived from human versus marine mammals (7). Human fecal matter contains only trace amounts of epicoprostanol. For Sydney inner-shelf sediments epicoprostanol was only a minor component (generally less than 10% of coprostanol), indicating that marine mammals are not major contributors in this environment.

The results indicate that between 1989 and 1992, after the commissioning of the ocean outfalls, most inner-shelf sediments showed an increase in the concentration of coprostanol (Table III). The areas of highest concentration were located in sediment adjacent to the deep ocean outfalls. Transport of sewage-derived matter further off-shore (and on-shore to beaches) is minimal. Rather this material may be moving in a north-south direction (14). However, the concentration of coprostanol measured in sediments from four cross-shelf sampling transects in the vicinity of the Bondi and

Table II. **Coprostanol content (ng/g) for sediments from selected environments**

Location	Coprostanol (ng/g)	Comment	Reference
Sydney, Australia			
Seven Mile and Palm Beach	-	no fecal pollution	17
Freshwater Beach	24	minor fecal pollution	17
Coastal sediments (24 sites)			
1989	7-1100	prior to ocean outfalls	17
1992	12-2350	after ocean outfalls	14
1993	30-1300	after ocean outfalls	-
Port Phillip Bay, Australia			
bay sediments	4-550	enclosed bay, secondary	20
influent creeks	50-2590	treatment effluent (sewage farm) with additional creek and storm water input	
Derwent River and Estuary, Australia			
pristine sediments	9	out of estuary	22
river sediments	360-1810	13 sewage treatment plants discharge to river	
Deep Water Dump site, North Atlantic			
dump site	80-820	185 km from coast	3
control site	8-10	approx. 100 km from dump site	
Venice, Italy			
canals	1000-41000	canals and lagoons, untreated waste	2
ref. station, Adriatic Sea	200		
Narragansett Bay			
surface sediments	130-39000	multiple treatment plants	4
Rhode Island control	200		
Antarctica			
Davis station	40-880	small station, treated effluent, 60 people	5
McMurdo, Cape Armitage	930	1000 people during summer untreated	6
Santa Monica Basin	380-3240	ocean outfall discharge	23
Boston Harbour	1000-16000	impacted by sewage	4
Ariake Sea, Japan	20-1770	impacted by sewage	24
Severn Estuary, England	900-3100	impacted by sewage	25
Clyde Estuary, Scotland	10-13600	impacted by sewage	8
Kaohsiung Harbour, Taiwan	580-128000	impacted by sewage	26

Malabar outfalls did not increase further after 1992 (Table III and unpublished data). The mean concentration of coprostanol generally was lower in sediments collected at the final sampling (1993) relative to the 1992 sampling (Table III; mean data for 24 sites, Figure 3). This observation is consistent with a decreased input after 1992 of sewage sludge through the deep ocean outfalls (AWT, personal communication).

The distribution of coprostanol in sediments provides an integrated picture of the variation of currents over extended periods. This variability may explain the spatial differences observed when comparing the 1992 and 1993 results. Over 9 days that RV Franklin was in the Sydney region during December 1993, surface (and midwater) currents were highly variable. Currents ranged from 1 knot southward, to weaker on-shore currents, then turned to northward at up to 0.7 knots for the final 2 days. Directional variability of the currents off Sydney suggest that sewage-derived material is not always swept out of this region. Rather some material may remain in the vicinity of the outfalls (or may move towards shore; 1993 results, Table III) ultimately being deposited in inner-shelf sediments as indicated by the coprostanol distribution.

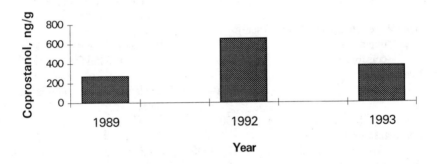

Mean sediment coprostanol concentration (4 repeat transects)

Figure 3.   Mean coprostanol concentration (ng/g) for sediments in Sydney coastal waters for inner-shelf sites sampled (0.5-5 nautical miles from shore; four transects). The coefficient of variation for replicate sediments from the same site has been estimated to be <25%.   Figure 1 shows the location of the Bondi and Malabar transects (see Table III), two of the transects analysed. The other two transects are approximately 2 nautical miles south of the Bondi and Malabar transects, respectively.

Coprostanol concentration data for sediments from a range of Australian (*14,17,20,22*) and other environments (*2-8,23-26*) are shown in Table II. Coprostanol concentrations >500 ng/g are equivalent to thermotolerant coliform abundances of >8000 cfu/100g (*14*) and are, we believe, indicative of significant sewage pollution. Results for Sydney inner-shelf sediments are comparable to those reported for sediments from other estuaries and coastal areas impacted by sewage. Concentrations over 50 000 ng/g clearly indicate extremely heavy inputs. The higher concentrations are usually associated with more enclosed environments such as bays and harbours where various mechanisms (e.g., prevailing ocean currents, tidal flushing or other) removing sewage-derived material are not as influential.

Table III. Coprostanol content of Sydney inner-shelf sediments

Location/distance offshore (nautical miles)	Coprostanol concentration (ng/g) December 1989	January 1992	December 1993
Bondi transect			
0.5	344	564	173
1	332	888	481
2	128	575	92
3	150	477	55
4	151	138	180
5	159	197	143
10	13	46	24
15	7	30	4
20	NS	NS	5
Malabar transect			
0.5	254	106	135
1	329	560	1910
2	531	2360	1350
3	126	1255	463
4	26	62	113
5	19	38	76
10	23	26	25
15	19	12	4
20	13	NS	58

SOURCE: 1989 and 1992 Malabar data from ref. 14. NS = not sampled.

Sewage sludge from the Sydney region contained around 10^6 ng/g coprostanol (*27*). The concentration of coprostanol in sediments within 0.5 km from the Malabar deep water outfall (1992, 2350 ng/g; 1993, 1300 ng/g) correspond to 1 in 400 (0.25% of sludge) and 1 in 750 (0.13%) dilutions of sludge. Coprostanol is mainly associated with fine sediments, as are most microbes, metals and organic molecules. This was the case for sediments in the Sydney region. In the 1992 and 1993 surveys, sites with the highest concentrations of organic carbon generally contained the highest concentrations of coprostanol and bacterial indicators (*14*, unpublished data).

Fine sediment was observed on filters obtained from bottom waters (1-5 m above sea floor) collected under the weather and current conditions encountered in December 1993 (mainly fine, wind speeds less than 20 knots). This observation indicates that transport of fine particles also occurred under milder weather conditions than in the intense storm events suggested earlier (*28*). Based on the coprostanol results, this would include under the influence of westerly currents. The implications of this finding in relation to potential mobilisation of microbes, including more persistent pathogens and viruses, that may be present in inner-shelf sediments remains to be determined. A better knowledge of the consequences of storm events with respect to transport of sewage-derived sedimentary matter will be useful to understand the ultimate fate of Sydney's sewage effluent.

Conclusions

The use of the fecal sterol molecular marker approach has provided an increased understanding of the source(s) and fate of sewage-derived material in the inland waters of Wyong and in the Sydney inner-shelf region, including both spatial and temporal resolution. For development of the molecular marker techniques to continue and be both widely accepted and routinely used, there is a need to: (i) further examine and validate differences in the fecal sterol profiles and the abundance of enteric bacteria within certain animal groups and, (ii) refine the precision and accuracy of the calculations used to apportion the various sources of fecal matter. Furthermore, we need to test laboratory results concerning the relationships between and relative persistence of fecal sterols and indicator microorganisms in the field to account for differences observed during wet and dry weather. Such findings and the application of these procedures will be of use to managers needing to distinguish sources of fecal pollution, and in studies that examine future temporal changes in the Sydney coastal environment with respect to possible changing effluent disposal practices.

Acknowledgements

We thank: the master and crew of Research Vessel Franklin; fellow CSIRO colleagues; AWT-Science and Environment (Nick Ashbolt and Andrew Ball), NSW Environmental Protection Authority and University of Western Sydney during field programs; George Cresswell for ongoing support; the Steroid Reference Collection for provision of sterol standards; Ed Butler, John Church, Peter Craig, Peter Franzmann, Carol Mancuso, Sarah Nespolo, Martin Riddle and others, particularly Bob Eganhouse of the United States Geological Survey and two anonymous reviewers, for assistance and comments on the manuscript. A portion of this work was funded by AWT and Wyong Shire Council.

Literature Cited

1. Vivian, C. M. G. *Sci. Total Environ.* **1986**, 53, 5-40.
2. Sherwin, M. R.; Van Vleet, E. S.; Fossato, V. U.; Dolchi, F. *Mar. Poll. Bull.* **1993**, 26, 501-507.
3. Takada, H.; Farrington, J. W.; Bothner, M. H.; Johnson, C. G; Tripp, B. W. *Environ. Sci. Tech.* **1994**, 28, 1062-1072.
4. LeBlanc, L. A.; Latimer, J. S.; Ellis, J. T.; Quinn, J. G. *Est. Coastal and Shelf Science.* **1992**, 34, 439-458.
5. Green, G.; Nichols, P. D *Ant. Sci.* **1995**, 7, 137-144.
6. Venkatesan, M. I.; Mirsadeghi, F. H. *Mar. Poll. Bull.* **1992**, 25, 328-333.
7. Venkatesan, M. I.; Ruth, E.; Kaplan, I. R. *Mar. Poll. Bull.*, **1986**, 17, 554-557.

8. Goodfellow, R. M.; Cardoso, J.; Eglinton, G.; Dawson, J. P.; Best, G. A. *Mar. Poll. Bull.* **1977**, 8, 272-276.

9. Grimalt, J. O.; Fernandez, P.; Bayona, J. M; Albaigés, J. *Environ. Sci. Tech.* **1990**, 24, 357-363.

10. Readman, J. W.; Preston, M. R.; Mantoura, R. F. C. *Mar. Poll. Bull.* **1986**, 17, 298-398.

11. Venkatesan, M. I.; Kaplan, I. R. *Environ. Sci. Tech.* **1990**, 24, 208-214.

12. Bethell, P. H; Goad, L. J.; Evershed, R. P. *J. Arch. Sci.* **1994**, 21, 619-632.

13. Evershed, R. P.; Bethell, P. H. Archaeological Chemistry. Organic, Inorganic and Biochemical Analysis. ACS Symposium Series 625, Washington D. C., Anaheim, California, 1995, pp 157-172.

14. Nichols, P. D.; Leeming, R.; Rayner, M. S.; Latham, V.; Ashbolt, N. J.; Turner, C. *J. Chromatog.* **1993**, 643, 189-195.

15. Cresswell, G. *Aust. J. Mar. Freshwater Res.* **1994**, 45, 677-691.

16. Bligh, E. G.; Dyer, W. M. *Can. J. Biochem. Physiol.* **1959**, 35, 911-917.

17. Nichols, P. D.; Leeming, R.; Rayner M. Latham, V. *J. Chromatog. A* **1996**, 733, 497-509.

18. Leeming, R.; Ball, A.; Ashbolt, N.; Nichols, P. *Water Res.* **1996**, 30, 2893-2900.

19. Leeming, R.; Nichols, P. *Water Res.*, **1996**, 30, 2997-3006.

20. Fagan, P.; Miskiewicz, A. G.; Tate, P. M. *Mar. Poll. Bull.* **1992**, 25, 172-180.

21. O'Leary, T.; Leeming, R; Nichols, P. D.; Volkman, J. K. **1994**, CSIRO Marine Laboratories Report No. PPB94-F, Hobart, Tasmania, 39 pp.

22. Leeming, R.; Nichols, P. D. *J. Mar. Freshwater Res.* (in press).

23. Venkatesan, M. I.; Santiago, C. A. *Mar. Biol.* **1989**, 102, 431-437.

24. Kanazawa, A.; Teshima, S. I. *Oceanologica Acta.* **1978**, 1, 39-44.

25. McCalley, D.V.; Cooke, M.; Nickless, G. *Water Res.* **1981**, 15, 1019-1025.

26. Jeng, W-L; Han, B-C. *Mar. Poll. Bull.* **1994**, 28, 494-499.

27. Nichols, P. D.; Espey, Q. I. *Aust. J. Mar. Freshwater Res.* **1991**, 42, 327-348.

28. Gordon, A. D.; Hoffman, J. G. Public Works Department, NSW, **1985**, Technical memo No. 85/2, Sydney, NSW, 33 pp.

Chapter 21

Questions Remain in the Use of Coprostanol and Epicoprostanol as Domestic Waste Markers: Examples from Coastal Florida

Paul M. Sherblom[1,2], Michael S. Henry[2], and Dan Kelly[2]

[1]Department of Environmental and Occupational Health, University of South Florida, 13201 Bruce B. Downs Boulevard, Tampa, FL 33617
[2]Mote Marine Laboratory, 1600 Thompson Parkway, Sarasota, FL 34236

Concentrations of the fecal sterol, coprostanol, and its epimer, epicoprostanol, were determined in sediments from three areas of coastal Florida. Also analyzed were feces from two species of marine mammals which frequent these waters. The sediments come from environments which may be impacted by diverse sources of contamination ranging from wild animal wastes to direct discharge of wastewater effluent. These areas displayed a wide concentration range for both epimers, with most samples having low concentrations (< 1 $\mu g/g$) of the fecal sterols. Sediment concentrations and the resulting epimer ratios could not always be related to proximity of known or potential sources. The results indicate that there remain questions concerning the usefulness of these compounds in differentiating marine and terrestrial sources when they are present at low concentrations. Additionally, the coprostanol to epicoprostanol ratio does not always allow unequivocal attribution of sediment contamination to domestic waste related, rather than natural sources.

Identifying the extent of environmental contamination related to a particular source is one object of the use of molecular markers. A common source of contaminants to the coastal environment is discharge of domestic wastewaters. The potential for public health and environmental impacts from these discharges has contributed to the evaluation of a number of compounds as potential molecular markers for domestic waste. This report focuses on the environmental distribution and potential sources of one type of these molecular markers, the coprostanols. These compounds have been widely used to delineate the contribution of domestic wastewater to environmental contaminant loadings. The history and use of coprostanol as a sewage marker has been reviewed by Walker *et al.* (*1*), while Vivian (*2*) discussed the use of sewage markers in general. Coprostanol (5β-cholestan-3β-ol) is produced by the anaerobic microbial hydrogenation of the double bond between carbons 5 and 6 in cholesterol (cholest-5-en-3β-ol). Under aerobic environmental conditions cholesterol is hydrogenated to cholestanol (5α-cholestan-3β-ol). Both coprostanol and cholestanol also have 3α epimers. While cholestanol has other sources in the environment (and is the thermodynamically favored product during hydrogenation), coprostanol is primarily produced by the enteric bacteria of higher animals, and there is reportedly little decay of this material in anaerobic sediments (*1*). There is some evidence that coprostanol can be produced *in situ* under anaerobic conditions in areas uncontaminated by mammalian fecal wastes (*3,4*). The relative magnitude and rate of this environmental production of coprostanol has yet to be determined, but is likely to be strongly controlled by sediment composition and the bacterial community present.

The preponderance of available literature regarding coprostanol in environmental samples can be considered monitoring data. Most coprostanol measurements have been made near known sources, and the presence of coprostanol is interpreted as indicating the extent of contamination by the domestic waste source. The adequacy of using coprostanol as an indicator of human waste inputs has been questioned by several investigators (*5-8*) due to its presence in other mammalian (animal) wastes. The work of Venkatesan *et al.* (*5*) and Venkatesan and Santiago (*6*) has shown that marine mammals may contribute to the fecal sterol burden of some coastal sediments. Venkatesan and coworkers (*5-7*) suggested that the ratio of coprostanol to epicoprostanol (c/e ratio) can be used to distinguish inputs of domestic wastes from material of non-human origin. They suggest that human related wastes will result in a ratio which is close to or larger than one (greater proportion of coprostanol). Lower values for the ratio would indicate the dominant inputs were from marine mammals. Grimalt *et al.* (*8*) have also reported that the presence of coprostanol does not necessarily mean that an area is contaminated with domestic waste. They utilized the relative amounts of the coprostanols (5β stanols) to the summation of the coprostanols and cholestanols (5β + 5α stanols) as a means of differentiating natural sources from domestic waste contributions. The main commonality among these reports is their evaluation of the relative abundance of the coprostanols to each other and/or other cholesterol derivatives present in the sample. In most coprostanol studies the relative abundance of the 3α and 3β epimers has not been determined, and the concentrations of coprostanone or the 24-methyl or ethyl coprostanols have not usually been investigated.

Environmental sources of human fecal material are dominated by municipal treatment plants and potential leaching from onsite disposal (septic) systems. Both coprostanol and epicoprostanol are produced during anaerobic microbial action on sewage sludge (*9*). Formation of epicoprostanol is apparently enhanced in the anaerobic digestion process, which leads to changes in the c/e ratio. This has been documented for two treatment works by McCalley *et al.* (*9*), and is supported by the observed c/e ratios (0.9 to 9.0) of sludges from six treatment plants which use various treatment methods (*10*). Reports of coprostanol and epicoprostanol concentrations in effluents, sludges, and environmental samples (*8-11*) have shown that the epimer ratio is extremely variable. From the work discussed above it can be seen that treatment method and depositional environment can affect both the absolute concentrations and the c/e ratio of fecal sterols from domestic wastes. These studies leave uncertainty regarding the significance of detecting low concentrations of these fecal sterols in environmental samples. We present here results from three sediment surveys conducted in coastal Florida which represent different degrees of direct human impact. Since marine mammal inputs are considered as a possible source to these sediments, fecal samples from each of the two types of marine mammals likely to impact these waters have also been collected and analyzed. The results are evaluated with regard to the specificity of the fecal sterols as molecular markers of domestic waste when they are present at low concentrations. The epimer (c/e) ratio is also evaluated for its usefulness in separating or identifying sources in these situations.

Experimental

Study Sites. The sediment surveys represent different levels of source strength and receiving environments. The Key West study area provides surficial sediment concentrations away from a known source (secondary wastewater effluent) discharging into a zone of high dilution. Sarasota Bay is a shallow, urban embayment with multiple potential sources including occasional direct wastewater discharge (advanced treatment, wet weather flow), septic tank leachate, recreational boating and potential impacts from resident and migrating marine mammal populations as well as stormwater runoff potentially contaminated with domestic or agricultural animal wastes. The Weeki Wachee river represents a system with more restricted sources. Specifically these include septic tank leachate and domestic animal wastes from a few housing developments on canals bordering the river, and the introduction of wastes from marine mammals inhabiting the river or terrestrial animals living in the undeveloped areas bounding large sections of the river.

Sampling. The locations of sediment sample stations within the three study areas are shown in Figure 1. In the Key West study, sampling transects were established North-South and East-West across the outfall, as well as along the shoreline. Three reference stations (not shown in Figure 1), well removed from potential human influence, were also sampled. The Sarasota Bay sampling sites were selected to provide a spatial characterization within the Bay, with special focus on locations thought to be impacted (either currently or historically) or of interest as wildlife habitat. Within each sampling site a suspected gradient in sediment quality was sampled in transects of 3 stations each, with 'A' being closest to shore and 'C' being furthest into the bay (gradient sampling stations not detailed in Figure 1). Duplicate grab samples of the upper 5 cm of sediment were collected using a petite ponar sampler. A slightly different sampling schedule was used in the Weeki Wachee river. Samples were collected from four river stations (numbered) above and below (two each) the influence of the canal systems and six stations located within the canal systems (stations A through F). Several of these stations (1, 3, A, D, E and F) were repetitively sampled over the course of nine months (spring to winter). Weeki Wachee and Key West Survey samples were collected from the top 3 cm using a stainless steel diver core (Key West samples collected and supplied by EPA personnel). Marine mammal fecal samples were obtained from wounded manatees (*Trichechus manati*, courtesy of Florida DEP) undergoing rehabilitation, or from stranded (deceased) dolphins (*Tursiops truncatus*) recovered by the Mote Marine Laboratory Marine Mammal Stranding Response Team. For

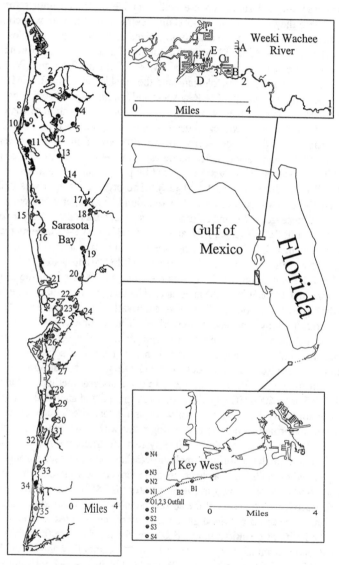

Figure 1. Maps of study sites showing sediment sampling stations.

comparison, sediments from a holding lagoon where a dolphin was undergoing rehabilitation prior to release were also collected and analyzed.

Sample processing. All samples were held over ice or frozen until returned to the laboratory, where they were stored at -5°C until extraction. Prior to extraction, samples were thawed at room temperature and homogenized by stirring. Subsamples were taken for analysis of moisture content and extraction for sterol analysis. Approximately 20 gram aliquots (wet) were extracted for sterol analysis. Sarasota Bay and Weeki Wachee sediments were extracted using cyclohexane and then dichloromethane (DCM) reflux with a Stark and Dean moisture trap. The Key West and holding lagoon sediments, as well as the marine mammal fecal samples, were extracted using a modification of EPA's SW-846 (*12*) method 3540B (Sohxlet extraction using acetone and hexane 1:1). Androstanol was added to all samples as a recovery surrogate, prior to extraction. Extracts were reduced in volume *via* rotary evaporation under gentle vacuum. When the extract was reduced to a few milliliters, residual water was eliminated by partitioning the extract between pre-extracted HPLC grade water and DCM (3x). The organic fraction was collected, re-concentrated and exchanged into hexane. Compound class separations were performed using combined silica gel - alumina columns. This procedure is essentially a combination of EPA (*12*) methods 3630B (silica gel), and 3611A (alumina). Adsorbents were cleaned by sonication in methanol, DCM and hexane, dried and activated at 210°C. Upon cooling, the adsorbents were deactivated (5% w/w) with HPLC grade water and stored under hexane. The 2:1 silica:alumina column was eluted using 20 ml of hexane, 20 ml of 5% DCM in hexane and 40 ml 30% DCM in hexane. The final fraction, 40 ml of ethyl acetate, contained the sterols and alcohols. These were derivatized to the corresponding acetates using anhydrous pyridine and acetic anhydride. The acetates were isolated from the derivatization solution by repeated (3x) evaporation of the solution to dryness and redissolving in DCM. The acetates were then taken up in a solution containing the quantitation standard (cholestane) and analyzed *via* gas chromatography with flame ionization detection (GC/FID) using the internal standard method. Analyses were performed on a Varian 6000 GC with data collected, stored and analyzed using PE/Nelson 2600 Chromatography Data System software and interfaces. Selected confirmatory analyses were performed using gas chromatography with mass spectrometry (GC/MS) using a Varian Saturn II ion trap GC/MS/DS system. Extraction efficiency and potential loss or alteration of analytes during sample processing were assessed through the use of spiked samples and the addition of androstanol as a recovery surrogate to the samples prior to extraction. Androstanol recovery averaged greater than 85%, and results have not been corrected for recovery. Analysis of the spiked samples demonstrated that the extraction and derivatization methods provided good recovery of the analytes with no experimental artifacts. The limits of quantitation and detection for the data are (dry wgt): Key West Sediment and marine mammal samples 10 and 1 ng/g; Sarasota Bay and Weeki Wachee 20 and 5 ng/g (LOQ and LOD respectively). Procedural blanks were consistently below the limit of quantitation.

Results and Discussion

Key West. Coprostanol and epicoprostanol concentrations (ng/g dry weight) of samples with detectable fecal sterols from the Key West (stations beginning with K) and Sarasota Bay (stations beginning with S) study areas are plotted in Figure 2 (note the logarithmic scale). For comparison, a line representing a c/e ratio of 1 is also plotted. Sediments from the Key West area normally plot above this line (except KS2 and KO 1, 2, 3); Sarasota Bay sediment stations all plot below this line (except S26A). The relevance of this will be discussed further below. Key West sediments had variable concentrations (30 to 340 ng/g dry) with c/e ratios of 0.2 to 9.0. Coprostanols were detected in sediments from the vicinity of the outfall as well as from a transect along the shore of Key West, but not in sediments collected from reference areas or those collected east and west of the outfall. Triplicate grab samples collected adjacent to the outfall (KO 1, 2, 3) were dominated by epicoprostanol and varied from each other by as much as a three fold difference in sterol concentrations with a two fold difference in the epimer ratio (Figure 2). This may reflect either variations in source material and/or deposition, or could indicate possible *in situ* alteration of the epimer ratio. The sediment transects north (KN) and south (KS) of the outfall were dominated by coprostanol relative to epicoprostanol with resulting c/e ratios being greater than 1. The variable sediment concentrations likely reflect differences in depositional activity but may also represent either variation of the epimer ratio in the effluent discharged or subsequent environmental alteration of the sterol signature. Samples from the beach transect (KB) had low concentrations of fecal sterols. These sediments may indicate transport of waste related material from offshore, or the fecal sterols may result from *in situ* production or inputs from runoff or septic leachate related sterols.

Sarasota Bay. Coprostanol concentrations in some Sarasota Bay sediments have been reported previously (*13*). The prior study, in 1982, focused on potential contamination by wastewater discharged near site 20. Concentrations ranged from ≤0.01 to 2.5 µg/g sediment from that portion of the bay (*13*). The summed coprostanol and epicoprostanol concentrations of sediments investigated here exhibit a similar range. Two of the 35 study sites (9 & 10), had fecal sterol concentrations below the limit of quantitation for all samples along the three station gradient. The remaining 33 study sites provided at least one grab sample with quantifiable concentrations of the fecal sterols. Station mean coprostanol and epicoprostanol concentrations are shown in Figure 2. As was stated above, most samples plot below the line indicating a c/e ratio of 1 reflecting the dominance of epicoprostanol in these sediments. Coprostanol was typically one-third of the epicoprostanol levels, with many samples having little or no coprostanol, resulting in many stations plotting at the limit of quantitation for coprostanol and, thus, overlapping on the plot. As stated previously, potential sources of the fecal sterols for the bay consist of recreational boat use, septic field leachate, surface runoff contaminated by domestic animal waste, marine mammal contributions and advanced wastewater treatment plant discharges. Evaluation of these results with regard to known or suspected sources of wastes to the bay does not fully explain the fecal sterol concentrations measured. While stations near the mouths of some tributaries entering the bay had higher fecal sterol concentrations (i.e. sites S12, S17, S18, S28), the measured concentrations varied widely within most transects away from shore

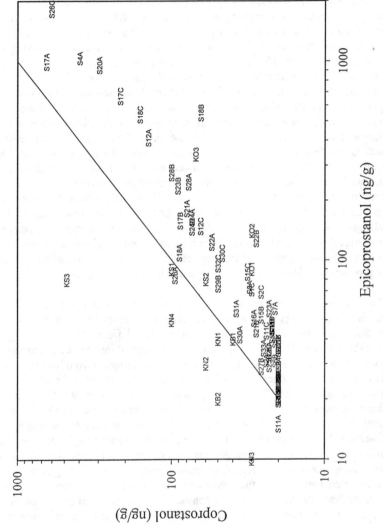

Figure 2. Coprostanol and epicoprostanol concentrations in sediment samples from Key West and Sarasota Bay Study sites.

(e.g. S17 A, B, C), as did the c/e ratio (e.g. S18 A. B, & C). Neither the absolute fecal sterol concentration nor the c/e ratio was consistently related to proximity to known shore-related sources of the fecal sterols. Station S26C, which is distant from known sources, had among the highest concentrations of both epimers as well as a c/e ratio of 0.4.

The c/e ratio of the Sarasota Bay samples ranged from 0.2 to 1.5, with 88% of the sediments having c/e ratios of 0.6 or below. The predominance of low c/e ratio values determined in these samples could suggest several processes for the fecal sterols in Sarasota Bay. These results could imply that: 1) all wastes, human or otherwise, entering the bay are very similar in source composition, 2) sediments within the bay are dominated by non-human sources for the coprostanols, 3) inputs from a variety of sources rapidly reach steady state conditions with respect to the relative proportions of these epimers through either rapid mixing or alteration of the sterol signature. This latter explaination could imply *in situ* coprostanol and epicoprostanol formation and/or degradation such that the epimer levels throughout the bay equilibrate, even with different source strengths and sediment concentrations. Considering these possible explanations in order: First, it is unlikely that the fecal sterol composition of each of the potential sources would be so similar. There is scant literature on the epimer ratio for terrestrial and domestic animal wastes, but it has been suggested (1,6) that dietary and species differences lead to large differences in both absolute and relative production of the fecal sterols. Second, given the population of the region and the urban and suburban nature of the drainage area of Sarasota Bay, it is unlikely that contributions by the relatively few resident marine mammals could dominate the mass flux of fecal sterols to the sediments. There is little agricultural activity remaining in the drainage basin for the bay. However, domestic animals could contribute to a runoff related source. This could be an important contributor to the flux of fecal sterols to this bay as well as to other coastal urban areas. The potential significance of this source needs to be elucidated before the presence of the fecal sterols can be attributed unequivocally to impact by domestic wastes. Third, it is possible that differences in epimer degradation during transport to, or burial in the Bay sediments could affect the epimer ratios observed. Domestic wastewater sources should have a greater contribution of coprostanol, but if the 3β epimer degraded more rapidly than the 3α epimer, or underwent isomerization, the c/e ratio would decline. If this happened, elevated fecal sterol concentrations would mean greater deposition of waste related material and higher c/e ratios may indicate more recent inputs of aerobically degraded or relatively untreated wastes.

Weeki Wachee River. All sediments analyzed during the Weeki Wachee River survey except those of river station 3 were found to contain fecal sterols on at least one sampling date (Table I). The highest concentrations were in sediments from the canals located in housing subdivisions. Low concentrations were found in sediments from the main river channel. Marine mammal and wild animal contributions should contribute more significantly to the main river than to these residential channels. These results indicate that the canal sediments may be impacted by septic tank leachate or surface runoff contaminated by domestic animal wastes. The sediment fecal sterol concentrations at a given station varied by as much as 65 fold over the nine month sampling period. Spatial variability of sediment concentrations might contribute to these differences, however, analysis of duplicate grab samples indicated that sediment concentrations at a sampling station varied

Table I. Weeki Wachee River: coprostanol and epicoprostanol concentrations (ng/g dry)

Station	Date Sampled[1]	Coprostanol	Epicoprostanol	c/e ratio
1	Mar. 19[2]	235	NA[2]	NA[2]
	Aug. 6	BLD[3]	BLD	- -
	Aug. 29	6	10	0.62
	Oct. 10	BLD	BLD	- -
	Nov. 18	17	14	1.20
	Dec. 16	11	9	1.24
2	All Dates	BLD	BLD	- -
3	Mar. 19	239	NA	NA
	Dec. 16	BLD	BLD	- -
4	Mar. 19	275	NA	NA
	Dec. 16	16	35	0.45
A	Mar. 19	4851	NA	NA
	Aug. 8	23	55	0.42
	Aug. 29	61	248	0.25
	Oct. 10	145	167	0.87
	Nov. 18	179	261	0.69
	Dec. 16	99	85	1.17
B	Mar. 19	2022	NA	NA
	Dec. 16	568	421	1.35
C	Mar. 19	1200	NA	NA
	Dec. 16	249	327	0.76
D	Mar. 19	1926	NA	NA
	Dec. 16	52	110	0.48
E	Aug. 8	67	102	0.65
	Aug. 29	248	1006	0.25
	Oct. 10	238	396	0.60
	Nov. 18	325	608	0.53
	Dec. 16	647	1434	0.45
F	Aug. 8	245	193	1.27
	Aug. 29	1558	1736	0.90
	Oct. 10	1005	619	1.62
	Nov. 18	268	384	0.70
	Dec. 16	733	894	0.82

1. All samples collected during 1991.
2. March samples were not derivatized prior to analysis, coprostanol and epicoprostanol would coelute and are reported as coprostanol. NA is Not Applicable. Dashes - no ratio determined.
3. BLD: below limit of detection (5 ng/g).

less than 20% at any one time. The c/e ratio was also highly variable over time (0.25 to 1.27 at a single station) and did not appear to be correlated with total fecal sterol concentrations. These data suggest that the sediment concentration as well as the c/e ratio may vary with season, either due to changes in source, movement or dilution of bottom sediments, or *in situ* processes such as biological activity.

Marine Mammals. Results from the marine mammal feces are shown in Table II. The two species studied are picivorous (dolphins) and herbiverous (manatees). The manatee feces provided the highest concentrations of fecal sterols (902 to 1577 µg/g) with c/e ratios in the range of 1.4 to 3.3. While the concentrations of the fecal sterols in manatee feces are lower than those observed in whale feces (*7*), the concentrations and c/e ratios determined are similar to those reported for domestic wastewaters.   In contrast to the report by Venkatesan and Santiago (*7*), who reported no detectable fecal sterols in a dolphin fecal sample, we found summed coprostanol and epicoprostanol concentrations of 1.1 to 15.4 µg/g for the dolphin feces with c/e ratios of 0.05 to 0.7 (Table II). The lowest concentration and highest ratio was determined in the fecal sample of a calf which was thought to still be nursing. The literature (*1*) suggests that in humans, the enteric bacterial population needed for the production of the fecal sterols does not develop until after the transition to solid food. Whether this represents a real difference between the species or that the calf had in fact started the transition to fish is unknown. The rehabilitation lagoon sediments had concentrations similar to those observed in the other sediments studied here (130-160 ng/g), with epimer ratios of 0.75 to 0.90. The differences in coprostanol concentrations and epimer ratios between these two species of marine mammals may be either dietary or physiologically based, both have been demonstrated to affect coprostanol production in many other animals, including humans (*1*).

Summary and Remaining Questions

Summed fecal sterol concentrations found in the sediments studied here (30 to 2437 ng/g) generally compare favorably with coprostanol concentrations observed in other coastal Florida sediments (*14-16*). Coprostanol concentrations ranging from 110 to 1250 ng/g in sediments from the Southeast Atlantic and Gulf Coasts were reported by the Benthic Surveillance Project of the NOAA Status and Trends Program (*14*). Sediment concentrations have been reported to range from < 10 to 4800 ng/g in Biscayne Bay (*15*), and to 3700 ng/g in the St. Johns River Basin (*16*). Since these earlier investigations did not separate the epimers the coprostanol concentrations reported are likely to represent a summation of both epimers.

As discussed above, the c/e ratio of Sarasota Bay sediments ranged from 0.2 to 1.5 with most values below 0.5. The Weeki Wachee sediments had c/e ratios of 0.25 to 1.35 and were highly variable at a given station over time. Sediments adjacent to the Key West wastewater outfall had c/e ratios of 0.2 to 0.4, whereas those collected further away had ratios which ranged from 0.7 to 9. If these data are interpreted based on the hypothesis forwarded by Venkatesan and Santiago (*7*) the conclusion would be that regardless of proximity to known sources, most of the fecal sterols found in these sediments were derived

Table II. Marine Mammal sterol concentrations (μg/gm dry weight)

Sample[1]	Coprostanol	Epicoprostanol	c/e ratio[2]
Manatees			
un-named	669	255	2.62
	2193	915	2.40
Gene	1208	368	3.28
Hugh	592	422	1.40
	544	358	1.52
	967	536	1.80
Dolphins			
92-05	4.38	11.1	0.40
92-04	0.37	6.95	0.05
92-01	0.47	0.66	0.71
Dolphin Holding Lagoon Sediment			
A	0.05	0.07	0.71
B1	0.07	0.09	0.78
B2	0.08	0.09	0.89

1. Marine mammal feces were not homogenized, duplicate analyses are
   listed separately when performed.
2. Coprostanol to epicoprostanol epimer ratio.

from marine mammal inputs (specifically dolphins) due to the low c/e ratios determined. However, as discussed earlier, this ratio can vary even between samples from wastewater treatment facilities. Eganhouse et al. (10) reported coprostanol and epicoprostanol concentrations in sludges from six treatment plants using different treatment methods. The c/e ratio for these six sludges ranged from 0.9 to 9.0. Wastewater effluent from one of these treatment plants was analyzed in 1988 (subsequent to the other study). While the anaerobically digested sludge had an c/e ratio of 0.9 in the study by Eganhouse et al. (10) the c/e ratio for solids isolated from 4 effluent samples from this plant ranged from 23 to 38, and the filtrate of these samples did not contain any measurable epicoprostanol (11). These studies demonstrate the variability of the c/e ratio even in domestic wastewaters.

Questions remain regarding the significance of the presence of low (ng/g dry weight) concentrations of the fecal sterols. Environmental factors, such as in situ biological activity, which could affect the persistence or conversion of the epimers need to be further elucidated. The potential contributions by stormwater runoff of domestic and agricultural animal wastes, to the fecal sterol sediment burdens, also requires further investigation. The importance of depositional zones, and potentially, the redox conditions in the impacted sediments is highlighted by the Sarasota Bay and Key West results. The Weeki Wachee River, and perhaps, the Sarasota Bay samples indicate that septic tank leachate or domestic animal related runoff may be a major source of fecal sterols in residential areas. The marine mammal data considered in conjunction with previous work (8,10,11) indicates that the epimer ratio does not provide a definitive method to separate domestic wastewater impacts from marine mammal wastes, particularly when the fecal sterols are present in low concentrations. The transport and fate of these compounds needs to de determined in order to evaluate their effectiveness as molecular markers for the presence of domestic waste. They may be useful in monitoring studies when the source is known (as in the Key West survey) or when the potential contributions from alternate sources can be evaluated (as in the Weeki Wachee River survey). However, without a clearer understanding of the cycling of these compounds, reliance on the presence of just coprostanol, or even both epimers, may lead to questionable conclusions regarding source contributions to a sediment (particularly at the concentrations observed in these studies).

Acknowledgments

The authors would like to thank Ms. Lisa Ford for assistance in extraction and analysis; L K Dixon, S Lowrey and the reviewers for useful suggestions concerning this report. Funding for these projects was provided by the EPA National Estuary Program, through the Sarasota Bay National Estuary Program, the EPA Office of Coastal Programs, the Southwest Florida Water Management District and the Mote Marine Laboratory Research Foundation.

Literature Cited

1. Walker, R. W.; Wun, C. K; Litsky W. *CRC Crit. Rev. in Environ. Control* **1982**. *10*, 91-112.
2. Vivian, C. M. G. *Sci. Total Environ.* **1986**. *53*, 5-40.
3. Taylor, C. D.; Smith, S. O.; Gagosian R. B. *Geochim. Cosmochim. Acta* **1981**. *45*, 2161-2168.
4. Toste, A. P. Ph.D. Dissertation, University of California at Berkeley. **1976**.
5. Venkatesan, M. I.; Ruth, E; Kaplan I. R. Mar. Pollut. Bull. **1986**. *17*, 554-557.
6. Venkatesan, M. I.; Santiago C. A. *Mar. Biol.* **1989**. *102*, 431-437.
7. Venkatesan, M. I.; Kaplan I. R. *Environ. Sci. Technol.* **1990**. *24*, 208-214.
8. Grimalt, J. O.; Fernández, P.; Bayona, J. M.; Albagés, J. *Environ. Sci. Technol.* **1990**. *24*, 357-363.
9. McCalley, D.V.; Cooke, M.; Nickless, G. *Water Res.* **1981**. *15*, 1019-1025.
10. Eganhouse, R.P.;Olaguer, D.P.; Gould, B.R.; Phinney, C.S. *Mar. Environ. Res.* **1988**. *25*, 1-22.
11. Sherblom, P.M. Ph.D. Dissertation, University of Massachusetts at Boston. **1990**.
12. EPA 530/SW-846 Test Methods for Evaluating Solid Waste: Physical/Chemical Methods 3rd ed. Final Update I, July **1992**.
13. Pierce, R.H.; Brown R.C. *Bull. Environ. Contam. Toxicol.* **1984**. *32*, 75-79.
14. NOAA. United States Department of Commerce. National Oceanic and Atmospheric Administration. National Status and Trends Program. Progress Report and Preliminary Assessment of Findings of the Benthic Surveillance Project **1984**. 81pp.
15. Pierce, R.H.; Brown R.C. A Survey of Coprostanol Concentrations in Biscayne Bay Sediments. Final Report to Dade County Environmental Resource Management, Miami, Florida. **1987**. 38pp
16. Pierce, R.H.; Dixon, L.K.; Brown, R.C.; Rodrick G. Characterization of Baseline Conditions of the Physical, Chemical and Microbiological Environments in the St. Johns River Estuary. Final Report to the Florida Department of Environmental Regulation. Contract # SP132. **1988**. 11pp.

Chapter 22

The Use of Azaarenes as Indicators of Combustion Sources to the Urban Atmosphere and Estuarine Systems

Martin R. Preston, Hung-Yu Chen, and Peter J. Osborne

Oceanography Laboratories, Department of Earth Sciences, University of Liverpool, Liverpool L69 3BX, United Kingdom

Azaarenes produced by the combustion of fossil fuels are present in the Liverpool urban atmosphere in particulate and vapour forms. They are also present in the sediments and suspended matter of the River Mersey estuary. The atmospheric partitioning of these chemicals is temperature dependent and well described by the Pankow and Bidleman (*1*) relationship. Both the total atmospheric concentrations and the vapour/particle partitioning show a strong seasonal variability. Laboratory studies reveal that a significant proportion of the particle associated azaarenes in the atmosphere are water soluble with the extent and rate of dissolution being dependent on pH and both increasing as the pH decreases. The consequences of this dissolution for the use of azaarenes as markers of combustion products are explored.

Azaarenes are polycyclic compounds containing a nitrogen atom within the ring structure. Examples (Figure 1) include quinoline and isoquinoline (2 ring), acridine, benzoquinolines and phenanthridine (3 ring) and azapyrene, azachrysene and benzacridine (4 ring). They are produced (either as the parent compounds or methylated derivatives) largely by the combustion of fossil fuels (*2,3*). They are also present in, for example, smoke contaminated food or cigarette smoke (*4*). Azaarenes have been detected in a number of environmental matrices including urban air samples (*5,6*) marine and freshwater sediments (*7-9*), stack gases from coal combustion sources and industrial effluents (*10*).

Azaarenes are also known to have significant toxicological properties including those of tumour initiators on mouse skin (*11-13*). In these respects they are similar to the structurally similar polycyclic aromatic hydrocarbons (PAH). Azaarenes differ however, in that they have a significantly greater range of volatility than PAH. They also have basic properties which render them much more soluble in aqueous (particularly acidic) solutions. They are therefore considerably more responsive to external environmental factors such as temperature changes and rainfall than PAH and this makes them potentially interesting as markers of combustion inputs to both atmospheric and aquatic environments.

Despite their relatively high toxicity azaarenes have received comparatively little scientific attention as environmental contaminants. However, as part of an ongoing programme of studies relating to the inputs of atmospheric organic pollutants to marine systems we have been conducting a detailed investigation of the atmospheric and estuarine variability of azaarenes. These field measurements have been augmented with some laboratory based studies of the solubility characteristics of azaarenes in urban aerosol samples.

Analytical methodology

Sampling. All suspended sediment and most sediment samples were collected from the Mersey Estuary (see Figure 2) during a number of exercises conducted in 1993 and 1994. A custom built sampling unit (*14*) or a hand held grab was deployed, using the National Rivers Authority boat *Sea Jet* as a sampling platform. A few sediment samples were collected from inter-tidal mud flats in places where these were accessible. Suspended sediment samples were collected in 2.5 litre glass Winchester bottles sealed with PTFE lined caps and pre-ashed aluminium foil. These samples were filtered as soon as possible onto pre-ashed (400 °C, 8 h) Whatman GF/F filters using a N_2 pressured (positive pressure) stainless steel filtration unit (Sartorius).

Sediments recovered with the grab (representing approximately the upper 5 cm) were emptied onto a tray and transferred into glass jars with a solvent rinsed metal spatula. The jars were covered with pre-ashed aluminium foil before the cap was screwed on. Both sediment and suspended sediment filters were then freeze dried before being taken on to the extraction stage. In the case of the sediments the <125 μm fraction was analysed and this was separated by dry, mechanical sieving on an automated sieve shaker and using solvent washed, all metal standard sieves.

Atmospheric samples were collected using a Sierra Anderson Hi-Vol sampling unit mounted on the roof of a 22 m high building within the campus of the University of Liverpool. Samples were collected at a flow rate of 0.8 m^3 min^{-1} onto pre-ashed glass fibre filters. The filters have a reported efficiency of 99.998 % for 0.3-0.4 μm particles. Sampling periods were generally around 48 h corresponding to a sampling volume of ca. 2300 m^3. A Sierra Anderson PUF (polyurethane foam) sampler unit was used for the collection of aerosol/vapour phase materials using a flow rate of 0.28 $m^3 \cdot min^{-1}$ with a sample volume of ca. 2400 m^3.

Glassware, filters and PUF plugs. All glassware was initially soaked overnight in Decon detergent. It was then rinsed with 2M hydrochloric acid, distilled water and then soaked in 2M sodium hydroxide overnight. Finally, the glassware used for water/suspended sediment sampling (2.5 litre Winchester bottles) was rinsed with Milli-Q water, sealed with pre-ashed aluminium foil and used within 48 h. Glass jars used for sediments were treated as above but were oven dried (150 °C, 2 h) and then baked in a muffle furnace (350 °C, 18 h). After this, the jars were sealed with pre-ashed aluminium

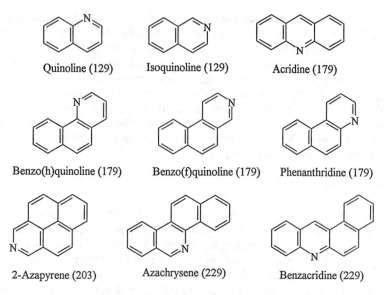

Quinoline (129)   Isoquinoline (129)   Acridine (179)

Benzo(h)quinoline (179)   Benzo(f)quinoline (179)   Phenanthridine (179)

2-Azapyrene (203)   Azachrysene (229)   Benzacridine (229)

Figure 1. The structure of some azaarenes.

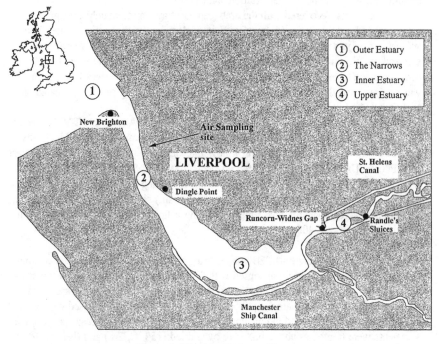

① Outer Estuary
② The Narrows
③ Inner Estuary
④ Upper Estuary

New Brighton

Air Sampling site

LIVERPOOL

St. Helens Canal

Dingle Point

Runcorn-Widnes Gap

Randle's Sluices

Manchester Ship Canal

Figure 2. The River Mersey Estuary.

foil until required. Laboratory glassware was washed as above but, in the final stage, was extracted with redistilled extraction solvent.

Glass fibre filters for air sampling were ashed in a muffle furnace (400 °C overnight) then wrapped in pre-ashed aluminium foil until required. PUF plugs were exhaustively extracted with dichloromethane in a large Soxhlet apparatus, dried, then stored in pre-ashed foil. All filters and PUF plugs were loaded into and removed from the holders in a laminar flow cupboard.

Solvents. HPLC grade solvents (Rathburn Chemicals Limited, Walkerburn, Scotland and Fisons Scientific Equipment, Loughborough) were used throughout the study and were redistilled through an all glass (1 m column) fractional distillation unit before use. Each solvent batch was tested before use by concentrating a 300 mL aliquot to 50 μL (initially by rotary evaporation to 1 mL followed by blow down to dryness using a gentle stream of purified nitrogen) and analysing the concentrate for impurities by gas chromatography-mass spectrometry (GC-MS).

Standards. Standard azaarenes were obtained from the following sources: 6-methylquinoline (Sigma Chemical Co. Ltd, UK) 8-methylquinoline, acridine, phenanthridine, 2,6,-di-*p*-tolylpyridine (Aldrich Chemical Co. Ltd, UK); 2,6-dimethylquinoline, 7,8-benzoquinoline, 9-methylacridine (Janssen Chimica, Cheshire, UK); 2-azapyrene, 2-azachrysene (NCI Chemical Carcinogen Repository, MRI Michigan, USA); d_9-acridine (MSD Isotopes, Cambrian Gases, K&K-Greef Ltd., , UK).

Gas chromatography (GC) and Gas Chromatography-Mass Spectrometry (GC-MS). All GC work was carried out on a Hewlett Packard HP5890A Series II instrument with on-column injection, flame ionisation and nitrogen phosphorus detectors and a HP3396A integrator. GC-MS was conducted on a similar system (HP5890A Series I GC) coupled to a VG250 double focusing, magnetic sector mass spectrometer using selective ion monitoring mode. Data processing was performed using a networked pair of VAX 2000 Workstations. The main chromatography columns used were 5% phenyl/methylsilicone; either HP-5 (25 m, 0.32 mm i.d., 0.52 μm phase thickness - Hewlett Packard) or DB5-MS (25 m, 0.25 mm i.d., 0.25 μm phase thickness - J&W Scientific). These were programmed from an initial 60 °C (1 min. hold) at 10 °C min^{-1} to 300 °C with a final 10 min hold giving a total analysis time of 35 min.

The compounds quantified (and the representative ions utilised were):-Quinoline/isoquinoline (m/z 129) methylquinolines/methylisoquinolines (6 isomers; m/z 143), dimethylquinolines (11 isomers; m/z 157), trimethylquinoline isomers (10 isomers; m/z 171), benzoquinolines (2 isomers; m/z 179), acridine/phenanthridine (m/z 179), methylated 3-ring azaarenes (7 isomers; m/z 193), azafluoranthene (2 isomers, m/z 203), azapyrene (2 isomers m/z 203), benzoazafluorene (2 isomers; m/z 217) benzacridine/ azachrysene (4 isomers; m/z 229). Total azaarenes were quantified as the sum of the above components using an internal standard method.

Sample extraction and cleanup: The main analytical procedures utilised for azaarene analyses were based on those described by Brumley *et al.*, (*15*) as modified by Osborne (*16*).

Dried samples (sediments or aerosol or suspended matter including filter) were extracted for 18 h with dichloromethane in a Soxhlet apparatus. An extraction rate of 12 cycles per hour was shown to be adequate to extract >92 % of the azaarenes present (16). The organic extract (ca. 250 mL) was then concentrated to ca. 1 mL by rotary evaporation and was quantitatively transferred to a pre-cleaned glass vial using 2x1 mL aliquots of dichloromethane for washing the flask. If the clean-up procedure was not to be carried out immediately then the vials were tightly capped and stored at -24 °C.

In the case of the leaching experiments, identical pairs of exposed high volume filters were first weighed and cut into equal sized pieces (6 in total) which were also weighed. One piece was retained unleached as a control and each of the others was then placed in a beaker containing a large excess (1 litre) of potassium dihydrogen phosphate buffer solution (adjusted to pH 6.04, 7.12 or 8.14 with sodium hydroxide) and allowed to soak for a specified time. The solutions were not stirred. After this period each filter piece was removed and any remaining particles dislodged from the filter were recovered from the buffer solution by filtration onto a pre-ashed glass fibre filter. The leached filter plus the recovered particles were then freeze-dried for 24 hours before being soxhlet extracted and treated as below.

The concentrated extracts were further processed by first adding an internal standard (2,6,-di-p-tolylpyridine or d_9-acridine) and then a 1 mL aliquot of 4M HCl. The mixture was shaken for 10 minutes on a IKA Vibrax-VXR Shaker and the two phases allowed to separate. The acidified layer was removed and placed in a clean vial whilst the organic layer was extracted with acid two further times. All acidic extracts were then combined and cooled in an ice bath for 15 minutes. Sufficient 6M NaOH was then added to raise the pH to 14, and this basic aqueous phase was back extracted into dichloromethane (3×) using the same method as for the initial acid extraction. The organic layers were combined, dried by passage through a short column of anhydrous sodium sulphate and evaporated to approximately 1 mL under a gentle stream of nitrogen while maintaining a low temperature. The samples were then either analysed immediately or stored at -24 °C until analysis could be performed. The coefficient of variation of 4 replicate sediment sample analyses was within the range 2.6 to 9.0 % with the exception of the most volatile methylquinolines where the coefficient of variation was 24 %. For 'total azaarenes' (see above) the coefficient of variation was 7.1 %. For 4 replicate aerosol filter samples the coefficient of variation of individual compounds was generally <10 %. Corrections for recovery were made on the basis of recovery of standard compounds relative to the internal standard (60-85 %).

Results and Discussion

Atmospheric Azaarenes.

Composition. In total, 152 aerosol samples were collected, and their azaarene concentration data are summarised in Table I. The composition of the aerosol is dominated by lower molecular weight species with methylated quinolines and isoquinolines representing around 60-65 % of the total (median value 63.5 %, range 11-86 %). In general, there are strong correlations between the Σ2 ring and Σ3 ring species and between the Σ3 ring and Σ4 ring species (best shown on a logarithmic scale, Figures 3a,b) with log-log correlations of r = 0.79 and 0.91 (p<0.01, n =152) respectively. A weaker correlation between $\Sigma 2$ ring and $\Sigma 4$ ring species (r = 0.57, p<0.01, n =152) was also observed.

Variability. There is a clear seasonal pattern of variability in both the azaarene concentrations in the aerosol (Figure 4a) and the vapour/particle partitioning (Figure 4b). There is also an indication from Figure 4a that there is an inverse relationship between the total quantity of azaarene in the aerosol and the ambient temperature, at least on a monthly mean basis.

There are clearly two possible explanations for this seasonal behaviour, and these are not mutually exclusive. The first is that there in a seasonal fluctuation in the flux of azaarenes to the atmosphere reflecting the different levels of usage of fossil fuels at different times of year. The second explanation is that the variability is due to changes in the partitioning behaviour with a much higher proportion of the azaarenes in the vapour phase in warmer weather. On the basis of other experiments that we have conducted with vapour/particle phase partitioning (see below) we believe that a change in source strength at different times of year is the dominating influence.

Table I Summary of azaarene concentrations in aerosol samples

Compounds	Concentration (ng m^{-3})		
	Mean	Median	Std Dev.
quinoline/isoquinoline	0.32	0.10	0.62
methylquinolines	0.41	0.16	0.79
dimethylquinolines	0.99	0.56	1.48
trimethylquinolines	0.39	0.16	0.63
Total Σ2-ring species	2.11	1.08	3.36
acridine/phenanthridine/ benzoquinolines	0.54	0.27	0.79
methylacridines/phenanthridine/ benzoquinolines	0.49	0.23	0.71
Total Σ3-ring species	1.03	0.53	1.51
azapyrenes/benzoazafluoranthenes	0.07	0.04	0.09
azachrysenes/benzacridines	0.02	0.01	0.05
Total Σ4-ring species	0.10	0.06	0.12
TOTAL AZAARENES	3.24	1.65	4.82

During the study period, ambient temperatures ranged between ~ -10 and +30 °C. Usage of coal and oil in domestic fires, boilers and the major local power station (2500 MW burning up to ~10^5 t coal per week) can be assumed to have

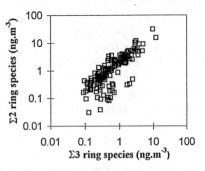

Figure 3(a). Concentrations of Σ2 ring azaarene species versus Σ3 ring species.

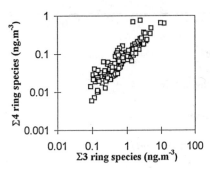

Figure 3(b). Concentrations of Σ4 ring azaarene species versus Σ3 ring species.

varied according to demands for heat and power. The first hypothesis can be tested using Figure 4(b) which clearly demonstrates the importance of the variable source strength by showing that there are significant changes in the total (particle + vapour) azaarene concentrations over the study period with the same pattern of high winter and low summer concentrations. Within this overall variability there are differences in the particle/vapour concentration ratios.

This partitioning relationship can be examined using the Pankow and Bidleman partitioning coefficient K_p (1). This is defined as:-

$$K_p = m_p \frac{F/[TSP]}{AT} + b_p \qquad (1)$$

where F (ng m^{-3}) (normalised to the total particle concentration TSP (μg m^{-3})) and A (ng m^{-3}) are the particulate and gas-phase concentrations respectively, T is the absolute temperature and m_p and b_p are the slope and intercept respectively of the plot of K_p against $1/T$. A plot of this type is shown in Figure 5.

Using plots of this type it is possible to calculate both the m_p and b_p parameters from equation 1 and these are shown in Table II together with some comparative values for selected PAH. It is clear from these data that there is a considerably greater range of partitioning behaviour for azaarenes than for the PAH which behave rather more as a single entity. This singularity of PAH behaviour has also been noted in estuarine suspended material and sediments and is attributed to the occlusion of PAH within particles ($18,19$). Our results indicate that azaarenes are less firmly associated with particulate matter than PAH and are possibly incorporated through surface adsorption. As azaarenes are apparently formed by the same combustion mechanisms as PAH the differences in observed behaviour are notable. The explanation may lie in the fact that the greater volatility/solubility of azaarenes as compared to PAH leads to greater lability within the particles.

Post-Depositional Behaviour of Aerosol Azaarenes. A number of laboratory experiments were conducted to examine the post-depositional behaviour of atmospheric aerosol particles introduced to water. It was strongly suspected that significant changes would occur because of the much more basic behaviour of azaarenes compared to PAH and their greater water solubility.

A series of high volume filter samples of the urban aerosol were collected. These filters were divided, and leached in buffer solutions as detailed in the Experimental section and the rates and extents of the dissolution of the azaarenes was determined The pH of the buffer solutions were within the range commonly encountered in estuarine and marine waters.

The results of these experiments are summarised in Figure 6 from which it can be seen that within the pH of many natural water (pH 6-8), azaarenes very rapidly dissolve with both the rate and extent of the dissolution being slightly greater at lower pH. Experiments were also conducted at pH<6 but in these cases the dissolution rates were too high to permit useful quantitation. The significance of these results is that under most conditions of dry deposition of aerosol to surface waters rapid removal of azaarenes is to be expected. In the case of wet deposition, the acidity of rainfall, particularly in relatively polluted environments, suggests that

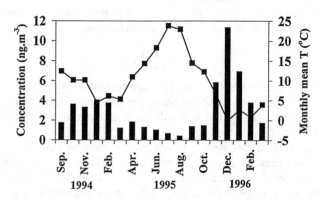

Figure 4(a). Monthly mean variations in total particulate azaarene concentrations (bars) and ambient temperature (solid line).

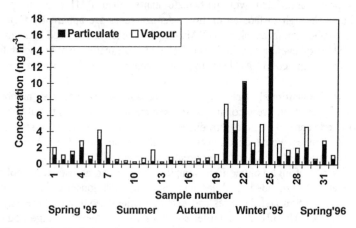

Figure 4(b). Variations in the proportions of particulate and vapour phase azaarenes

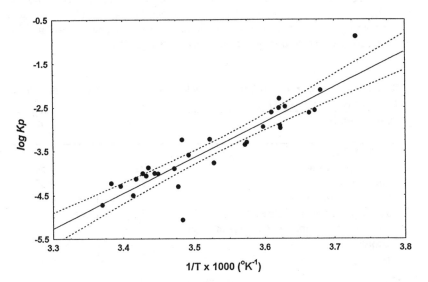

Figure 5. log *Kp* (equation 1) plotted against inverse absolute temperature for the acridine, benzoquinoline, phenanthridine group (m/z 179). The dotted lines indicate 95% confidence limits.

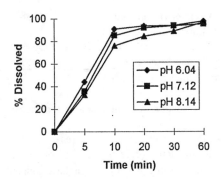

Figure 6. Dissolution rates of total azaarenes from aerosol samples as a function of pH.

most, if not all, of the azaarenes initially present in association with atmospheric particles will reach the water surface in dissolved form.

One unquantified parameter influencing the rate of dissolution was temperature. Dissolution experiments were performed at ambient laboratory temperatures and it is to be expected that rates will be significantly slower at the lower temperatures encountered in the natural environment, particularly in winter when azaarene concentrations in the atmosphere are highest.

TABLE II m_p, b_p and correlation coefficient (r)values for azaarenes (present work) and PAH (1)

Compound	m_p	b_p	$r^{(1)}$
AZAARENES			
methylquinolines	2294	-9.89	0.56
dimethylquinolines	2800	-11.36	0.59
trimethylquinolines	3847	-14.60	0.78
benzoquinolines/acridine/phenanthridine	3512	-13.88	0.91
methylacridines/methylbenzoquinolines	3914	-15.05	0.91
azapyrenes	5613	-20.59	0.87
azachrysenes	5694	-20.35	0.80
PAH			
phenanthrene/anthracene	4095	-18.38	
Me-phenanthrene/Me-anthracene	3359	-15.44	
fluoranthene	4402	-18.44	
pyrene	4167	-17.50	
benzo(a)pyrene/benzo(b)fluorene)	4538	-18.44	
chrysene/benz(a)anthracene/triphenylene	5806	-21.83	
benzo(b)fluoranthene/benzo(k)fluoranthene	5677	-20.19	
benzo(a)pyrene/benzo(b)pyrene	4867	-17.00	

[1] values are significant at p<0.01

Azaarenes in estuarine suspended material and sediments

Extensive studies of the concentrations of azaarenes in both suspended matter and sediments have been conducted (16,19). Concentrations of total azaarenes vary between low ng g^{-1} and low μg g^{-1} with the highest concentrations (≤ 2.5 μg g^{-1}) being found in cohesive, muddy sediments in local canals. The very limited water exchange in the local canal system means that these waterways act to some extent as traps for azaarenes in both runoff and from atmospheric deposition. Interestingly, the slope of the relationship between trimethyl and dimethyl quinolines in both the canal sediments and atmospheric aerosol material is very similar (19), suggesting that either there is a common source to both reservoirs or that the main source to the canal sediments is through atmospheric deposition.

Within the main body of the estuary the tidal range is high (≥10 m) with very high currents which tend to scour fine-grained deposited material from the sediments. Concentrations of total azaarenes in the suspended matter fall within the range of 50-250 ng g⁻¹. However, there are changes in the composition of the suspended matter as it progresses down the estuary which modify the relative abundance of azaarene components. These are summarised in Figure 7 from which it can be seen that the proportion of lower molecular weight, two-ring species decrease fairly rapidly with increasing distance from the tidal limit suggesting that selective leaching of these species is occurring. It therefore follows that the azaarenes exported from the estuarine system in particulate form will be dominated by higher molecular weight species, and it is these that are most likely to enter the food chain through the actions of detritus feeding organisms.

The pH dependence of solubility behaviour revealed in the experiments discussed above also have implications for estuarine cycling of azaarenes. In sediments, particularly oxygen depleted sediments, the pH of pore waters drops and this will then cause sedimentary azaarenes to dissolve. If, at a later stage, tidal resuspension of these sediments injects sedimentary pore waters into the water column then it is conceivable that azaarenes may then tend to reassociate with particulate matter. Given that lower molecular weight species tend to dissolve more rapidly, this combination of pH variation and sediment resuspension could well lead to the selective transport of lower molecular weight as opposed to higher molecular weight compounds. The importance of this cycling and fractionation remains to be investigated.

Conclusions

There is clear evidence that azaarenes are widely distributed within both the aquatic and atmospheric environments. These compounds display more variation in their behaviour than do PAH, and their distributions are strongly influenced by both temperature and pH. The vapour/particulate phase partitioning is well described by the Pankow and Bidleman approach to a linear Langmuir isotherm, allowing the reliable prediction of partitioning behaviour to be achieved. Seasonal variations in atmospheric azaarene concentrations are predominantly due to variations in the strength of the source signal. Recent work (*20*) indicates that there is a significant contribution of anthropogenically derived organic compounds dissolved in rainwater to the overall nitrogen flux to the oceans. The fact that azaarenes dissolve rapidly in natural waters at representative pH values indicates that these compounds may make a measurable contribution to global nitrogen fluxes. It is also possible that this behaviour has consequences for human health where the dissolution of inhaled aerosol particles containing azaarenes could lead to locally high concentrations of potentially damaging chemicals.

In estuarine systems there is evidence of similarities between atmospheric and sedimentary azaarene distribution suggesting common sources (16, 19). Once present in the wider estuarine system however, fractionation of particulate azaarenes occurs with a rapid diminution of the relative amounts of lower molecular weight azaarene species.

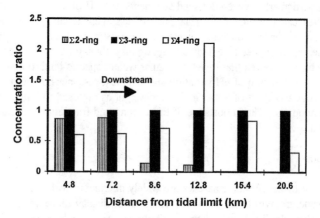

Figure 7. The relative amounts of 2,3, and 4 ring azaarenes in Mersey Estuary suspended matter normalised to total 3 ring species.

Acknowledgments

We should like to acknowledge the support of the Association of Commonwealth Universities (PJO) and Chiu-Mao Chen (H-Y C) for financial support during the course of this study. We are also very grateful to Dr Peter Jones and his colleagues from the National Rivers Authority for their assistance with the fieldwork and to Dr Anu Thompson for assistance with GC-MS analysis.

Literature Cited

1.   Pankow, J.F.; Bidleman, T.F. *Atmospheric Environment* **1991**, *25*, 2241-2249.
2.   Sawicki, E.; Meeker, J.E.; Morgan, M.J. *Archiv. Env. Health* **1965**, *11*, 773-775.
3.   Sawicki, E.; Meeker, J.E.; Morgan, M.J. *Int. J. Air Wat. Poll.* **1965**, *9*, 291-298.
4.   Grimmer, G.; Naujack, K.W.; Dettbarn, G. *Toxicol. Lett.* **1987**, *35*, 117-124.
5.   Dong, M.W.; Locke, D.C.; Hoffmann, D. *Env. Sci. Technol.* **1977**, *11*, 612-618.
6.   Warzecha, L. *Chemia Analilyczna* **1993**, *38*, 571-583.
7.   Wakeham, S.G. *Env. Sci. Technol.* **1979**, *13*, 1118-1123.
8.   Furlong, E.T.; Carpenter, R. *Geochim. Cosmochim. Acta* **1982**, *46*, 1385-1396.
9.   Onsuka, T.I.; Terry, K.A. *J. High Res. Chrom.* **1989**, *12*, 362-367.
10.   Catallo, W.J.; Hoover, D.G.; Vargas, D. *Aquatic Toxicology* **1994**, *29*, 291-303.
11.   IARC, IARC Monographs on the Evaluation of the Carcinogenic Risk of Chemicals to Humans: Polynuclear Aromatic Compounds, Part 1, Chemical, Environmental and Experimental Data, World Health Organization, France **1983** Vol. 32; pp. 123-134,
12.   Vo-Dinh, T. In *Chemical Analysis of Polycyclic Aromatic Compounds* Vo-Dinh, T. Ed.; John Wiley & Sons: New York. **1989**, pp. 1-30.
13.   Matzner, R.; Bales, R.C. *Chemosphere* **1994**, *29*, 1755-1773.
14.   Preston, M.R.; Al-Omran, L.A. *Mar. Poll. Bull.* **1986**, *17*, 548-553
15.   Brumley, W.C.; Brownrigg, C.M.; Brilis, G.M. *J. Chromatog.* **1991**, *558*, 223-233.
16.   Osborne, P.J. *Azaarenes in Some Merseyside Sediments.* Ph.D. Thesis, University of Liverpool, **1995**, 257pp.
17.   Readman, J.W.; Mantoura, R.F.C.; Rhead, M.M. *Fresenius Z. Anal. Chem.* **1984**, *319*, 126.
18.   Readman, J.W.; Mantoura, R.F.C.; Preston, M.R.; Llewllyn, C.A.; Reeves, A.D. *Int. J. Environ. Anal. Chem.* **1987**, *27*, 29-54.
19.   Osborne, P.J.; Preston, M.R.; Chen, H.-Y. *Marine Chemistry* (submitted).
20.   Cornell, S.; Rendell, A.; Jickells, T. *Nature* **1995**, *376*, 243-246.

Chapter 23

Toxaphene Residue Composition as an Indicator of Degradation Pathways

H. Parlar[1], G. Fingerling[1], D. Angerhöfer[1], G. Christ[1], and M. Coelhan[2]

[1]Department of Chemical Technical Analysis, Technical University Munich,
85350 Freising-Weihenstephan, Germany
[2]Department of Analytical Chemistry, University of Kassel,
Heinrich-Plett-Strasse 40, 34109 Kassel, Germany

Toxaphene, a mixture of chlorinated bornanes and camphenes with a high proportion of enantiomers, can be used as an indicator for the investigation of environmental conditions which lead to the compositional pattern of toxaphene in environmental samples. Due to its persistence and long range transport, toxaphene is distributed throughout the world. The changes in composition after biotic or abiotic transformation are visible in the gas chromatograms of different environmental samples. These are important clues to translocation and degradation pathways. The results of enantioselective residue analysis and of degradation experiments with single toxaphene components suggest the dominance of photolysis over other abiotic or biotic processes in air, water, and fish.

After being released into the environment, chemicals are normally changed by abiotic and/or biotic processes with the result that residue analysis sometimes reveals complex patterns of the parent compounds and numerous products of variable transformation pathways. As different processes often lead to the same products, it is sometimes difficult to decide which of several possible mechanisms is responsible. The main question is whether a given derivative identified by an analytical method is a metabolite, that is, a product of enzymatic transformation, or a product of abiotic processes, such as photolysis, thermolysis, hydrolysis, or radical induced reactions. This problem can be solved analytically, if either the parent compound or the product is optically active (Figure 1). If the parent compound (A) is a racemate with an enantiomeric ratio of 1:1, the ratio should change during biodegradation, as enzymatic processes normally are enantioselective resulting in different rate constants ($k_+ \neq k_-$). After abiotic degradation, the enantiomer ratio remains unchanged. The same should hold with symmetrical parent compounds which are converted to asymmetrical products (+C and –C). Enantiomer selective conversion has been investigated especially with α-HCH (1-3), chlordane (4), and, recently, with toxaphene (5).

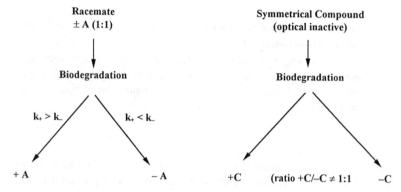

Figure 1. Possible routes of enantioselective degradation of environmentally relevant chemicals

Enantiomer selective residue analysis provides an opportunity to distinguish between these two main pathways, provided, a suitable chemical is at hand. Suitable substances are those where either parent compounds or products are asymmetrical, which can be found in different environmental compartments (air, water, soil, and organisms) in concentrations high enough to be easily quantified, and where analytical methods for enantioselective separation are available. In this respect, toxaphene is interesting for the following reasons:

1. Toxaphene has been one of the most widely used insecticides in the world; usage from 1950 to 1993 has been estimated to be about 1,330,000 tons (6).

2. Due to its high persistence and mobility, toxaphene is distributed throughout the world (7-15).

3. Toxaphene is a complex mixture of more than 200 compounds with 98% of them being racemates (5).

4. Some of the components are easily degraded in various environmental systems while others are stable. This results in a highly variable gas chromatographic patterns depending on the nature and origin of the sample (16-21).

Experimental

In this section, experimental procedures are given only for work that has been executed in our laboratory (residue analysis in fish and fish products, degradation in soil, and photolysis). For background on other analytical data see special references.

Residue Analysis. Technical toxaphene standard (Camphechlor) and standard toxaphene components were obtained from Ehrenstorfer, Germany. Organic solvents used were of purity grade for residue analysis. Na_2SO_4 and H_2SO_4 (95-97%) were from Merck, Germany. The analyzed fish and fish products were kept under $-12°C$ until use. 15-30g fish or caviar tissue were minced with 60-120g anhydrous Na_2SO_4 to fine powder and Soxhlet extracted with cyclohexane/dichloromethane (2:1) for 4h. The extracts were reduced to about 30mL by a rotary evaporator. Lipids were treated with concentrated H_2SO_4 until the H_2SO_4 phase remained colorless. The organic phases were collected, washed with tridistilled water until they were neutral, dried with 10g anhydrous Na_2SO_4 overnight, reduced to 0.5mL, and stored at -12°C until further fractionation on silica gel (20, 51). Oil samples (2g) were dissolved in 50mL n-hexane followed by DMF/n-hexane partitioning to separate the fat. In a second clean-up step, the n-hexane phase was treated with concentrated H_2SO_4. A pre-separation was carried out by silica gel fractionation with an overpressure of 0.3 bar N_2. The columns were prepared with 4.3g silica gel 60 (70-230 mesh, activated at 140°C for 25h and then deactivated with 5% water). The fractionation was done first with n-hexane (including 0.5mL of the sample) and then with 9 mL of n-hexane/dichloromethane (2:1). The third of the resulting fractions contained the toxaphene. This fraction was reduced to approximately 1mL and stored at -12°C. Prior to high resolution gas chromatography-electron capture detection (HRGC-ECD) or HRGC-EC negative ion mass spectrometry (NI-MS) analysis, the extracts were reduced to ca. 200 μL under N_2 (*20*).

Standards and samples were analyzed on a Hewlett-Packard 5890/5988A GC-MS system equipped with a 25m x 0.2mm capillary column (HP-5, film thickness 0.33μm) with helium as carrier gas (ca. 2mL/min). The conditions were as follows: splitless injection; temperature program: 140°C (1min), 250°C (4°C/min); injection port and transfer lines: 280°C; ion source temperature: 100°C; moderating gas: CH_4; emission current: ca. 200μA.

HRGC-ECD measurements were performed with a Varian 3300 system equipped with a 25m x 0.2mm capillary column (HP-5, film thickness 0.5μm) with N_2 as carrier gas (ca. 2mL/min). The conditions were as follows: splitless injection; temperature program: 150°C (0min), 250°C (5°C/min); injection port: 250°C; detector (ECD): 280°C (*20*).

HRGC-ECD-on column measurements were performed with a Varian 3400 system equipped with a 60m x 0.25mm capillary column (HP-5, film thickness 0.25μm) with N_2 as carrier gas (ca. 1.5mL/min). The conditions were as follows: injector oven temperature: 100°C (0min), 230°C (180°C/min), 50min; column temperature: 120°C (0min), 200°C (20°C/min), 250°C (5°C/min), 36min; detector: 280 °C (*51*).

Photolysis: UV-irradiated toxaphene has been prepared in the following way: 1 g of technical toxaphene (62.7% Camphechlor of Ehrenstorfer, Augsburg, Germany) was dissolved in 1.5 L degassed *n*-hexane, irradiated for 2 h with a low pressure Hg lamp (Vycor 250 mA/500 V, Fa. Gräntzel, Karlsruhe, Germany, emission maximum at 254 nm), concentrated, and cleaned by column chromatography on silica gel 60 (70-230 mesh) with *n*-hexane (*51*). Single substances were irradiated under the same conditions, but with a total irradiation time of 12 h, during which samples were taken after 3, 6, and 9 h (*21*).

Degradation in Soil: Technical toxaphene standard (Camphechlor) was obtained from Ehrenstorfer, Germany. Single toxaphene components were isolated from technical toxaphene as previously described (*31, 32*). The soil, a loamy silt (pH 6.7, 1.8% organic carbon), was collected from a field in the surroundings of Kassel. It had been chosen because of ist lack of any detectable organochlorine contaminants. The soil was air-dried and passed through a 2-mm sieve prior to use.

Incubation was done with 80g portions of soil each placed in a 200mL Erlenmeyer flask and fortified with 1mL of acetone containing 400μg of the toxaphene component. After adding 150mL of sterile, distilled water to each flask, they were shaken, and the dissolved O_2 was removed with a stream of nitrogen for 30min. Then the flasks were tightly capped with Teflon-coated stoppers and kept in the dark at ca. 30°C. Two samples as well as two blanks were prepared in this way for each component. In the case of technical toxaphene, a total of 800μg (10μg/g of soil) was added to the soil. One series of the spiked soil samples was sterilized by autoclaving (121°C, 15psi) for two 1h periods at intervals of 24h before adding the components. Samples for analysis were taken weekly during the first two months and for the rest of the time – a total of 4 and 6 months, respectively – in intervals of 2 weeks. The flasks were shaken and then opened under a stream of N_2, and 10mL aliquots of the suspension were taken. Each sample was acidified with H_2SO_4 to pH ca. 1 and extracted with a mixture of 5mL of petroleum ether (45-65°C)/acetone (1:1) in an ultrasonic bath for 30min. The petroleum ether layer was separated, and the aqueous layer was reextracted twice with 2 x 5mL of petroleum ether (45-65°C). Finally, the organic phases were combined, dried over Na_2SO_4, and concentrated under N_2 to ca. 1mL. These extracts were directly used for GC analysis.

HRGC-ECD measurements were performed with a Varian 3300 gas chromatograph. The conditions were as follows: injector: splitless (0.5 min)/split, 230°C; column: DB-5 30m x 0.25mm, film thickness 0.5µm; carrier gas: N_2 (2 mL/min); temp. program: 120°C (0 min), 200°C (20°C/min), 230°C (5°C/min), 250°C (1.5°C/min), 15 min; detector (ECD): 280°C (32).

HRGC-ECNI-MS measurements were performed on a Hewlett-Packard 5890/5988A system under the following conditions: column: HP-5 25m x 0.2mm, film thickness 0.33µm; carrier gas: helium (ca. 2mL/min); injection: splitless(0.5min)/split; temperature program: 140°C (1min), 250°C (4°C/min); injection port and transfer lines: 280°C; ion source temperature: 100°C; moderating gas: CH_4; emission current: ca. 200µA.

Toxaphene in the Environment

Results of Toxaphene Residue Analysis. Several problems have had to be resolved before residue analysis could be used for the discrimination of biotic and abiotic degradation of different components of technical toxaphene. Analytical methods for separation of the complex mixture of technical toxaphene were developed. The main components of technical toxaphene and the most abundant toxaphene components in environmental samples were isolated and structurally identified, so that data on toxaphene composition as well as residue levels in different environmental samples could be acquired. Today the structure of more than 60 environmentally relevant toxaphene components are known (5, 22-34). Toxaphene has been estimated to consist of about 76% chlorobornanes, 18% chlorobornenes, 2% chlorobornadienes, 1% chlorinated hydrocarbons, and 3% nonchlorinated hydrocarbons (35), although the only unsaturated components isolated to date are chlorinated camphenes. The percentage of the identified components cannot be given exactly because of coelution, even when using high resolution capillary columns, and because response factors are known only for some of the components. These points aside, the peak area percentage of all identified components (measured with ECD) amounts to 50% of the total technical toxaphene (30, 36, 37).

Of these more than 60 known components, only about 25 are regularly found in environmental samples. While most of the nona- and decachlorobornanes are relatively rapidly degraded, many of the hexa- and heptachlorobornanes as well as some of the octa- and nonachlorobornanes (Figure 2) are persistent and, therefore, undergo long range transport. This results in contamination of especially aquatic biota collected from nearly all parts of the northern hemisphere. Atmospheric transport of toxaphene to the Arctic and North Atlantic has been demonstrated by residue analysis of air and water samples from different regions (14, 15, 38-42). Sometimes, similarities in the gas chromatograms are found consisting of a shift to shorter retention times together with the dominance of the same peaks or peak groups, which can be seen in chromatograms

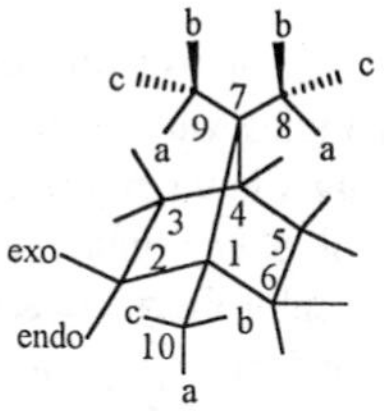

No. 26

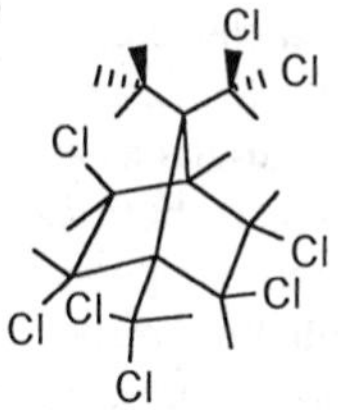

(±)-2-endo,3-exo,5-endo,6-
exo,8b,8c,10a,10c-
octachlorobornane

No. 40

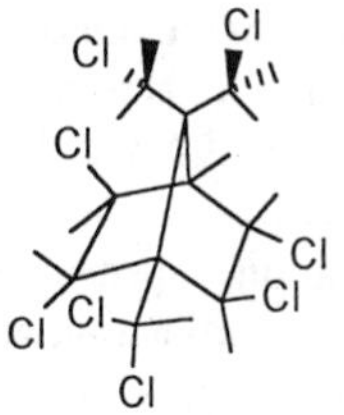

(±)-2-endo,3-exo,5-endo,6-
exo,8b,9c,10a,10c-
octachlorobornane

No. 41

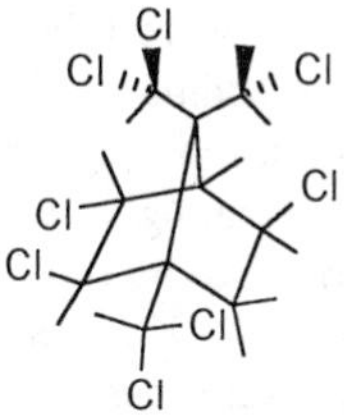

(±)-2-exo,3-endo,5-
exo,8b,9b,9c,10a,10b-
octachlorobornane

No. 44

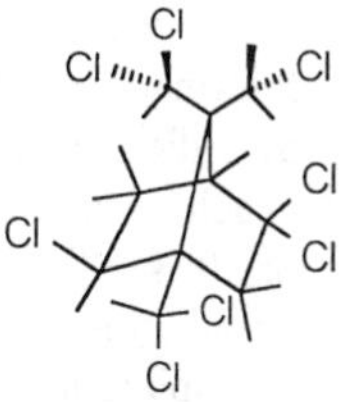

(±)-2-
exo,5,5,8c,9b,9c,10a,10b-
octachlorobornane

No. 50

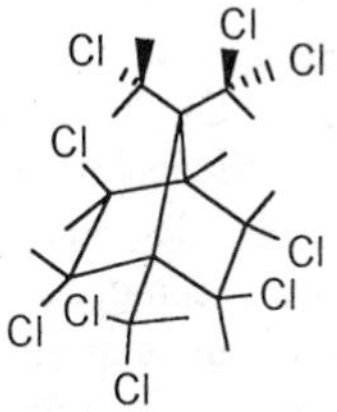

(±)-2-endo,3-exo,5-endo,6-
exo,8b,8c,9c,10a,10c-
nonachlorobornane

No. 62

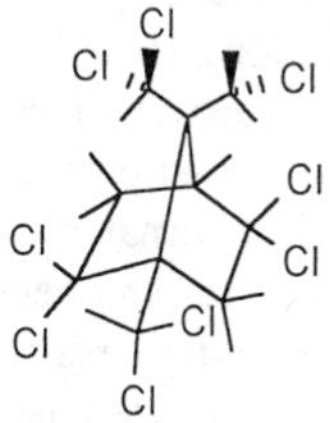

(±)-2,2,5,5,8c,9b,9c,10a,10b-
nonachlorobornane

Figure 2. Structures of toxaphene components frequently detected in environmental samples. The numbers given are the respective Parlar numbers (*13*).

from samples of air, water, and fish or fish products (Figure 3). More commonly, the patterns of different kinds of samples vary widely. Tissues of some animals, such as marine mammals, may contain only a few components in rather high concentrations (43-46). In one case, beluga whale blubber contained 1300 ng/g lipid weight of 2-exo,3-endo,5-exo,6-endo,8,8,10,10-octachlorobornane and 2350 ng/g lipid weight of 2-exo,3-endo,5-exo,6-endo,8,8,9,10,10-nonachlorobornane with total toxaphene contents of 4130 ng/g lipid weight (29).

UV-Degradation of Toxaphene. Irradiation of toxaphene in solution with wavelengths above 290 nm only leads to a slight degradation of the mixture (17). After irradiation at wavelengths below 290 nm, the same peak pattern results as at wavelenghts above 290 nm, while the reaction time is shorter (Figure 4A). In both cases, dechlorination and dehydrochlorination, especially of higher chlorinated components, take place. This can be seen in the gas chromatogram by a shift in distribution to shorter retention times (18). The mechanism proceeds *via* photolytic loss of one chlorine atom, preferentially in the C_2-position, followed by abstraction of hydrogen from the solvent. Irradiation of single toxaphene components results in dechlorination products with reaction rates depending on the structure of parent compounds (Figure 4B) (18, 21, 28, 30, 35). Generally, during irradiation in solvents, the bornane structure is preserved. The degradation rate shows a positive correlation to the chlorine substitution at the bridge head (21). Photolability seems to depend on the presence of a geminal dichloro group at the C_2-position. The dechlorination rate is enhanced by an additional chlorine atom at the C_3-position but not by a dichloro group in the C_5-position. Components with only a single chlorine atom at each secondary ring atom in alternating orientation (such as Parlar numbers 26, 40, or 50, see Figure 2) have been found to be extremely photostable (21). Contrary to this, irradiation of toxaphene adsorbed to silica gel at wavelengths above 290 nm results in complete mineralization. Surface water has no significant influence on the degradation process (17).

Degradation of Toxaphene Components in Soil. Under aerobic conditions, the degradation of toxaphene appears to proceed rather slowly, as residue analysis of aerated soil samples from a former toxaphene manufacturing plant exhibited chromatograms indistinguishable from those of technical toxaphene (21). Under anaerobic conditions, toxaphene is more easily degraded by microorganisms (21, 35, 47-50). Gas chromatography-electron capture detector (GC-ECD) analysis has shown that the highest chlorinated components (i.e. nona- and decachlorobornanes) as well as part of the octachlorobornanes are nearly completely degraded to lower chlorinated bornanes, while hexa- and heptachlorobornanes are accumulated (49, 50). It is not clear whether the remaining hexa- and heptachloro compounds are produced by degradation of the higher chlorinated isomers or whether they are original constituents of the technical mixture. In laboratory experiments with technical toxaphene in a loamy soil under anaerobic conditions, a significant shift of the GC-ECD peak pattern of toxaphene towards lower retention times was observed after only one week (Figure 5) (21). After three months, an additional decrease of detectable components could be observed, the main components being two hepta- and one hexachlorobornane. The peak pattern became still simpler after 5 months. The main component was 2-exo,3-endo,6-

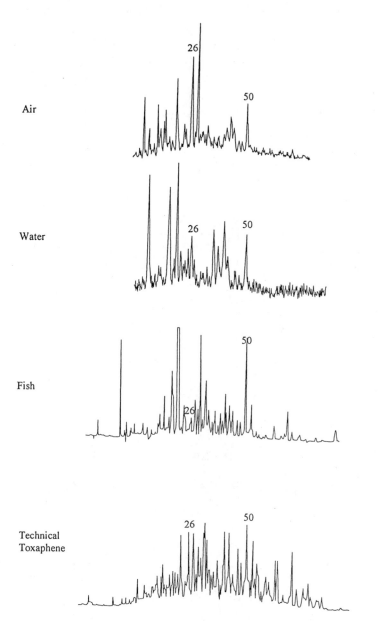

Figure 3. GC-ECD peak patterns of some environmental samples (*39, 54*) compared to that of technical toxaphene (*13*)

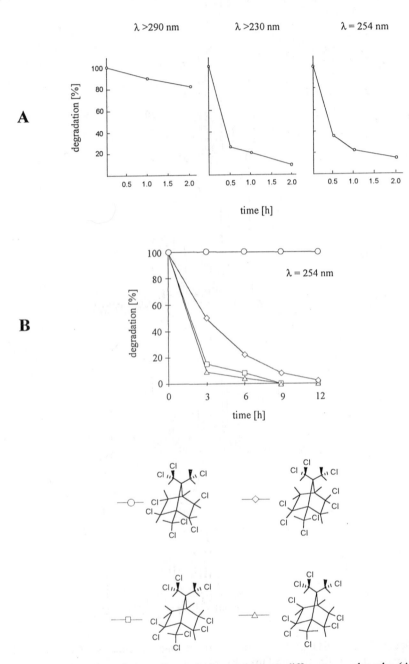

Figure 4. Photodegradation of technical toxaphene at different wavelengths (A) (*28*) and of four single components at 254 nm (B) (*32*)

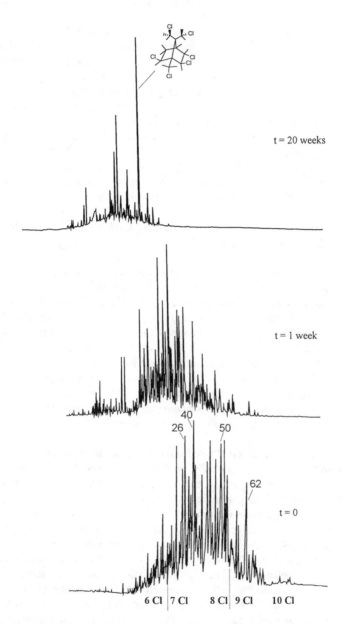

Figure 5. GC-ECD peak pattern of toxaphene degraded in a loamy soil under sulfate-reducing conditions (*21*)

exo,8,9,10-hexachlorobornane, but hexachlorobornenes and a tetrachlorocamphene-2-one were also found.

Studies of the degradation of single toxaphene components in a loamy soil under anaerobic conditions have confirmed the dependence of microbial dechlorination rates on the chlorine substitution pattern of the six-membered ring (21). Components with only one chlorine atom at each C-atom in alternating orientation were found to be highly persistent, while components having geminal dichloro groups on the ring were rather labile, especially when the dichloro group was localized at the C_2-atom (Figure 6). As in the case of photolytic reactions, microbial degradation of all compounds investigated was enhanced by the presence of a second geminal dichloro group. The dead-end metabolite, 2-exo,3-endo,6-exo,8c,9b,10a-hexachlorobornane, has also been found in residue analysis of sediment (51) and fish (52). Whether this is due to accumulation of this component or due to a high proportion of components in technical toxaphene with a chlorine substitution pattern that leads to this product is unknown.

Application of Enantiomer Selective Residue Analysis to the Discrimination of Toxaphene Degradation Pathways

Results of Toxaphene Degradation Studies and Residue Analysis. The studies on the degradation of single toxaphene components as well as technical toxaphene in a loamy soil under anaerobic conditions all gave (1S,2R,3R,4R,6R),2-exo,3-endo,6-exo,8,9,10-hexachlorobornane as the main metabolite (21). This enantiomer was the only one that could be isolated even though all parent compounds in toxaphene occur as racemates. It is not clear whether this is an artifact or the result of enantioselective degradation. Stereoselective conversion is known for many enzymes. Therefore, it may be that the different enantiomers of toxaphene components are degraded differentially. By contrast, enantioselective residue analysis of different samples of fish and fish products showed no significant discrimination of toxaphene enantiomers (Table I). The enantiomer ratio of chlorobornanes in all samples is about 1:1.

Comparison of Toxaphene GC Peak Patterns. The similarity of the GC-peak patterns of some kinds of samples, especially of fish samples and UV-irradiated toxaphene (Figure 7), leads to the hypothesis that the same or similar degradation pathway is responsable for the resultant pattern in both cases. Previously, it has been suggested that the mechanisms of toxaphene metabolism in fish should lead to the same products as photolysis (21, 53, 54). However, because no discrimination of enantiomers has been found in fish samples, it may be that in fish no significant metabolism occurs at all. If this is correct, then all toxaphene conversion products found in fish may be the result of unselective accumulation of mainly photostable toxaphene components after transformation predominantly in the atmosphere. Selective uptake or accumulation of toxaphene by fish has as yet not been confirmed, but experiments with bluegill fish showed little metabolism (55). Although the resistance of some individual toxaphene components to microbial and photodegradation seems to be comparable, differences between the chromatograms of samples of fish tissues, on the one hand, and samples of soil or tissues of marine mammals, on the other hand, show that, even without consideration of enantiomeric compositions, these two processes give rise to different toxaphene residue compositions. Therefore, the similar distribution of part of the

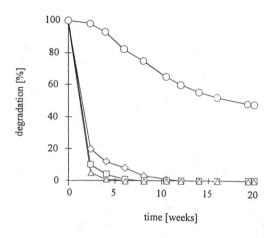

Figure 6. Degradation of four single toxaphene components in a loamy soil under sulfate-reducing conditions (*32*)

Table I. Enantiomer ratios (+/−) of the toxaphene indicator components No. 26, 31, 40 + 41, 44, 50, 58, and 62 determined by GC-ECNI-MS-SIM. Data combined from ref. 5, 59, and 60.

Sample	Component						
	26	31	40+41	44	50	58	62
Cod liver oil (Germany)	0.97	n.d.	0.95	0.96	1.08	1.00	0.96
Cod liver oil (France)	1.01	1.07	1.10	0.94	1.07	1.05	1.00
Cod liver oil (Iceland)	1.03	1.03	0.98	0.98	1.08	n.d.	1.03
Herring (Baltic Sea)	0.96	1.00	0.91	0.93	1.08	1.05	0.98
Halibut (North Atlantic)	0.98	n.d.	0.91	1.01	1.03	n.d.	1.05
Caviar substitute 1 (North Sea)	1.00	n.d.	1.04	1.07	0.98	0.97	0.99
Caviar substitute 2 (North Sea)	1.00	n.d.	1.04	1.07	0.98	0.97	0.99
Redfish (North Atlantic)	1.01	1.00	1.03	1.08	1.07	1.03	1.00

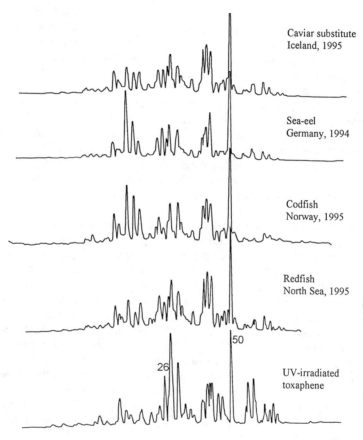

Figure 7. ECNI-MS-SIM chromatograms of different aquatic samples from the North Atlantic and North Sea (*41*)

toxaphene components in air, water, and fish indicates that in this case most of the degradation may be due to photolytic processes in air.

Degradation Processes of Toxaphene in the Troposphere

Relative Photostability of Toxaphene Components under Environmental Conditions. The chlorobornanes in technical toxaphene mainly belong to three groups: 1) those with only one chlorine atom at the ring C-atoms (e. g. Parlar No. 26, 40, and 50), 2) those with a geminal dichloro group at the ring and one chlorine atom in the α-position (Parlar No. 39 and 58), and 3) those with geminal dichloro groups, but without a directly adjacent chlorine atom (e. g. Parlar No. 44 and 62) (Figure 8). The first of these groups is extremely photostable, while the second is easily degraded by intramolecular photoelimination of Cl_2. This process leads to chlorobornenes which are also easily degraded either by UV-light or by reactive oxygen species giving polar conversion products. This second group amounts to more than 70% of the chlorobornanes in technical toxaphene. The third group is intermediate, being rather stable except when a second dichloro group exists in position 10. In this case, chlorocamphenes can be formed *via* Wagner-Meerwein rearrangement. Dechlorination cannot take place in the air on a large scale, however, because of the lack of sufficient concentrations of proton donors, such as alkanes, alkenes, or terpenes.

Other Mechanisms Possibly Relevant for Degradation of Toxaphene in Air. Though chlorobornanes cannot react with O_3, they are rather easily attacked by OH-radicals. Because of the low vapour pressure of chlorobornanes gas phase experiments cannot be performed, but the rate constants estimated for different components using the formula of Atkinson (56) are comparable ($k = 3 \cdot 10^{-12}$ to $6 \cdot 10^{-12}$ $cm^{-3} \cdot molecule^{-1} \cdot s^{-1}$). Therefore, it can be assumed that the reaction with OH-radicals in the troposphere is nonselective relative to the structure of the chlorobornanes and should lead to a general decrease in toxaphene concentration in the air without a significant change in composition. Irradiation of the technical mixture was also performed in fully deuterated *n*-hexane because of the low solubility of toxaphene in D_2O. The results show that none of the products which had deuterium incorporated is identical with those components that accumulate in the environment (Figure 2). (57). This means that the toxaphene components found in the environment are not derived from higher chlorinated compounds of technical toxaphene, but are stable constituents of the original mixture that have been selectively preserved.

Conclusions

Residue analysis data of toxaphene in different environmental samples, as well as from degradation experiments, point to abiotic processes being the main removal mechanism for toxaphene in the troposphere. It appears that toxaphene compositions in fish are directly the result of uptake of toxaphene partly degraded in air and water, because enantioselective analysis shows that metabolism in fish is probably negligible. The reaction of toxaphene with OH-radicals is too nonselective to yield the highly specific patterns characteristic of samples from air, water, or fish, in which the photostable

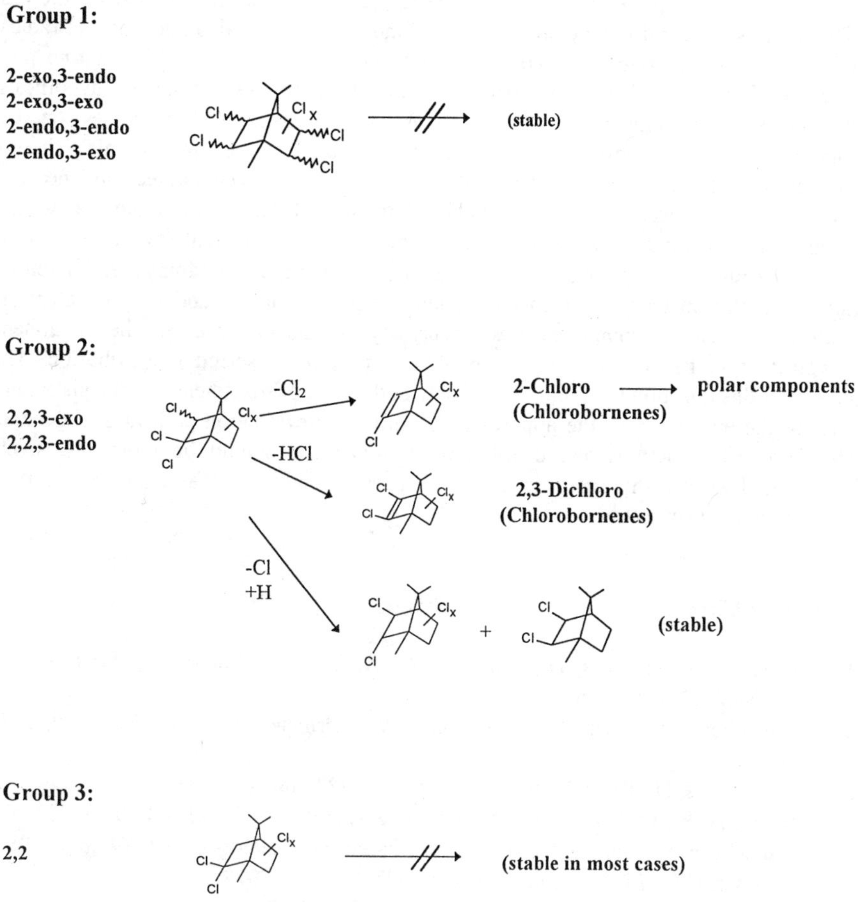
Group 1:
2-exo,3-endo
2-exo,3-exo
2-endo,3-endo
2-endo,3-exo
Cl
Cl
Cl x
Cl
Cl
(stable)
Group 2:
2,2,3-exo
2,2,3-endo
-Cl2
-HCl
-Cl
+H
2-Chloro
(Chlorobornenes)
polar components
2,3-Dichloro
(Chlorobornenes)
+
(stable)
Group 3:
2,2
Clx
(stable in most cases)

Figure 8. Groups of chlorobornanes in technical toxaphene and their degradation pathways under varying environmental conditions

components dominate. In all probability, these patterns originate from photolysis. UV radiation is the only abiotic process which is able to compete with OH-radical induced degradation. As photodegradation of toxaphene at wavelengths above 290 nm is rather slow, UV-radiation in the troposphere seems to contribute   more to toxaphene photolysis than assumed before now. This is in contradiction with the assumption that the photolytic effect of shorter wavelengths in the troposphere is insignificant. The abiotic degradation patterns of toxaphene reflected in environmental samples is the first hint that this assumption may be incorrect. A more thorough calculation of OH-radical induced degradation rates of toxaphene in the troposphere combined with quantum yield data of single toxaphene components should help to test the new hypothesis. Nevertheless, the influence of phase transfer and distribution processes on the composition of toxaphene residues cannot be estimated at the present time. Though equilibration of the technical mixture with air and water has been studied (58), no such experiments have been done on single chlorobornanes. Differences in molecular weight alone cannot account for the toxaphene composition in fish residues where also some higher chlorinated components, such as Parlar No. 50 and 62, predominate. Therefore, further careful analysis of toxaphene residues especially in air and ocean water and studies with single components are necessary to decide whether the toxaphene composition in fish owns more to photolysis or more to selective distribution. The answer to this question is of interest not only in the field of toxaphene ecochemistry, but also for the evaluation of the importance of different degradation pathways in general. Therefore, the reactions and distribution of toxaphene should be more intensively investigated also with regard to the atmospheric conditions that lead to observed environmental distribution.

Literature Cited

1.   Faller, J.; Hühnerfuss, H.; König, W.A.; Krebber, R.; Ludwig, P. *Environ. Sci. Technol.* **1991** *25*, 676-678.
2.   Kallenborn, R.; Hühnerfuss, H.; König, W.A. *Angew. Chem. Int. Ed. Engl.* **1991** *30*, 320-321.
3.   Hühnerfuss, H.; Kallenborn, R. *J. Chromat.* **1992** *580*, 191-124.
4.   Buser, H.-R.; Müller, M.D. *Environ. Sci. Techn.* **1993** *21*, 1211-1220.
5.   Kallenborn, R.; Oehme, M.; Vetter, W.; Parlar, H. *Chemosphere* **1994** *28*, 89-98.
6.   Voldner, E.C.; Li, Y.F. *Chemosphere* **1993** *27*, 2073-2078.
7.   Bidleman, T.F.; Olney, C.E. *Nature* **1975** *257*, 475-477.
8.   Villeneuve, J.-P.; Cattini, C. *Chemosphere* **1986** *15*, 115-120.
9.   Bidleman, T.F.; Patton, G.W.; Walla, M.D.; Hargrave, B.T.; Vass, W.P.; Erickson, P.; Fowler, B.; Scott, V.; Gregor, D.J. *Arctic* **1989** *42*, 307-313.
10.   Muir, D.C.G.; Grift, N.P.; Ford, C.A.; Reiger, A.W.; Hendzel, M.R.; Lockhart, W.L. In *Long Range Transport of Pesticides*, Kurtz, D.A., Ed.; Lewis: Chelsea, Michigan, **1990**.
11.   Luckas, B.; Vetter, W.; Fischer, P.; Heidemann, G.; Plötz, J. *Chemosphere* **1990** *21*, 13-19.

12. Hargrave, B.T.; Harding, G.C.; Vass, W.P.; Erickson, P.E.; Fowler, B.R.; Scott, V. *Arch. Environ. Contam. Toxicol.* **1992** *22*, 41-54.
13. Calero, S., Fomsgaard, I.; Lacayo, M.L.; Martinez, V.; Rugama, R. *Chemosphere* **1992** *24*, 1413-1419.
14. Hoff, M.; Muir, D.C.G.; Grift, N.P. *Environ. Sci. Technol.* **1992** *26*, 276-283.
15. Hoff, R.M.; Muir, D.C.G.; Grift, N.P.; Brice, K.A. *Chemosphere* **1993** *27*, 2057-2062.
16. Parlar, H.; Gäb, S.; Nitz, S.; Korte, F. *Chemosphere* **1976** *5*, 333-338.
17. Becker, F. *Thesis*; Technical University Munich, Germany, **1987**.
18. Parlar, H.; Kotzias, D.; Korte, F. *Chemosphere* **1983** *12*, 1453-1458.
19. Saleh, M.A.; Casida, J.E. *J. Agr. Food Chem.* **1978** *26*, 583-590.
20. Burhenne, J. *Thesis*; University of Kassel, Germany, **1993**.
21. Fingerling, G. *Thesis*; University of Kassel, Germany, **1995**.
22. Black, D.K. *16th ACS National Meeting*, Atlantic City, N. S., Sept. **1974**.
23. Anagnostopoulos, M.L.; Parlar, H.; Korte, F. *Chemosphere* **1974** *3*, 65-70.
24. Landrum, P.; Pollock, G.; Seiber, J. *Chemosphere* **1976** *5*, 63-69.
25. Saleh, M.A.; Turner, W.V.; Casida, J.E. *Science* **1977** *198*, 1256-1258.
26. Turner, W.V.; Engel, J.L.; Casida, J.E. *J. Agr. Food Chem.* **1977** *25*, 1394-1401.
27. Chandurkar, P.; Matsumura, F. *Bull. Environ. Contam. Toxicol.* **1979** *21*, 539-547.
28. Parlar, H. *Chemosphere* **1988** *17*, 2141-2150.
29. Stern, G.A.; Muir, D.C.G.; Grift, N.P.; Dewailly, E.; Bidleman, T.F.; Walla, M.D. *Environ. Sci. Technol.* **1992** *26*, 1838-1840.
30. Burhenne, J. *Thesis*, Kassel, FRG, **1993**.
31. Hainzl, D.; Burhenne, J.; Parlar, H. *Chemosphere* **1993** *27*, 1857-1863.
32. Hainzl, D.; Burhenne, J.; Barlas, H.; Parlar, H. *Fresenius J. Anal. Chem.* **1994** *351*, 271-285.
33. Vetter, W.; Scherer, G.; Schlabach, M.; Lukas, B.; Oehme, M. *Fresenius J. Anal. Chem.* **1994** *349*, 552-558.
34. Nikiforov, V.A.; Tribulovich, V.G.; Karavan, V.S. *Organohalogen Compounds* **1995** *26*, 393-396.
35. Saleh, M.A. *J. Agr. Food Chem.* **1983** *31*, 748-751.
36. Parlar, H.; Becker, F.; Müller, R.; Lach, G. *Fresenius Z. Anal. Chem.* **1988** *331*, 804-810.
37. Hainzl, D. *Thesis*, University of Kassel, **1994**.
38. Bidleman, T.F.; Widequist, U.; Jansson, B.; Söderlund, R. *Atmos. Environ.* **1987** *21*, 641-654.
39. Bidleman, T.F.; Falconer, R.L.; Walla, M.D. *Sci. Total Environ.* **1995** *160/161*, 55-63.
40. Arthur, R.D.; Cain, J.D.; Barrentine, B.F. *Bull. Environ. Contam. Toxicol.* **1976** *15*, 129-134.
41. Bidleman, T.F.; Zaranski, M.T.; Walla, M.D. In *Toxic Contamination in Large Lakes*; Schmidke, N.W., Ed.; Lewis: Chelsea, Michigan, **1988**; Vol. 1; pp 257-284.
42. Hoff, R.M.; Bidleman, T.F.; Eisenreich, S.J. *Chemosphere* **1993** *27*, 2047-2055.

43. Muir, D.C.G; Norstrom, R.J.; Simon, M. *Environ. Sci. Technol.* **1988** *22* 1071-1079.

44. Muir, D.C.G.; Ford, C.A.; Grift, N.P.; Steward, R.E.A.; Bidleman, T.F. *Envir. Pollut.* **1992** *75*, 307-316.

45. Bidleman, T.F.; Walla, M.D.; Muir, D.C.G.; Stern, G.A. *Environ. Toxicol. Chem.* **1993** *12*, 701-709.

46. Muir, D.C.G.; Ford, C.A.; Stewart, R.E.A.; Smith, T.G.; Addison, R.F.; Zinck, M.E.; Béland, P. *Can. Bull. Fish. Aquat. Sci.* **1991** *224*, 165-190.

47. Parr, J.F.; Smith, S. *Soil Sci.* **1976** *121*, 52-57.

48. Clark, J.M.; Matsumura, F. *Arch. Environ. Toxicol.* **1979** *8*, 285-298.

49. Murthy, N.; Lushy, W.; Oliver, J.; Kearney, P. *J. Nuclear Agr. Biol.* **1984** *13*, 16-17.

50. Mirsatari, S.; McChesney, M.; Craigmill, A.; Winterlin, W.; Seiber, S. *J. Environ. Sci. Hlth* **1987** *B22*, 663-690.

51. Stern, G.A.; Loewen, M.D.; Miskimmin, B.M.; Muir, D.C.G.; Westmore, L.B. *Environ. Sci. Technol.* **1996** *30*, 2251-2258.

52. Bartha, R.; Vetter, W.; Luckas, B. *Fresenius J. Anal. Chem.*, submitted.

53. Lach, G. *Thesis*, University of Kassel, **1990**.

54. Xu, L. *Thesis*, University of Kassel, **1994**.

55. Isensee, A.R.; Jones, G.E.; McCann, J.A.; Pitcher, F.G. *J. Agr. Food Chem.* **1979** *27*, 1041-1046

56. Atkinson, R. *Chem. Rev.* **1986** *86*, 69-201.

57. Behr, H. *Diploma Thesis*, University of Kassel, Germany, 1992.

58. Murphy, T.J.; Mullin, M.D.; Meyer, J.A. *Environ. Sci. Technol.* **1987** *21*, 155-162

59. Karlsson, H.; Oehme, M. *Organohalogen Compounds* **1996** *28*, 369-374

60. Alder, L.; Palavinskas, R.; Andrews, P. *Organohalogen Compounds* **1996** *28*, 410-415

Chapter 24

Soot as a Strong Partition Medium for Polycyclic Aromatic Hydrocarbons in Aquatic Systems

Örjan Gustafsson and Philip M. Gschwend

R. M. Parsons Laboratory, MIT 48–415, Department of Civil and Environmental
Engineering, Massachusetts Institute of Technology, Cambridge, MA 02139

Abstract. The spatial distributions of polycyclic aromatic
hydrocarbons (PAHs) in many aquatic environments appear to be
dictated by partitioning with soot as opposed to with bulk organic
matter. Recent field-observations of the solid-water distribution
coefficients of members of this contaminant assemblage are
consistently elevated compared to expectations from organic-matter-
based partition models. Increasing hydrophobicity across the PAH
assemblage is seen to affect relative distributions. Using PAHs as
molecular markers of sorption suggests that an active exchange of such
planar molecules between a strong sorbent, such as soot, and the
surrounding water is taking place. Quantification of PAHs, organic
carbon, and soot carbon in surficial continental shelf sediments off
New England revealed that the distribution of PAHs was highly
correlated with soot carbon. Estimates of the soot-water partition
coefficient for several PAHs, assuming sorbate-soot association is
thermodynamically similar to sorbate fusion, are found to agree
reasonably well with published aqueous sorption constants to activated
carbon.

Several polycyclic aromatic hydrocarbons (PAHs) are carcinogenic and have been
classified by the US Environmental Protection Agency as "Priority Pollutants" (1). In
fact, PAHs have been identified as the principal human cell mutagens in extracts of a
particular natural sediment, clearly illustrating the need for environmental chemists to
emphasize studies of this compound class (2). Furthermore, large amounts of these
organic contaminants continue to be released into the environment, primarily from
combustion of fossil and wood fuel, but also from direct petroleum spills.
　A useful starting point for anticipating these and other chemicals' dispersal and
effects in the environment is to consider their "phase"-distribution. This physico-
chemical speciation largely governs the extent to which a given compound may
undergo transport and transformation. PAHs are both easy to analyze and span a
range of well-known thermodynamic properties, making them suitable as molecular
markers of environmental partitioning. Elucidation of the behavior of a few PAHs

thus affords extrapolation to many other compounds of similar structure through linear free energy relationships. Due to their high aqueous activity coefficients, PAHs and many other xenobiotic chemicals have a high affinity for solid phases. In the widely accepted solid-water partition model for such hydrophobic compounds, developed by Karickhoff and Chiou (3, 4), particulate PAH concentrations are normalized to the total organic carbon content of the solid, believed to be the sorbent property that dictates the tendencies of hydrophobic sorbates to associate with the solid. This classical organic-matter-based partition model has shown its ability to predict, from thermodynamic properties of the system, the chemical speciation in hundreds of laboratory-based sorption studies where a wide array of soils and sediments have been spiked with organic chemicals like PAHs (e.g., 5, 6).

However, many recent field observations have reported *in situ* distribution coefficients ($(K_{oc})_{obs}$) of PAHs that are orders of magnitude above the predictions of the classical model (Figure 1). These enhanced particle-associations for PAHs appear to be widespread as they have been documented in such diverse environmental compartments as surface waters of lakes (7), rivers (8), estuaries (9), and the ocean (10, 11), as well as in sediment porewater (12) and rainwater (13). In contrast, $(K_{oc})_{obs}$ of similarly hydrophobic polychlorinated biphenyls (PCBs) are typically lower and agree better with predictions from the Karickhoff-Chiou model (e.g., 13-16).

An association between PAHs and various pyrogenic carbon phases, all henceforth referred to as soot, has been previously postulated to control the distribution and bioavailability of these compounds (e.g., 9, 12, 17-25). However, only in two previous studies has a direct link between soot and particulate PAHs been examined through simultaneous quantification in natural sediments (18, 23). Broman and co-workers performed microscopy-counting of coarse (> 5-10 μm) charcoal particles in sediment-trap collected material and found a correlation between such particle counts and PAH concentrations. However, because most PAH-carrying soot particles in the atmosphere are much smaller than can be detected with optical microscopy (e.g., 26, 27), it is not possible to interpret quantitatively their results in terms of a PAH partition model. Recently, a thermal-oxidation method which determines the total mass of sedimentary soot carbon, SC, was developed (18). Our earlier work demonstrated that the vertical distribution of SC in a lacustrine core correlated with the PAH pattern, while being decoupled from the organic carbon profile. Measurements of total SC in Boston Harbor sediments of a few parts per thousand (by dry mass) were also used to rationalize previously reported (12) elevated PAH $(K_{oc})_{obs}$ in two sediment-porewater systems (18).

Obviously, inclusion of soot-bound PAHs in the quantified particulate fraction of field studies (e.g., those of Figure 1) would yield elevated $(K_{oc})_{obs}$. However, inspection of the entire contaminant assemblage exhibiting elevated distribution coefficients reveals that there is still a pronounced effect of hydrophobicity on the relative solid-solution distributions. This observation strongly suggests that an active exchange between the, presumed, soot-associated PAHs and their dissolved counterparts is taking place. Hence, a significant fraction of pyrogenic PAHs appears to be not "permanently occluded" (e.g., 9, 12, 22) but rather to partition between what seems to be a strongly sorbed state and the surrounding water (18). In the absence of controlled sorption studies in soot-water systems, we estimated PAH partition coefficients between activated carbon (as an analog for soot) and water from the linear portion of sorption isotherms developed by others (28). Such a pyrogenic phase is a significantly better sorbent than natural organic matter for PAHs on a carbon-mass basis. Such observations have lead to the hypothesis that for planar aromatic molecules such as PAHs, we need to expand the classical hydrophobic partion model to include sorption, not only into natural organic carbon, but also with soot carbon phases (18):

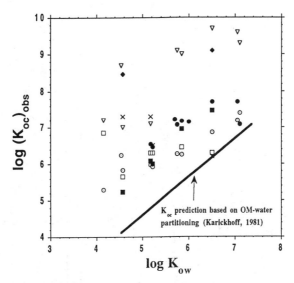

Figure 1. Enhanced organic-carbon normalized *in situ* partition coefficients $(K_{oc})_{obs}$ for PAHs observed in diverse environmental regimes: surface waters of lakes (open squares = ref. 7), rivers (filled squares = ref. 8), estuaries (filled diamonds = ref. 9), and the ocean (filled circles = ref. 10; dotted open circle = ref. 11), as well as in sediment porewater (x = ref. 12) and rainwater (inverted open triangles = 13). K_{ow} values are from (3, 80, 81) and the model-predicted regression with K_{oc} is from (79).

$$K_d = f_{oc}K_{oc} + f_{sc}K_{sc} \qquad (1)$$

where f_{oc} and f_{sc} are the non-soot organic carbon and soot carbon mass-fractions of the solid matrix, respectively, and K_{oc} and K_{sc} are the organic carbon and soot carbon normalized equilibrium distribution coefficients, respectively.

The first objective of this paper is to demonstrate that the concentrations of PAHs in marine sediments are better correlated with soot carbon content (f_{sc}) than with organic carbon content (f_{oc}), implying that PAH sorption to soot controls PAH cycling and effects. We further evaluate the relevant physico-chemical sorbent properties of soot, discuss potential sorption mechanisms, and derive an estimate of the magnitudes of soot-water equilibrium partitioning (K_{sc}) for typical PAHs.

Methods.

Sediment Sampling. Large-volume box-cores were obtained from different locations in the Gulf of Maine during several cruises (Figure 2). In 1990, two stations in Boston Harbor: at the mouth of Fort Point Channel (FPC) and off Spectacle Island (SI) were cored. In May 1994, sediments in Portland Harbor (PH), outer Casco Bay (CB), and a station in the open Gulf of Maine, Wilkinson Basin (WB) were sampled, while in July 1996, sediments were collected from two sites in Cape Cod Bay (CCB1 and CCB2), Massachusetts Bay (MB), and near Stellwagen Bank (SB). All samples were taken with a Sandia-Hessler type MK3 (Ocean Instruments, San Diego, CA) corer with a 0.25 m² by 0.7 m deep box. Where the sediment-water interface appeared undisturbed, we used acrylic liners (ca. 13 cm diameter) to extract sub-cores. The sub-cores were immediately extruded on board and trimmed. Sections for

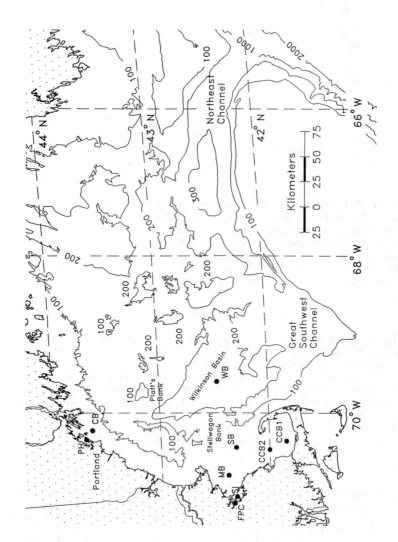

Figure 2. Locations of sediment sampling stations in the Gulf of Maine.

organic compound, OC, and SC determination were stored frozen in solvent-rinsed amber glass bottles with Al-foil lined Teflon caps.

Quantification of Sedimentary Total Organic and Soot Carbon. Our analytical method for measuring OC and SC in complex sedimentary matrices has been detailed and validated elsewhere (18). Briefly, the OC+SC content of dried sediment is obtained by carbon elemental analysis (PE 2400 CHN; Perkin Elmer Corp., Norwalk, CT) after destruction of carbonates through mild acidification of the sample in Ag capsules (Elemental Microanalysis Ltd., Manchester, NH). For SC, dried sediment is thermally oxidized at 375°C for 24 h, followed by acidification and carbon elemental analysis. Concentration of OC is then calculated as the difference between OC+SC and SC.

Quantification of Sedimentary PAHs. Concentrations of PAHs in FPC and SI were taken from published estimates in the same cores (12). No PAH data exist for CCB1 and CCB2 sediments. In all other samples, about 10 g of the wet sediments were transferred to a soxhlet extractor and spiked with four deuterated PAH recovery standards, spanning a range in aqueous solubilities. The sediments were extracted for 48 h in a 9:1 mixture of methylene chloride and methanol. Extracts were concentrated using Kuderna-Danish or rotary evaporation and were then charged to a fully activated silica gel gravity column containing $NaSO_4$ (anh.) and activated copper (prepared according to ref. 29) to remove residual water and elemental sulfur, respectively. The PAHs were eluted in the third fraction with 30 mL 3:1 hexane - toluene and quantified by gas chromatography - mass spectrometry (Hewlett-Packard 5995B). Recoveries of PAH internal standards for both samples and blanks were 65-95% for the entire procedure, and reported concentrations are corrected for these recoveries. PAH contents of the samples were several orders of magnitude above those of the blanks.

Results and Discussion.

Source-Diagnostic PAH Ratios. The relative abundances of alkylated-to-unsubstituted parent PAHs contain information useful in elucidating the origin of environmental PAHs (e.g., 17, 30, 31). The alkyl homolog distribution of petroleum generally exhibits an increasing abundance with increasing carbon number over the first four-six homologs (e.g., 17) while the unsubstituted congeners are the most abundant in PAH assemblages from combustion processes (e.g., 17, 32).

Since the form in which PAHs are introduced to the environment may affect their dispersal and transformations, we have consolidated source-diagnostic ratios of the 178 (phenanthrene) and 202 (pyrene) series from a large literature data set (about 20 independent studies representing over one-hundred observations: refs. 19, 21, 23-25, 32-44, Gustafsson-Gschwend *et al.*, unpublished data from Gulf of Maine) to elucidate the relative importance of petroleum and pyrogenic sources of PAHs to a variety of aquatic environments (Figure 3). The reported ratios support the contention by LaFlamme and Hites (30) that the global distribution of PAHs in contemporary aquatic environments is dominated by an input from combustion processes (Figure 3). The historical sedimentary record of PAHs follows the industrial usage of fossil fuel (e.g., 18-19, 45-48), and we may thus conclude that the predominantly pyrogenic PAHs in today's environment are from anthropogenic combustion as opposed to natural forest fires.

The Sedimentary Distribution of Organic and Soot Carbon. Examining nine sediments from diverse environments in detail, we found organic carbon concentrations ranging from 30-50 mg OC per gram dry weight sediment (gdw sed.) in rapidly accumulating harbor regimes to 10-20 mg OC/gdw sed. at offshore continental shelf locations, to a value of 1.5 mg OC/ gdw sed. at a highly winnowed site with coarse sand (Table I). The sedimentary soot carbon concentrations varied

Diagnostic PAH Source Ratios - 178 series -

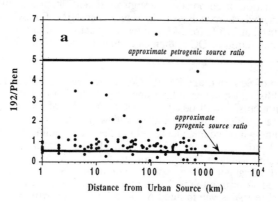

Diagnostic PAH Source Ratios   - 202 series -

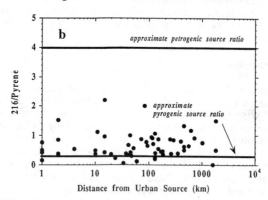

Figure 3. The PAH source-diagnostic ratios of (a) the sum of methyl-phenanthrenes and -anthracenes to phenanthrene and (b) the methyl-pyrenes and -fluoranthenes to pyrene from a wide range of atmospheric, water-column, and sedimentary environments (19, 21, 23-25, 32-44, Gustafsson-Gschwend *et al.*, unpublished data). The solid lines refer to typical ratios found in petrogenic and pyrogenic sources of PAHs (17, 31, 32).

even more widely (factor of 15), decreasing with increasing distance from urban sources. SC concentrations ranged from 2-7 mg SC/gdw sed. in urban harbors to 0.1-0.7 mgSC /gdw sed. further offshore (Table I). These results suggest that 3-13% of reduced carbon in modern continental shelf sediments is anthropogenic soot, with the highest fractions being found in or near urban harbors.

Table I. Organic and Soot Carbon Concentrations in Gulf of Maine Sediments

Sample depth (cm)		Lattitude (N)	Longitude (W)	Organic Carbon (mg OC/ gdw sed.)	Soot Carbon (mg SC/ gdw sed.)
FPC	7-9	42°21'22"	71°02'41"	50.9±0.7	6.61±0.84
SI	0-1	42°19'46"	70°59'34"	28.8±2.0	2.72±0.40
PH	2-3	43°39'63"	70°14'32"	37.0±4.3	1.76±0.56
CB	2-3	43°36'94"	70°09'31"	12.5±3.5	0.72±0.10
WB	2-3	42°38'00"	69°36'26"	22.5±0.5	0.71±0.30
CCB1	2-3	41°51'18"	70°17'54"	11.1±0.2	0.34±0.13
CCB2	2-3	42°03'30"	70°24'36"	13.0±0.2	0.42±0.04
MB	2-3	42°22'00"	70°54'00"	1.50±0.7	0.11±0.07
SB	2-3	42°20'33"	70°23'30"	22.3±0.4	0.75±0.01

Regression of PAHs with OC and SC. Our data indicate a significantly higher degree of correlation for PAHs with SC than with OC (Table II and Figure 4). Regressions were performed for three PAHs with 3-, 4-, and 5-ringed systems, respectively, and consequently spanning a range in hydrophobicities. In addition to higher correlation coefficients ("R^2"), a greater degree of significance for each of the PAH-soot relations was calculated ("p") (Table II). We tested the Null hypothesis that SC and PAH are not correlated and found that its probability to be valid was less than 1%. However, the statistical analysis indicated quite significant probabilites for OC and PAH to not be correlated (Table II).

These strong correlations between field measurements of SC and PAHs lend credence to the applicability of the analytical soot carbon method. More importantly, the SC data strongly suggest that the spatial distribution of PAHs are chiefly dictated by interactions with soot as opposed to with bulk organic matter. Since both amorphous organic and soot carbon are likely to be largely associated with the "fines" fraction of accumulating sediments, some correlation between OC and SC may also be observed. However, since grain size is not the relevant intrinsic sorbent property, and since the statistical analysis yields significantly better correlation for PAHs with SC than with OC, future work on the phase-speciation of PAHs should focus on the soot carbon content of sediments. One exception is obviously near oil spills. These results suggest that it is necessary to revise the fundamental assumption of bulk organic matter partitioning currently made in fugacity and bioavailablity models when applied to ambient PAHs (49-52). A strong soot sorbent would mean significantly lower dissolved fractions and correspondingly lower predictions of bioavailabilities.

Table II. Correlations of PAHs with OC and SC in Seven Marine Sediments

Compound	Organic Carbon		Soot Carbon	
	Correlation Coefficient - R^2	Probability Level - p	Correlation Coefficient R^2	Probability Level p
Phenanthrene	0.74	< 0.10	0.99	< 0.01
Pyrene	0.59	< 0.10	0.97	< 0.01
Benzo[a]pyrene	0.63	> 0.50	0.98	< 0.01

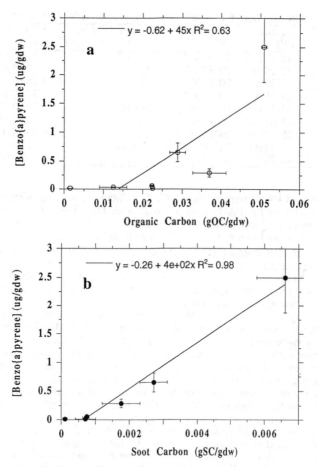

Figure 4. Regressions (n = 7) between sedimentary concentrations of benzo[a]pyrene with (a) organic carbon and (b) soot carbon, respectively, illustrating the better correlation of PAHs with soot carbon (see Table II).

Physico-Chemical Properties of Soot. To develop an understanding of the underlying mechanisms and driving force behind the association of PAHs with soot, it is helpful to consider both macroscopic features and molecular details of the soot matrix.

Soot Particles - Formation and Physical Properties. Earlier microscopy-based approaches to quantifying sedimentary soot have revealed some of the physical properties of these pyrogenic carbon phases found in aquatic environments. Grossly, soot particles may be divided into two groups, where different formation processes render distinct macroscopic properties. The few-to-tens of μm-sized coarse "charcoal" particles (e.g., 53-57), resulting from incomplete combustion of fuel particles, often occur as single spheroidal forms in sediments. The fine "soot" particles (e.g., 58), occuring as individual spheres of 3-30 nm diameter, are found aggregated into grape-like clusters. Such fine soot is formed from vapor-phase condensation processes where free-radical reactions between polyaromatic and acetylenic species play a key role during both inception and surface growth (e.g., 59-62). Hence, in addition to being a dilute fraction of the entire sedimentary matrix, the fact that soot particles may exhibit very different morphologies and span several orders of magnitude in size renders microscopy-based approaches to quantify the entire pool of PAH-carrying sedimentary soot less attractive than bulk chemical analysis (18).

However, there are physical similarities (e.g., density, surface area, and surface chemical structures) among soots from different sources whether furnace flames, diesel and gasoline engines, or by commercial processes (62, 63). Specific densities are near 1.8-2.1 g/cm^3, similar to the density of pure graphite (2.1-2.2 g/cm^3; 63, 64). Brunauer-Emmet-Teller (BET) surface areas of soot appear to be around 100 m^2/g, not much greater than two-dimensional (exterior) surface areas calculated from the SEM-derived diameters of the spheres and assuming the above density (e.g., 63-66). The similarity in three- and two-dimensional surface areas suggest that soots' pores are generally too narrow for penetration by the adsorbing species. High-resolution transmission electron microscopy (HRTEM) images have recently been presented with reported interlayer spacing for diesel soot on the order of 4 Å (67), suggesting that the pore-width is at most a few Å.

Soot Particles - Chemical Properties. Recent TEM studies, in combination with spectroscopic results, confirm the picture of soot as a multi-layered aromatic system as was hypothesized long ago (e.g., 68). Large C:H and C:O elemental ratios suggest that soot must possess a highly condensed and conjugated structure (e.g., 62, 66). The most prominent feature in FTIR spectra of soot is the absorption band at 1590 cm^{-1}, corresponding to C=C stretching mode of polyaromatic systems (66, 69). This is consistent with earlier findings of aromatic stretching modes of graphitic planes (e.g., 70, 71). Minor bands in FTIR spectra suggest the presence of some C=O and C-O bonds in soot (66). The same dominant 1590 cm^{-1} absorption band was reported also in the Raman spectrum of soot, indicating lattice vibrations of graphitic structure (i.e., crystal symmetry; ref. 66). With dimensions for crystallites of carbon black (72), Akhter and co-workers calculated that there should be nearly 30 aromatic rings per layer of this overall aromatic network (66). Cross-polarization magic-angle-spinning (CP/MAS; solid state) ^{13}C NMR results indicate that 90-100% of the total organic carbon in soot is aromatic, supporting the overall condensed backbone picture (66). Hence, soot may be thought of as particles of multi-layered macro-PAHs (conjugated blocks of about 30 rings), structurally held together with some carbon-bridges and ether-linkages, and containing some substitution of conjugated carbonyls and acid anhydrides (a hypothetical structure is shown in Fig. 16 in ref. 73).

Soot-Water Partitioning of PAHs. Strong interaction between PAH molecules and the condensed aromatic structure of soot may be anticipated. The positioning of sorbed PAHs on or in the soot matrix may have an effect on the ease whereby the molecules may exchange with the surrounding media. Urban atmospheric soot is commonly described as containing an organic film of liquid compounds (e.g., 60, 63). PAHs are often found to make up a significant fraction of the extractable fraction (e.g., 60, 74). However in aquatic environments, this liquid layer is likely to either dissolve in the surrounding water or degrade. In fact, our sedimentary PAH-SC regressions (e.g., Figure 3) suggest that the extractable PAHs contribute less than one per cent of the SC mass. Hence, in aquatic sedimentary environments, PAHs may be anticipated to interact directly with the soot backbone structure. There, the PAHs can exist, for instance, in an aromatic layer "trapped" between macro-PAH blocks. The few-Å-"thick" PAHs may also be sorbed on the exterior layer of the soot, or in the interplanar spacing between two layers.

It is easy to envision the favorable π-cloud overlap between the planar PAHs and the flat "macro-PAH" surface of soot. This close-up contact would likely lead to greatly reduced rotational and translational freedom of the PAH molecules. This sorbed state may hence be best modeled as a fusion (solid precipitation) onto the planar soot surface. Note that the strength of dispersion forces is highly dependent on the distance of separation between the soot surface and the center of the sorbent molecule (e.g., ref. 75). As a result, other non-planar aromatic molecules, such as *ortho*-substituted PCBs, may not experience fully reduced rotation and, thus, not be "locked" in a solid-like state. It is noteworthy that non-*ortho* substituted (coplanar) PCBs exhibit enhanced affinity to urban aerosols compared to more non-planar congeners despite similar subcooled liquid vapor pressures (76), suggesting the importance of a close and favorable alignment between sorbate and soot.

Estimation of PAH Soot-Water Partitioning. To explore potential mechanisms and the physico-chemical driving forces of soot-water partitioning of PAHs, it is useful to attempt to conceptualize the process and isolate thermodynamic measures of each part's importance. We will explore the partitioning mechanism envisioned above, where PAHs are either dissolved in the aqueous phase or fused on the semi-crystalline soot surface. The latter would lead to greatly reduced rotational and translational freedom of the molecules. This sorbed state is assumed to best be modeled as a solid precipitation.

For a PAH distributing itself between the soot and the surrounding aqueous phase, at equilibrium, its fugacities (f_i) in the two media are equal:

$$f_{\text{soot-surf}} \equiv f_{\text{water}} \qquad (2)$$

These fugacities may be expressed relative to the fugacities of a PAH in its pure solid and liquid forms, respectively:

$$\gamma_{\text{soot-surf}} \; x_{\text{soot-surf}} \, P_{\text{soot-surf}} \, (s) \; \equiv \; \gamma_w \, x_w \, P^\circ_w(l) \qquad (3)$$

where γ, and x are the PAH activity coefficients and mole fractions on the soot surface and in the water, respectively, and P is the vapor pressure. The formulation on the left hand side follows the description for solid solutions of Prausnitz (ref. 77; p. 403), with the assumption that sorption to the surface of the soot is thermodynamically similar to sorption onto a pure PAH solid. If $P°_w(l)$ is chosen as the reference state, one may use the Prausnitz' (ref. 77; p. 390) approximation to relate the solid vapor pressure of a compound to its subcooled liquid vapor pressure for the change in free energy associated with the phase transition (ΔG_{fusion}), and recast Equation 3:

$$\gamma_s \, x_s \, \exp[-(\Delta S_\phi/R)((T_m/T) - 1)] \, P°_w(l) \equiv \gamma_w \, x_w \, P°_w(l) \tag{4}$$

where ΔS_ϕ is the change in entropy resulting from the phase-transition, T_m is the melting point of the solid PAH, and T is the absolute temperature of the system.

A soot-water equilibrium distribution constant (K_{sw}) may be defined as:

$$K_{sw} \, (L/m^2) \equiv \frac{C_s}{C_w} = \frac{x_s}{A_s} \frac{V_w}{x_w} \tag{5}$$

where C_s is the concentration of the compound on the soot surface (mol/m^2) and C_w is its concentration in the aqueous phase (mol/L). A_s is the specific area of the soot (m^2/mol) and V_w is the specific volume of the aqueous solution (L/mol); these terms are constant for dilute conditions. From above (equation 5) we now substitute for the ratio x_s/x_w:

$$K_{sw} \, (L/m^2) = \frac{\gamma_w \, V_w}{\gamma_s \, \exp[-(\Delta S_\phi/R)((T_m/T) - 1)] \, A_s} \tag{6}$$

with the liquid fugacities canceling out. This equation can be manipulated to separate terms:

$$\log K_{sw} = \log \gamma_w - \log \gamma_s + \log \left(\frac{V_w}{A_s}\right) + \frac{\Delta S_\phi}{2.3 \, R}\left(\frac{T_m}{T} - 1\right) \tag{7}$$

As is apparent from Equation 7, the magnitude of K_{sw} is dependent on the relative (in)compatibilities of a PAH with the aqueous and soot phases (i.e., the aqueous and solid activity coefficients), and the free energy costs of phase transition.

If the terms in Equation 7 could be estimated, we would be able to *a priori* predict K_{sw}. We start with the first term, using anthracene to illustrate the relative magnitudes of the terms. The aqueous activity coefficient is a measure of the non-ideality of the solution (relative to pure subcooled liquid anthracene). Since the solubility of anthracene is low ($C_w{}^{sat} = 10^{-4.40}$ M), γ_w must be very close to $\gamma_w{}^{sat}$. Since the specific volume of a dilute aqueous solution is known (0.018 L/mol), γ_w may be estimated from:

$$\gamma_w \approx \gamma_w{}^{sat} = \frac{1}{x_w{}^{sat}} = \frac{1}{C_w{}^{sat} V_w} = \frac{1}{10^{-4.40} \, 0.018} = 10^{6.1} \tag{8}$$

For soot association, in the infinite dilution case (Henry's law region; where sorption isotherm is linear during initial low coverage), the same assumption of no interaction between sorbed molecules may be made. Following the above assumption that PAH sorption onto soot is thermodynamically similar to associating with a pure PAH crystal, the activity coefficient of soot-sorbed PAHs can be approximated by:

$$\gamma_s \approx \gamma_s^{sat} = \frac{1}{x_s^{sat}} = 1 \tag{9}$$

Considering the relative interactions with the surrounding media, it is not a surprise that PAHs are much more compatible with soot than with water as these relative activity coefficients indicate. In the infinite dilution case, the surfaces of the aqueous soot particles are covered by water molecules. Hence, the specific surface of soot in an aqueous system is:

$$A_s = A_{H2O} N_A = 0.5 \ \text{Å}^2 \times 6.023 \ 10^{23} \ \text{mol}^{-1} = 3000 \ \text{m}^2/\text{mol} \tag{10}$$

where A_{H2O} is the two-dimensional surface area of a water molecule and N_A is Avogadro's number. The entropy of liquid-solid phase transfer (ΔS_m) at the melting point for rigid molecules is relatively invariant and may be estimated from (78):

$$\Delta S_m(T_m) \approx 56.5 \qquad (\text{J/mol K}) \tag{11}$$

The total entropy-change upon fusion of PAHs is made up of approximately 15 J/mol K due to translational restriction and 42 J/mol K due to loss of rotational freedom (78). As noted above, we suspect that interaction of PAHs with an aromatic soot surface leads to reduced rotational and translational freedom compared to the liquid state of the molecule. However, because a PAH molecule parallel to the soot plane is restricted in only one of three translational planes (orthogonal to surface), and in two of three rotational dimensions, this close-up surface interaction is likely less restricting than formation of a three-dimensional solid crystal. Given only one-third reduction in translation (1/3 x 15 J/mol K) and two-thirds reduction in rotation (2/3 x 42 J/mol K), we estimate ΔS_Φ of PAH soot sorption to be approximately 33 J/mol K. Since for anthracene $T_m = 489$ K,

$$\frac{\Delta S_\phi}{2.3 \ R}(\frac{T_m}{T} - 1) \approx \frac{33}{2.3 \times 8.3}(\frac{489}{298} - 1) = 0.8 \tag{12}$$

Substituting the estimated values of the terms making up K_{sw} back into Equation 7 results in a thermodynamically-based *a priori* estimate of the soot-water distribution for anthracene of:

$$\log K_{sw} \ (\text{anthracene}) \ \approx 1.7 \ L_w/\text{m}^2_{soot} \tag{13}$$

Since the surface area of soot particles such as diesel particulate matter, flame soot, and graphitized carbon blacks typically cluster around 10^5 m^2/kg (e.g., 63, 64, 73), the equilibrium distribution coefficient may be presented in sorbent-mass based units, fascilitating comparison:

$$\log K_{sw} \text{ (ant)} \approx 6.7 \text{ L}_w/\text{kg}_{soot} \tag{14}$$

This estimate may be compared to values for partitioning of anthracene in an activated-carbon - water system ($\log K_{ac\text{-}w}$) of 6.5 (ref. 82) and 6.8 (calculated in ref. 18, from the Henry's law region of the sorption isotherms in ref. 28). In contrast, the organic carbon - water partition constant ($\log K_{oc}$) for anthracene (≈ 4.1) is much smaller (e.g., 79). Theoretical K_{sw} partition coefficients were calculated for a set of PAHs and compared with measured and/or estimated partitioning to activated carbon and natural organic carbon matrices, respectively (Table III).

Table III. Comparison of Estimated K_{sw} with $K_{ac\text{-}w}$ and K_{oc} Values (L_w/kg_C)

Compound	T_m (K)	$\log \gamma_w$	$\log K_{sw}$ estimated	$\log K_{ac\text{-}w}$ measured		$\log K_{oc}$
	ref. 80	ref. 80		ref.28	ref.82	ref.79
Acenaphthene	369	5.59	5.8	7.2	6.2	3.5
Fluorene	389	5.86	6.2	7.1	6.3	3.8
Anthracene	489	6.14	6.7	6.8	6.5	4.1
Phenanthrene	374	6.34	6.4	7.0	6.5	4.2
Pyrene	429	6.62	7.0	7.7		4.8
Fluoranthene	384	6.78	6.9	7.8	6.9	4.8
Chrysene	528	7.52	8.2	8.5		5.4
Benz[a]-anthracene	435	7.62	7.8	7.4		5.5
Benzo[a]pyrene	448	8.08	8.4		7.8	6.1
Dibenzo[a,h]-anthracene	539	8.09	9.3		8.2	6.8

The good match between our theoretically deduced soot-water equilibrium distribution and empirically obtained data from an assumed related sorbent matrix (activated carbon) suggests that the conceptualized mechanism and parameterization of the physico-chemical driving forces may be representative of the actual sorption process.

Conclusions.

The dominant form of PAHs in many aquatic environments is to be associated with soot. Inspection of the entire PAH assemblage reveals the importance of hydrophobicity on the relative solid-water distributions. This characteristic implies that an active exchange between soot-associated PAHs and their dissolved counterparts is taking place, albeit on an unknown timescale. Another implication of this study is that PAHs are good markers for the presence of soot and other soot-associated contaminants.

A much better correlation for PAHs was found with soot carbon compared with bulk organic carbon, despite the fact that soot carbon is only present at a few per cent of the non-soot organic carbon levels. Soot carbon could account for 97-99% of the sedimentary distributions observed for phenanthrene, pyrene, and benzo[a]pyrene; three PAHs of quite varying physico-chemical properties, whereas there was no significant correlation between PAHs and organic carbon at the 95% confidence limit.

A thermodynamic evaluation of the soot-water partition process resulted in theoretical estimation of PAH partition coefficients which were substantially larger than traditional organic-carbon normalized partition constants, but which agreed reasonably well with results from sorption onto activated carbon. The significantly enhanced affinity to highly condensed and aromatic surfaces is suggested to be a result of geometrically efficient π-cloud overlapping.

There are far-reaching implications of soot (de)sorption of PAHs dominating the environmental phase-speciation of these, and by inference many other physico-chemically similar, contaminants. Estimated K_{sw} together with measurements of soot carbon concentrations in coastal sediments suggest that the dissolved fraction of these compounds that are available for biological uptake and homogeneous phase reactions and transport are much lower than previously thought. This insight about PAH behavior may be extrapolatable to structurally similar contaminant assemblages such as coplanar PCBs and polychlorinated dioxins, and to lesser extents also to other hydrophobic pollutants. These findings should inspire controlled soot-water sorption studies of PAHs and other contaminants to further constrain the magnitude and kinetics of this partition process, as well as elucidating what fractions of soot-associated PAHs are available for desorption on different time scales.

Acknowledgments.

The captains and crew of R/V Asterias, R/V Argo Maine, and R/V Diane G. are thanked for their assistance during sediment sampling. John Farrington is thanked for lending us his MK-3 box corer. The able coring skills of Hovey Clifford, Chris Swartz, Chris Long, John MacFarlane, and Tom Ravens are recognized. John MacFarlane, Allsion MacKay, Keong Kim, and Rachel Adams are acknowledged for support and assistance during sample analysis. Inspiring discussions with John Farrington, Bob Eganhouse, Shige Takada, Johan Axelman, Dag Broman, and Adel Sarofim contributed to this work. Shige Takada is especially thanked for contributing a pre-publication data set to Figure 1. Financial support came from the Office of Naval Research (grant # NOOO14-93-1-0883), National Oceanic and Atmospheric Administration (# NA36RM044-UM-S242), National Institute of Environmental Health Sciences (# 2-P30-ESO-2109-11), and Massachusetts Water Resources Authority. The views herein are those of the authors and do not necessarily reflect the views of NOAA or any of its subagencies.

Literature Cited.

1. *Evaluation and Estimation of Potential Carcinogenic Risks of Polynuclear Aromatic Hydrocarbons* (Office of Health and Environmental Assessment, Office of Research and Development, US Environmental Protection Agency, Washington, DC, 1985).
2. Durant, J. L.; Thilly, W. G.; Hemond, H. F.; LaFleur, A. L. *Environ. Sci. Technol.* **1994**, *28*, 2033.
3. Karickhoff, S. W.; Brown, D. S.; Scott, T. A. *Water Res.* **1979**, *13*, 241.
4. Chiou, C. T.; Porter, P. E.; Schmedding, D. W. *Environ. Sci. Technol.* **1979**, *206*, 831.
5. Schwarzenbach, R. P.; Gschwend, P. M.; Imboden, D. M. *Environmental Organic Chemistry*. Wiley-Interscience: New York, NY, 1993, 269-276.

6. Kile, D. E., Chiou, C. T.; Zhou, H.; Li, H.; Xu, O. *Environ. Sci.Technol.* **1995**, *29*, 1401.
7. Baker, J. E.; Eisenreich, S. J.; Eadie, B. J. *Environ. Sci. Technol.* **1991**, *25*, 500.
8. Takada, H.; Kumata, H.; Satoh, F. 212th American Chemical Society National Meeting. Division of Environmental Chemistry extended abstracts, pp. 295-297.
9. Readman, J. W.; Mantoura, R. F. C.; Rhead, M. M. *Sci. Tot. Env.* **1987**, *66*, 73.
10. Broman, D.; Näf, C.; Rolff, C.; Zebühr, Y. *Environ. Sci. Technol.* **1991**, *25*, 1850.
11. Ko, F.-C.; Baker, J. E. *Mar. Chem.* **1996**, *49*, 171.
12. McGroddy, S. E.; Farrington, J. W. *Environ. Sci. Technol.* **1995**, *29*, 1542
13. Poster, D. L.; Baker; J. E. *Environ. Sci. Technol.* **1996**, *30*, 341.
14. McGroddy, S. E.; Farrington, J. W.; Gschwend, P. M. *Environ. Sci. Technol.* **1996**, *30*, 172.
15. Brownawell, B. J.; Farrington, J. W. *Geochim. Cosmochim. Acta* **1986**, *50*, 157.
16. Burgess, R. M.; McKinney, R. A.; Brown, W. A. *Environ. Sci. Technol.* **1996**, *30*, 2556.
17. Youngblood, W. W.; Blumer, M. *Geochim. Cosmochim. Acta* **1975**, *39*, 1303.
18. Gustafsson, Ö.; Haghseta, F.; Chan, C.; MacFarlane, J.; Gschwend, P. *Environ. Sci. Technol.* **1997**, *31*, 203.
19. Gschwend, P. M.; Hites, R. A. *Geochim. Cosmochim. Acta* **1981**, *45*, 2359.
20. Farrington, J. W.; Goldberg, E. D.; Risebrough, R. W.; Martin, J. H.; Bowen, V. T. *Environ. Sci. Technol.* **1983**, *17*, 490.
21. Prahl, F. G.; Carpenter, R. *Geochim. Cosmochim. Acta* **1983**, *47*, 1013.
22. Jones, D. M.; Rowland, S. J.; Douglas, A. G.; Howells, S. *Intern. J. Environ. Anal. Chem.* **1986**, *24*, 227.
23. Broman, D.; Näf, C.; Wik, M.; Renberg, I. *Chemosphere* **1990**, *21*, 69.
24. Bouloubassi, I.; Saliot, A. *Oceanologica Acta* **1993**, *16*, 145.
25. Wakeham, S. G. *Mar. Chem.* **1996**, *53*, 187.
26. Venkataraman, C.; Friedlander, S. K. *Environ. Sci. Technol.* **1994**, *28*, 563.
27. Allen, J. O.; Dookeran, N. M.; Smith, K. A.; Taghizadeh, K.; Lafleur, A. L. *Environ. Sci. Technol.* **1996**, *30*, 1023.
28. Walters, R. W.; Luthy, R. G. *Environ. Sci. Technol.* **1984**, *18*, 395.
29. Blumer, M. *Anal. Chem.* **1957**, *29*, 1039.
30. LaFlamme, R. E.; Hites, R. A. *Geochim. Cosmochim. Acta* **1978**, *42*, 289.
31. Sporstøl, S.; Gjøs, N.; Lichtenthaler, G.; Gustavsen, K. O.; Urdal, K.; Oreld, F.; Skei, J. *Environ. Sci. Technol.* **1983**, *17*, 282.
32. Lee, M. L.; Prado, G. P.; Howard, J. B.; Hites, R. A. *Biomed. Mass Spectrom.* **1977**, *4*, 182.
33. Takada, H.; Farrington, J. W.; Bothner, M. H.; Johnson, C. G.; Tripp, B. W. *Environ. Sci. Technol.* **1994**, *28*, 1062.
34. Lipiatou, E.; Saliot, A. *Mar. Chem.* **1991**, *32*, 51.
35. Lipiatou, E.; Marty, J.-C.; Saliot, A. *Mar. Chem.* **1993**, *44*, 43.
36. Bates, T. S.; Hamilton, S. E.; Cline, J. D. *Environ. Sci. Technol.* **1984**, *18*, 299.
37. Broman, D.; Colmsjö, A.; Näf, C. *Bull. Environ. Contam. Toxicol.* **1987**, *38*, 1020.
38. Dachs, J.; Bayona, J. M.; Fowler, S. W.; Miquel, J.-C.; Albaigés, J. *Mar. Chem.* **1996**, *52*, 75.
39. Barrick, R. C.; Prahl, F. G. *Estuarine Coastal Shelf Sci.* **1987**, *25*, 175.
40. Windsor Jr, J. G.; Hites, R. A. *Geochim. Cosmochim. Acta* **1979**, *43*, 27.

41. Hites, R. A.; LaFlamme, R. E.; Windsor Jr, J. G. In *Petroleum in the Marine Environment*; Petrakis, L.; Weiss, F. T., Eds.; Adv. in Chem. Ser. 185; American Chemical Society: Washington, DC, 1980; pp289-311.
42. Simo, R.; Colom-Altés, M.; Grimalt, J. O.; Albaigés, J. *Atmos. Env.* **1991**, *25A*, 1463.
43. Bjørseth, A.; Lunde, G.; Lindskog, A. *Atmos. Env.* **1979**, *13*, 45.
44. Ehrhardt, M.; Petrick, G. *Mar. Chem.* **1993**, *42*, 57.
45. Grimmer, G.; Böhnke, H. *Cancer Lett.* **1975**, *1*, 75.
46. Hites, R. A.; LaFlamme, R. E.; Farrington, J. W. *Science* **1977**, *198*, 829.
47. Prahl, F. G.; Carpenter R. *Geochim. Cosmochim. Acta* **1979**, *43*, 1959.
48. Hites, R. A.; LaFlamme, R. E.; Windsor Jr, J. G.; Farrington, J. W.; Deuser, W. G. *Geochim. Cosmochim. Acta* **1980**, *44*, 873.
49. Mackay, D.; Paterson, S. *Environ. Sci. Technol.* **1981**, *15*, 1006.
50. Mackay, D.; Paterson, S. *Environ. Sci. Technol.* **1991**, *25*, 427.
51. Boehm, P. D.; Farrington, J. W. *Environ. Sci. Technol.* **1984**, *18*, 840.
52. U.S. Environmental Protection Agency *Sediment quality criteria for the protection of benthic organisms: Fluoranthene*; EPA 822-R-93-012; Offices of Water, Research, and Development and, Science and Technology; Washington, DC, 1993.
53. Griffin, J. J.; Goldberg, E. D. *Science* **1979**, *206*, 563.
54. Griffin, J. J.; Goldberg, E. D. *Geochim. Cosmochim. Acta* **1981**, *45*, 763.
55. Goldberg, E. D.; Hodge, V. F.; Griffin, j. J.; Koide, M.; Edgington, D. N. *Environ. Sci. Technol.* **1981**, *15*, 466.
56. Renberg, I.; Wik, M. *Ecol. Bull.* **1985**, *37*, 53.
57. Wik, M; Natkanski, J. *Phil. Trans. R. Soc. Lond.* **1990**, *B327*, 319.
58. Griffin, J. J.; Goldberg, E. D. *Environ. Sci. Technol.* **1983**, *17*, 244.
59. Prado, G.; Lahaye, J. In *Mobile Source Emissions Including Polycyclic Organic Species*; Rondia, D.; Cooke, M.; Haroz, R. K., Eds.; NATO Advanced Research Workshop Series; Reidel: Liege, Belgium, 1983; pp. 259-275.
60. Longwell, J. P. In *Soot in Combustion Systems and its Toxic Properties*; Lahaye, J., Prado, G., Eds.; NATO/Plenum Press: New York, 1981; pp. 37-56.
61. Lahaye, J. *Polymer Degrad. Stabil.* **1990**, *30*, 111.
62. Hamins, A. In *Environmental Implications of Combustion Processes*, Puri, I. K., Ed.; CRC Press: Boca Raton, 1993; pp 71-95.
63. Risby, T. H.; Sehnert, S. S. *Environ. Health Perspec.* **1988**, *77*, 131.
64. Ross, M. M.; Risby, T. H.; Steele, W. A.; Lestz, S. S.; Yasbin, R. E. *Colloids and Surfaces* **1982**, *5*, 17
65. Ross, M. M.; Risby, T. H.; Lestz, S. S.; Yasbin, R. E. *Environ. Sci. Technol.* **1982**, *16*, 75.
66. Akhter, M. S.; Chughtai, A. R.; Smith, D. M. *Appl. Spectrosc.* **1985**, *39*, 143.
67. Rainey, L.; Palotas, A.; Bolsaitis, P.; Vander Sande, J. B.; Sarofim, A. F. *Appl. Occup. Environ. Hyg.* **1996**, *11*, 777.
68. Donnet, J. B.; Schultz, J.; Eckhardt, A. *Carbon* **1968**, 6, 781 (in French).
69. Smith, D. M.; Griffin, J. J.; Goldberg, E. D. *Anal. Chem.* **1975**, *47*, 233.
70. Tuinstra, F.; Koenig, J. L. *J. Chem. Phys.* **1970**, *53*, 1126.
71. Friedel, R. A.; Carlson, G. L. *J. Phys. Chem.* **1971**, *75*, 1149.
72. Avgul, N. N.; Kiselev, A. V. *Chemistry and Physics of Carbon*; Dekker: New York, NY, 1970.
73. Akhter, M. S.; Chughtai, A. R.; Smith, D. M. *Appl. Spectrosc.* **1985**, *39*, 154.
74. Lee, M. L.; Bartle, K. D. In *Particulate Carbon Formation During Combustion*; Siegla, D. C.; Smith, G. W., Eds.; Plenum 'ew York, NY, 1981, 91-104.

75. Israelachvili, J. *Intermolecular and Surface Forces*; Academic Press: San Diego, CA; 1992, 2nd Ed.
76. Falconer, R. L.; Bidleman, T. F.; Cotham, W.E. *Environ. Sci. Technol.* **1995**, *29*, 1666.
77. Prausnitz, J. M. *Molecular Thermodynamics of Fluid-Phase Equilibria*; Prentice-Hall: Englewood Cliffs, NJ, 1969.
78. Yalkowsky, S. H.; Valvani, S. C. *J. Chem. Eng. Data* **1979**, *24*, 127.
79. Karickhoff, S. W. *Chemosphere* **1981**, *10*, 833.
80. Miller, M. M.; Wassik, S. P.; Huang, G.-L.; Shiu, W.-Y.; Mackay, D. *Environ. Sci.Technol.* **1985**, *19*, 522.
81. Ruepert, C.; Grinwis, A.; Govers, H. *Chemosphere* **1985**, 14, 279.
82. Luehrs, D. C.; Hickey, J. P.; Nilsen, P. E.; Godbole, K. A.; Rogers, T. N. *Environ. Sci. Technol.* **1996**, *30*, 143

Chapter 25

Polychlorinated Biphenyls as Probes of Biogeochemical Processes in Rivers

S. A. Fitzgerald and J. J. Steuer

U.S. Geological Survey, 6417 Normandy Lane, Madison, WI 53719

A field study was conducted to investigate the use of PCB (polychlorinated biphenyl) congener and homolog assemblages as tracers of biogeochemical processes in the Milwaukee and Manitowoc Rivers in southeastern Wisconsin from 1993 to 1995. PCB congeners in the dissolved and suspended particle phases, along with various algal indicators (algal carbon and pigments), were quantitated in the water seasonally. In addition, PCB congener assemblages were determined seasonally in surficial bed sediments. Biogeochemical processes investigated included: determination of the source of suspended particles and bottom sediments by comparison with known Aroclor mixtures, water-solid partitioning, and algal uptake of PCBs. Seasonal differences among the PCB assemblages were observed mainly in the dissolved phase, somewhat less in the suspended particulate phase, and not at all in the bed sediments.

Despite being banned since the 1970's, polychlorinated biphenyls (PCBs) are ubiquitous contaminants, present not only in industrial areas where they were manufactured and used in cutting oils, sealants, hydraulic fluids and pesticides, but also in remote locales such as the polar regions due to atmospheric transport and deposition *(1)*. PCBs are a set of 209 related chlorinated organic compounds, some of which have demonstrated toxicity *(2)*. Being relatively hydrophobic and lipophilic, these compounds tend to adsorb onto clay surfaces or be associated with lipids present in algae or other aquatic organisms. Thus, when present, PCB distributions in aquatic environments such as rivers can potentially be used as tracers of various biogeochemical processes. Some of these processes include: partitioning to suspended and bottom sediments, and passive PCB uptake by algae. In addition, because PCBs tend to be refractory in most aquatic environments, it is often possible to determine the particular commercial mixtures of PCBs, termed Aroclors, that were released to the river. One exception is when an Aroclor mixture has been extensively 'weathered' by selective solubilization, volatilization, and/or microbially-mediated decomposition.

These biogeochemical processes were investigated as part of a larger study of PCB distribution and transport in the Milwaukee and Manitowoc Rivers in southeastern Wisconsin. The study was co-funded by the U.S. Geological Survey (USGS), including both the Federal/State Cooperative Program and the National Water Quality Assessment (NAWQA) Program, and the Wisconsin Department of Natural Resources. Concentrations and relative abundances of dissolved and adsorbed PCB congeners and homolog groups were determined seasonally. In addition, total suspended components of organic matter was partitioned into living (largely algae) and detrital (dead organic matter). Living and detrital organic matter differ significantly in both total organic carbon and lipid content -- two characteristics that determine the sorption capacity for PCBs. Uptake of PCBs by algae is a significant route of introduction into aquatic food webs. Algae from a wide spectrum of environments including the Arctic Ocean, have been shown to accumulate PCBs from water *(3-5)*. Particle-associated PCBs increased with primary production in large experimental mesocosms *(6)*. Concentrations of PCBs in surficial sediments of the Lagoon of Venice were correlated with the growth, senescence, and decomposition of the resident algal population *(7)*. PCB concentrations in sediments increased by more than an order of magnitude upon deposition of the algal biomass and subsequently decreased when the algae decomposed.

The objective of the present study was to use PCB assemblages of the dissolved and suspended particles in the water and those in bottom sediments as probes of various biogeochemical processes in these two rivers.

Methods

Field Sampling: Water, suspended solids, and bottom sediments were sampled at four sites in southeast Wisconsin, three on the Milwaukee River (Pioneer Road, Thiensville, and Estabrook Park) and one on the Manitowoc River, about 70 km north of the Milwaukee River sites (Figure 1) during four seasons near the middle of the month: summer (August '93), fall (November '93), winter (February '94), and spring (May '94). All water samples were collected at the USGS stations located downstream from the impoundments from which the bottom sediments were collected. The one exception was Pioneer Road where sediment samples were collected in the immediate area of the gage station. Water for all analyses was collected at four equally spaced points across the river. Samples for suspended organic carbon (SOC), chlorophyll-a, and total suspended solids (TSS) were collected in 1-L clean glass bottles that were first rinsed with water from the site and then submerged vertically through the entire water column. These water samples were composited in a precleaned Teflon churn splitter and subsampled according to standard USGS procedures *(8)*. Samples for PCBs ("dissolved" and "particulate" phases), and phytoplankton were taken at the same locations using different sampling methods (see below).

Bottom sediments were generally collected with an Ekman corer. However, during summer, sediments from the Thiensville and Manitowoc stations were obtained with a 3-inch acrylic gravity core liner. The Ekman corer was used because a large quantity of surficial (0 to 2 cm) sediment (~ 2 liters of wet sediment) was needed at each site to ensure that an adequate amount of sediment would be available for all the planned analyses (see below). Only cores with an undisturbed sediment/water

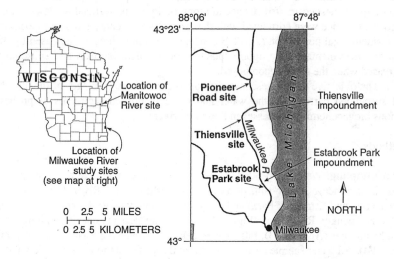

Figure 1. Location map for the Milwaukee River study sites.

interface as determined by lack of resuspended sediment in the overlying water of the core were used. The water was carefully removed with a plastic syringe prior to removing the top 2-cm of sediment in the Ekman corer with a clean glass jar. The top 2-cm of the gravity cores were collected *via* hydraulic extrusion into a short piece of acrylic core liner with the same internal dimension. All samples from Pioneer Road were collected by hand with an open-ended Teflon tube (held horizontally) that was used to collect the upper 2 cm of soft sediment (except the winter sample when an Ekman corer was used through the ice). Several of these upper 2-cm sections from a given site were combined and mixed in a clean glass bowl prior to subsampling for various constituents.

Dissolved and Particulate PCBs in Water. Bulk water samples (80 L) were collected in four 20-L pre-cleaned (HPLC-grade acetone and water) stainless steel canisters. These were first rinsed with native water from the site. Separation of "dissolved" and "particulate" phases occurred in the field immediately after sample collection. The water was first filtered through four in-line precombusted glass fiber filters (GF/F, 293 mm, 0.7 µM nominal pore size) at a maximum pressure of 5 psi into 20-L pre-cleaned (rinsed with HPLC-grade acetone and water) glass carboys. Pump tubing consisted of Teflon except for a short length of silicone plastic tubing (~15 cm) which was in the pump head. The pump tubing was rinsed with HPLC-grade acetone and water between samples. The filters were wrapped in aluminum foil, sealed in a plastic bag, and chilled (4°C) prior to analysis. The filtrate was passed through cleaned (9) Amberlite XAD-2 (20-60 mesh) resin columns (flow rate not exceeding 1 L·min^{-1}), and the volume was recorded prior to discarding the eluent. Resin columns were chilled (4°C) en route to the laboratory.

PCBs on the resins and filters were extracted separately (acetone:hexane::50:50 for 16 hours) in a Soxhlet apparatus. All PCB samples were analyzed on a congener-specific basis using capillary column gas chromatography with electron capture detection (HP 5890-II Gas Chromatograph with a 60-m DB5 column) at the State Laboratory of Hygiene (SLOH), Madison, Wisconsin (9). This method can determine up to 100 congeners (with 26 co-elutions). Quantitation was done using single point calibration standards consisting of a dilution of stock solutions of Aroclors 1232, 1248, and 1262 at 183 µg·mL^{-1}. Response factors were generated daily using single-point calibration of the diluted standard at a concentration of 0.549 µg·mL^{-1}. Surrogate standards (Congeners #14, #65, and #166) were added (at nominal concentrations of 20, 5, and 5 ng·L^{-1}) to each sample and blanks prior to extraction to monitor recovery. Matrix spike solutions consisted of the following Aroclors (nominal concentration): 1232 (0.25 mg·L^{-1}), 1248 (0.18 mg·L^{-1}), and 1262 (0.18 mg·L^{-1}). The average recovery of these surrogates and matrix spikes for all samples analyzed during the period coincident with the analysis of the samples from the present study is shown in Table I. Congeners #30 and #167 were used as retention time reference peaks and as internal standards for quantitation. Congeners eluting prior to and including #77/110 use congener #30 as an internal standard, whereas congeners eluting after #77/100 use congener #204 as the internal standard. Concentrations of ΣPCB were not corrected for percent recovery of either the surrogates or matrix spikes.

Field replicates of the "dissolved" and "particulate" fractions varied by 11% and 14%, respectively (one standard deviation). The detection limit for individual PCB congeners ranged from 0.02 to 0.09 ng·L^{-1} for both the dissolved and particulate phases. The concentration of dissolved ΣPCB in six blanks were as follows: below detection limit, 0.67, 0.58, 0.24, 0.07, and 1.3 ng·L^{-1}. Less than five percent of all dissolved ΣPCB samples had concentrations equal to, or less than the highest measured blank (1.3 ng·L^{-1}), and all samples had measured dissolved ΣPCB greater than the average of all six blanks (0.48 ng·L^{-1}). Concentrations of particle-associated ΣPCB in six blanks were as follows: 5 below detection limit and 0.45 ng·L^{-1}. The average of all particle-associated ΣPCB blanks (0.075 ng·L^{-1}) was at least an order of magnitude lower than the concentration in all samples, and the one blank detection was still a factor of two less than the lowest sample concentration.

Table I. Average of all PCB surrogate and matrix spike recoveries

	Dissolved (n)	Suspended Particles (n)	Bottom Sediment (n)
Surrogates:			
#14	77 (317)	82 (306)	86 (414)
#65	91 (329)	89 (306)	91 (416)
#166	95 (326)	104 (311)	102 (420)
Matrix Spikes (average of all congeners)	94 (22)	92 (21)	93 (60)

Algal Carbon in Water. One liter samples of bulk water were collected in plastic bottles submerged through the water column at the same locations and times as all other water column samples. Samples were preserved with glutaraldehyde to a final concentration of 0.2% (summer) and 0.5% (all other dates). The bottles were shipped within a few days to either the Milwaukee Metropolitan Sewerage District (MMSD), Milwaukee, Wis. (all summer samples) or Chadwick & Associates, Inc., (C&A) Littleton, Colorado (all other samples), for taxonomic and biovolume (only C&A samples) determinations according to standard procedures *(10)*. For the C&A samples, only live cells were enumerated and biovolume was determined on at least twenty individual cells per sample. MMSD diatom counts included live and dead cells and were corrected for live cells using the live to dead cell ratio from preliminary counts.

Because MMSD analyzed the summer samples, no biovolume measurements were available for these samples. Instead, estimated total biovolume values were calculated by using the average biovolume/cell determined for each of the major algal Divisions determined on samples from the other seasons by C&A. These average biovolume/cell values were then multiplied by the summer cells/mL values to produce estimated total biovolume values for the summer algal carbon calculations. Algal carbon was calculated using published relations between biovolume and algal carbon for diatoms and all other algal species separately *(11)*. Replicate algal samples from the water column that had been sent to both labs (C&A and MMSD) for the fall sampling from two sites (Estabrook Park and Thiensville) differed substantially. The different total phytoplankton densities were accounted for by the failure of MMSD to identify any cyanophytes. However, both labs reported similar total diatom densities and ratios of centric to pennate diatoms. Because of this discrepancy, C&A counts were used for all but the summer sampling dates, where only MMSD data were available. Thus, the algal carbon values for the summer samples might be low by an unknown amount equivalent to the cyanophytes. Historical summer (August) data exists for the Estabrook Park site in years 1974 through 1981 in the National Water Information System (NWIS) database of the USGS. Cyanophytes accounted for 0%-52% of the total number of algal cells, although during the last five years of that period, cyanophytes accounted for an average of 8% or less of the total cells. Moreover, no cyanophytes were found during three of those last five years (1976, 1977, and 1981) years. Thus, it is possible that there were in fact no cyanophytes in the Milwaukee River samples during summer.

Suspended Organic Carbon (SOC) in Water. A stainless steel Gelman filtering apparatus with 0.45 µM silver filters (Osmonics, Inc., Minnetonka, Minnesota) was used to collect SOC samples. The filters were first rinsed with 10 mL of HPLC-grade water. Suspended particles were filtered under nitrogen at a maximum pressure of 15 psi. The filters were carefully folded in half, placed in plastic petri dishes, and chilled (4°C) during shipment to the USGS's National Water Quality Laboratory (NWQL) for determination of "suspended organic carbon" (SOC) using acidification, wet oxidation and nondispersive infrared spectroscopy *(12)*. Average sampling precision for SOC was 29% determined on replicate field samples.

Chlorophyll-a in Water. For the chlorophyll-a analysis, a known volume of water was passed through mixed acetate and nitrate cellulose ester membrane filters (5.0 µM pore size). The filters were folded in half, placed in plastic petri dishes, wrapped in aluminum foil and frozen prior to analysis at SLOH *(10)*. Pigments were extracted in 90% acetone, and chlorophyll-a was determined spectrophotometrically. Average variation between replicate samples was 5%.

PCBs in Sediment. Subsamples of sediment to be analyzed for PCBs were placed in combusted glass bottles supplied by the contract laboratory (SLOH) and chilled (4°C). A 10 to 25 g subsample was analyzed for moisture content (103°C for at least 10 hours), a factor which was later used in the quantitation to report values on a dry-weight basis. Sediment samples were allowed to air dry (to ≤ 30% moisture by weight) and were then sieved (#10 mesh). PCBs absorbed to sediment were isolated by Soxhlet extraction and analyzed *via* capillary column gas chromatography with electron capture detection (same as for water column PCBs - see above)*(9)*. This method can determine up to 85 congeners (with 23 co-elutions). Quality assurance and control procedures for PCB congeners in sediments were identical to those for dissolved and particle-associated ΣPCB (see above) with a few exceptions. First, the matrix spikes were added to sediment samples in the Soxhlet apparatus, and the solvent was allowed to evaporate prior to extraction. Also, the surrogate standards were added to the Soxhlet thimble before extraction of PCB congeners. Recoveries of all matrix spikes and surrogates for all samples run by the lab during a time coincident with the analysis of the samples from the present study are shown in Table I. Quantitation was similar to that for dissolved and particle-associated PCBs except that masses of individual congeners were reported per mass dry weight of sediment ($\mu g \cdot g^{-1}$). The detection limit for individual PCB congeners in sediments ranged from 0.2-1.4 ng· g^{-1}. Concentrations of ΣPCB were not corrected for percent recovery of either the surrogates or matrix spikes.

Results and Discussion

Source of PCBs in Bed Sediments. Because they are relatively hydrophobic, PCBs in aqueous environments are expected to be largely associated with sediments. For example, more than 99.9% of Aroclors 1242, 1254, and 1260 were found to be associated with bottom sediments from the Duluth-Superior Harbor of Lake Superior with the remainder associated with some pore water phase (13). Because of this, absent bacterially-mediated degradation or selective loss due to physical processes such as dissolution, PCB congener assemblages in bed sediments might be expected to closely resemble known Aroclor mixtures. For example, PCB assemblages (averaged for all seasons at each site because they did not vary much) from the Manitowoc River and Estabrook Park sites resembled Aroclors 1254 and 1242, respectively (Figure 2). Unfortunately, no industrial purchase, use or release information on Aroclors exists for these sites. In contrast, records of industrial purchases of Aroclor 1260 and 1242 do exist for PCB point sources located upstream of the Pioneer Road and Thiensville sites (14). Congener assemblages at these two sites most resemble Aroclor 1260. However, the sediments also contain some lower chlorinated congeners that might be considered similar to Aroclor 1242, albeit at a smaller concentration than Aroclor 1260 (Figure 2). The similarity of PCB assemblages in these river sediments with known Aroclor mixtures suggests that microbially-mediated decomposition and/or dissolution has not significantly modified the PCB assemblages in these sediments.

Solid-Water Partitioning. In general, partitioning of PCBs from water to particles is expected to increase with increasing hydrophobicity (15,16). The expected PCB distribution is a prevalence of the higher chlorinated homolog groups in the suspended particle and bottom sediment phases and a prevalence of the less chlorinated homolog groups in the dissolved phase due to the relatively higher solubility of the less chlorinated congeners (17). This pattern was indeed observed at all four sites during all four seasons (Figure 3). Partitioning of the less chlorinated congeners to the dissolved phase leaves the suspended particles relatively enriched in the highly chlorinated homolog groups compared to bottom sediments. A common pattern observed was that the suspended particles have relatively less tri-, tetra-, and pentachlorobiphenyls and relatively more hexa-, hepta-, and octachlorobiphenyls compared to the bottom sediments, shown here for the Estabrook Park site (Figure 4). However, the suspended particle population consists not only of resuspended bottom sediments but also of various biogenic particles including algae and zooplankton. These biogenic particles might have different sorption characteristics compared to resuspended sediments, and if abundant, might control PCB sorption (see section on Algal Uptake of PCBs).

Algal Uptake. There is some evidence that PCBs sorbed to suspended particles in these rivers were associated with live algal cells as opposed to detritus during all

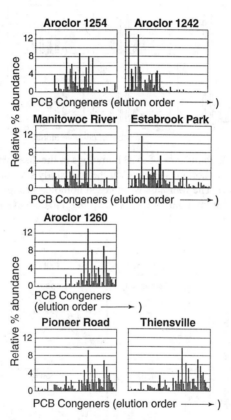

Figure 2. Relative % abundance of PCB congeners in Aroclor mixtures and bottom sediments at the four study sites (all seasons averaged).

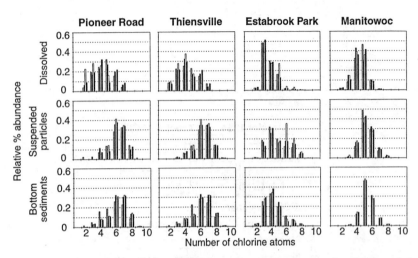

Figure 3. Relative % abundance of PCB homologs in the dissolved and suspended particulate phases and bottom sediments at the four study sites during all four seasons (1993-1994). The four bars within each homolog grouping, from right to left, correspond to summer, fall, winter, and spring. No data available for summer at Manitowoc River.

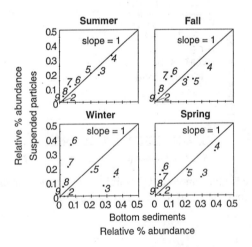

Figure 4. Relative % abundance of the homolog groups in the bottom sediments *versus* the suspended particulate phase during all four seasons (1993-1994) at Estabrook Park. Numbers correspond to the number of chlorine atoms. The solid line has a slope of unity.

seasons. For example, particle-associated ΣPCB concentrations were positively correlated with chlorophyll-a content during all seasons but only at the three impounded sites: Thiensville, Estabrook Park, and Manitowoc the River site ($r^2 = 0.87$, 0.88, and 0.94, respectively)(Figure 5 and Table II). If a positive correlation between these two parameters implies conditions close to equilibrium, then the larger residence time associated with impoundments might be necessary for the algal biomass to reach equilibrium with the dissolved PCB pool. In addition, the impoundments have large deposits of PCB-laden fine sediments which may supply dissolved PCBs to the water column. Pioneer Road had a much weaker ($r^2 = 0.25$), though still positive, correlation coefficient in addition to a smaller calculated slope. Concentrations of both particle-associated PCBs and chlorophyll-a were relatively low at this site compared to other sites. This, combined with the small number of data points and the relatively large error in the calculated slope make interpretation difficult. The slope of this relation at the Manitowoc River site was an order of magnitude higher that at the Milwaukee River sites due to the larger mass of ΣPCB in the bottom sediments, the presumed source of PCBs in all other phases (Table III). The higher slope is expected because algal uptake of PCBs and other hydrophobic organic compounds is through passive diffusion driven by the fugacity gradient between the truly dissolved phase and the hydrophobic biomass *(18)*. A positive correlation between chlorophyll-a and particle-associated ΣPCB concentrations has also been noted previously in the Fox River, WI. *(19)*.

The concentration of algal carbon and chlorophyll-a were positively correlated with particle-associated ΣPCB concentration ($r^2 = 0.59\text{-}0.98$)(Figure 5 and Table II), lending confidence in the algal-carbon calculations which relied on several assumptions (see methods). However, the relation between suspended particle-associated ΣPCB and algal carbon at the three Milwaukee River sites ($r^2 = 0.012$, 0.68,and 0.42 for Pioneer Road, Thiensville, and Estabrook Park, respectively) was not as good as that between particle-associated ΣPCB and chlorophyll-a (Figure 5 and Table II). This was probably due to the larger error associated with the algal carbon calculations as compared to the relatively smaller sampling error for chlorophyll-a analyses. Again, little correlation was observed between these two parameters at Pioneer Road. The slope of the particle-associated ΣPCB *versus* algal carbon was two-three fold higher at the Manitowoc site compared to the Milwaukee River sites, again in keeping with the larger concentration of dissolved ΣPCB at that site (Table III).

Algal blooms and their attendant relatively high organic carbon content, have been shown to influence the distribution of hydrophobic organic contaminants between the dissolved and particle-associated phases *(21)*. Algal carbon as a percentage of SOC was highest in spring (average = $50 \pm 16\%$), and lowest in summer (average = $15 \pm 10\%$) averaged for all sites. In contrast, algal carbon accounted for intermediate percentages in fall and winter ($39 \pm 21\%$ and $38 \pm 14\%$, respectively). Algal carbon as a percentage of SOC at the three Milwaukee River sites was approximately double that for the Manitowoc River site. Estabrook Park had the highest average percentage ($44 \pm 18\%$) and Manitowoc had the lowest ($20 \pm 7\%$) averaged for all seasons. Pioneer Road and Estabrook Park had intermediate (and similar) average percentages of $39 \pm 19\%$ and $38 \pm 21\%$, respectively. The range in seasonal values (15%-50%) was somewhat larger than the range of values among all the sites (20%-44%) and especially among the three Milwaukee River sites (39%-44%), suggesting that seasonal effects

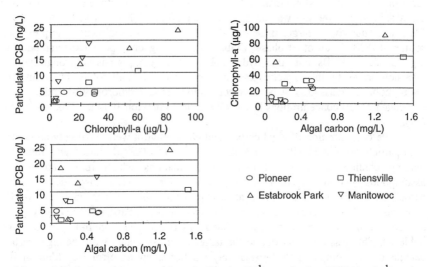

Figure 5. Relations between chlorophyll-a ($\mu g \cdot L^{-1}$), particulate PCB ($ng \cdot L^{-1}$) and algal carbon ($mg \cdot L^{-1}$) at Pioneer Road (●), Thiensville (■), Estabrook Park (▲), and the Manitowoc River (▼) during all fours seasons (1993-1994).

Table II. Summary statistics between chlorophyll-a, algal carbon, and PCBs in suspended particles

	Chlorophyll-a vs Particulate PCBs				Algal carbon vs Chlorophyll-a				Algal carbon vs Particulate PCBs			
	Pioneer Road	Thiensville	Estabrook Park	Manitowoc River	Pioneer Road	Thiensville	Estabrook Park	Manitowoc River	Pioneer Road	Thiensville	Estabrook Park	Manitowoc River
slope (m)	0.053	0.16	0.24	6.6	42	34	51	43	0.57	5.2	11	270
error (m)	0.064	0.045	0.060	1.2	20	9.5	30	6.2	3.7	2.6	9.0	55
r^2	0.25	0.87	0.88	0.94	0.69	0.87	0.59	0.98	0.012	0.68	0.42	0.96
n	4	4	4	3	4	4	4	3	4	4	4	3

Table III. Average concentrations (all seasons) of ΣPCB in all phases

	Dissolved ΣPCB (ng·L⁻¹)	Suspended Particle ΣPCB (ng·L⁻¹)	Bottom Sediment ΣPCB (μg·g⁻¹)
Pioneer Road	2.4	2.9	0.55
Thiensville	2.7	5.6	0.29
Estabrook Park	8.1	14	2.1
Manitowoc River	30	110	140

(temperature, spring addition of nutrients, etc.) were more important than site-specific characteristics in controlling algal growth in these rivers.

Despite these seasonal differences in algal carbon mass, PCB homolog assemblages were remarkably similar between the spring bloom and non-bloom seasons (average of other three seasons) at all four sites (data not shown). Based on this, if algae were quantitatively important as a sorption medium for dissolved PCBs, live algae and detritus appear to behave similarly with respect to sorption of all congeners. This was in agreement with the observation that the distribution of several hydrophobic organic compounds in Lake Michigan was remarkably constant despite large changes in both the concentration and composition of suspended particles (21).

However, averaging over all seasons may obscure more subtle relations seen in individual data (22). For example, when individual samples were considered at each site, those rich in algal carbon (spring) showed a prevalence of particular PCB homolog groups compared to samples low in algal carbon for the three sites where algal carbon and PCBs were reasonably well correlated (Figure 6). However, no consistent pattern was observed among the three sites. Recent kinetic modeling of algal uptake of PCBs predicted that the 'super' hydrophobic congeners (Log $K_{ow} \geq$ 6.5), which include the hexa-, hepta-, and octachlorobiphenyls, should be relatively depleted in algal biomass. This results from the relatively long time to reach equilibrium for these highly chlorinated congeners and should be most pronounced during spring and summer, seasons marked by high algal growth rates (23). Thus the enrichment of the octachlorobiphenyls at Thiensville and the hexa- and heptachlorobiphenyls at Estabrook Park in the algal-rich samples was not expected. It is possible that the relative difference in algal carbon between the highest and lowest values at each site (43%, 16% and 50% at Thiensville, Estabrook Park and Manitowoc River, respectively) was not large enough to see patterns predicted by theory.

Conclusions

PCB congener and homolog assemblages, coupled with theoretical predictions of chemical behavior, were used as probes of various biogeochemical processes. Solid-water partitioning produced the expected enrichment of less chlorinated congeners in the dissolved phase and heavily chlorinated congeners in the suspended particle phase. This pattern would be expected if the system was close to equilibrium with respect to

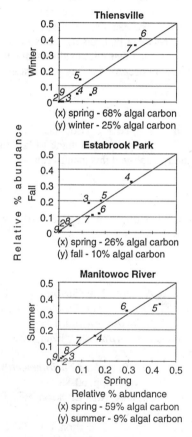

Figure 6. Relative % abundance of the homolog groups in the suspended particulate phase in algal-rich samples (spring) *versus* algal-poor samples (winter/ Thiensville, fall/Estabrook Park, and summer/Manitowoc River site).

partitioning from water to suspended particles. Chlorophyll-a, a proxy for algae, and calculated algal carbon were both reasonably well correlated with suspended particle-associated PCBs only at the three impounded sites. Thus, the residence time of water and algae in the system is related to algal uptake of PCBs in rivers. There was little agreement between the observed and theoretically predicted congener abundances during high algal growth (spring) periods. If the theory is right, further efforts must be undertaken to separate this algal fraction from detritus in order to better understand the sorption behavior of both phases for PCBs and other hydrophobic contaminants. PCB congener and homolog distributions can also be used to rule out other processes. For example, the high similarity between PCB distributions in bed sediments and known Aroclor mixtures is strong evidence that microbially-mediated decomposition (of the less chlorinated congeners) is not occurring to any measurable extent.

Acknowledgments

We would like to thank Judy Wierl, Nick Hanson, and Dave Housner for conducting and/or assisting in the field sampling. We also thank Mike Murray and the reviewers whose comments substantially improved this manuscript.

Note

Use of brand names is for identification purposes only and does not constitute endorsement by the U.S. Geological Survey.

Literature Cited

(1) Muir, D.C.G.; Omelchenko, A.; Grift, N.P.; Savoie, D.A.; Lockhart, W.L.; Wilkinson, P.; Brunskill, G.J. *Environ. Sci. Technol.* **1996**, *30*, 3609-3617.

(2) McFarland, V.A.; Clarke, J.U. *Environ. Health.* **1989**, *81*, 225-239.

(3) Mahanty, H.K. In *PCBs and the Environment; Waid, J.S., Ed.*; CRC Press: Boca Raton, Fla., 1986, Vol. 2; pp. 1-8.

(4) Hamdy, M.K.; Gooch, J.A. In *PCBs and the Environment; Waid, J.S., Ed.*; CRC Press: Boca Raton, Fla., 1986, Vol. 2; pp. 63-88.

(5) Hargrave, B.T.; Harding, G.C.; Vass, W.P.; Erickson, P.E.; Fowler, B.R.; Scott, V. *Environ. Contam. Toxicol.* **1992**, *22*, 41-54.

(6) Millard, E.S.; Halfon, E.; Minns, C.K.;Charlton, C.C. *Environ. Toxicol. Chem.* **1993**, *12*, 931-946.

(7) Pavoni, B.; Calvo, C.; Sfriso, A.; Orio, A.A. *Sci. Total Env.* **1990**, *91*, 13-21.

(8) Ward, J.R.; Harr, C.A. United States Geological Survey (USGS), Open File Report #90-140.

(9) *Methods and Quality Control for the Organic Chemistry Unit of the Wisconsin State Laboratory of Hygiene*; Degenhardt, D., Ed.; Wisconsin State Laboratory of Hygiene: Madison, WI, 1996, 1910 p.

(10) *Standard Methods for the Examination of Water and Wastewater. 18th ed.*; Clesceri, L.S.; Greenberg, A.E.; Trussell, R.R. Eds., American Public Health Association: Washington, D.C., 1992.

(11) Strathmann, R.R. *Limnol. Oceanogr.* **1966**, *11*, 411-418.

(12) Wershaw, R.L.; Fishman, M.J.; Grabbe, R.R.; and Lowe, L.E. United States Geological Survey (USGS), Open File Report #82-1004.

(13) Capel, P.D.: Eisenreich, S.J. *J. Great Lakes Res.* **1990**, *16*, 245-257.

(14) Westenbroek, S. Wisconsin Department of Natural Resources, Cedar Creek PCB Mass Balance - Final Draft, 1993, 156 p.

(15) Karickhoff, S.W.; Brown, D.S.; Scott, T.A. *Water Res.* **1979**, *13*, 241-248.

(16) Chiou, C.T.; Porter, P.E.; Schmedding, D.W. *Environ. Sci. Technol.* **1983**, *17*, 227-231.

(17) Miller, M.M.; Ghodbane, S.; Wasik, S.P.; Tewarl, Y.B.; Martire, D.E. *J. Chem. Eng. Data* **1984**, *29*, 184-190.

(18) Swackhamer, D.L.; Skoglund, R.S. *Environ. Toxicol. Chem.* **1993**, *12*, 831-838.

(19) Steuer, J.J.; Jaeger, S.; Patterson, D. Wisconsin Department of Natural Resources, Publ. WR 389-95, 1995.

(20) Koelmans, A.A.; Lijklema, L. *Wat. Res.*, **1992**, *26*, 327-337.

(21) Eadie, B.J.; Morehead, N.R.; Landrum, P.F. *Chemosphere* **1990**, *20*, 161-178.

(22) Stow, C.A.; Carpenter, S.R. *Environ. Sci. Technol.* **1994**, *28*, 1543-1549.

(23) Skoglund, R.S.; Stange, K; Swackhamer, D.L. *Environ. Sci. Technol.* **1996**, *30*, 2113-2120.

Chapter 26

Transport and Degradation of Semivolatile Hydrocarbons in a Petroleum-Contaminated Aquifer, Bemidji, Minnesota

E. T. Furlong[1], J. C. Koleis[2], and G. R. Aiken[2]

[1]National Water Quality Laboratory, U.S. Geological Survey, 5293 Ward Road, Arvada, CO 80002
[2]National Research Program, U.S. Geological Survey, 3215 Marine Street, Boulder, CO 80303

Polycyclic aromatic hydrocarbons (PAH) were used as probes to identify the processes controlling the transport and fate of aqueous semivolatile hydrocarbons (SVHCs) in a petroleum-contaminated aquifer near Bemidji, Minnesota. PAH and other SVHCs were isolated from ground water by field solid-phase extraction and analyzed using gas chromatography/mass spectrometry. Close to the oil body, aqueous aliphatic hydrocarbon compositions are substantially different from the parent oil, suggesting microbial alteration prior to or during dissolution. Aqueous PAH concentrations are elevated above oil-water equilibrium concentrations directly beneath the oil and decrease dramatically at distances ranging from the 25 to 65 m downgradient from the leading edge of the oil body. Variations in downgradient distributions of naphthalene, fluorene and phenanthrene, coupled with their biodegradation, partitioning and volatility characteristics, suggest that the PAH are useful probes for distinguishing between the biogeochemical processes affecting SVHC transport and persistence in ground water.

If physical properties alone are considered, semivolatile hydrocarbons (SVHCs) are not expected to travel significant distances in ground water due to rapid sorption onto aquifer material. However, in several instances, SVHCs have been transported in ground water further than a simple two-component transport model would predict. This enhanced or "facilitated" transport is often attributed to enhanced cosolvation of hydrophobic compounds in the mobile aqueous phase by *in situ* dissolved organic carbon, or DOC. (*1-5*). DOC contains both colloidal and dissolved organic carbon (*6*) and has been extensively investigated as a transport facilitating agent (*6-9*).

The effect of DOC on transport of SVHCs has been extensively studied in batch and column experiments (*1, 4-5, 8-9*) often using individual polycyclic aromatic hydrocarbons (PAH) as model compounds. Isolating the effects of DOC from the effects of other variables that influence the transport of SVHCs is difficult in field settings (*6*). Microbial degradation of many SVHCs can attenuate the contaminant signal in a plume, suggesting that individual SVHCs must be persistent if they are to be useful probes of DOC-SVHC interactions and other variables affecting transport. Temporal and spatial variability in hydrologic transport also complicates the interpretation of aqueous SVHC distributions. The effects of DOC and other variables enhance or alter SVHC transport. These effects are best identified in simple hydrologic environments, where a discrete source of persistent SVHCs occurs producing detectable contamination downgradient from the source over multiple field seasons.

In this study, we used disk solid-phase extraction to isolate PAH and alkyl-substituted PAH down to low parts-per-trillion concentrations in ground water samples. Samples were collected in 1995 from wells downgradient of an oil body in a petroleum-contaminated, sand-and-gravel aquifer. In addition to PAH and alkyl-substituted PAH, we evaluated the relative abundance of degraded aliphatic hydrocarbons, often described as a chromatographically unresolved complex mixture or UCM. PAH composition and concentration changes and changes in the composition of the UCM, as reflected by chromatogram shape, are used as probes for identifying processes affecting transport of SVHCs in ground water in the presence of oil-derived DOC.

Field Site Setting.

On August 17, 1979, a pressurized underground oil-pipeline ruptured approximately 16 km northwest of Bemidji, Minnesota. Approximately 1,670 m^3 of a paraffinic Alberta crude oil was injected directly into the sand and gravel aquifer or was sprayed over the soil surface, infiltrating the aquifer (*10*). After cleanup, some fraction of the 410 m^3 of the oil not recovered was dispersed across the 7-8 m unsaturated zone in an approximately 1 meter thick oil lens that sits on top of the water table. The approximate location of the oil lens is shown in Figure 1, as are the locations of oil and water sampling wells used in this study. The hydrology, inorganic geochemistry, biogeochemistry, and aspects of the organic geochemistry have been described (*11-13*).

Several distinct zones of ground water chemistry occur at this site. Downgradient of the oil lens, along the axis of ground water flow, a plume of contaminated water has developed, containing elevated DOC and oil component concentrations (*12*). From the center of the oil lens, the contamination plume extends approximately 110 m downgradient and varies in thickness from 3 to 5 m, extending down from the water table. The zones in the plume change from anoxic (zone III) to suboxic (zone IV) to oxic (zone V). Elevated DOC concentrations in the these zones, coupled with the odor and sheen of petroleum detected in water collected from some wells at these sites, suggest that petroleum and microbially degraded petroleum components are being transported downgradient from the spilled oil. Adjacent to the oil, benzene and alkylated monoaromatic compounds reflect a balance of dissolution from the oil and removal from the aqueous phase (*14*). Transport of these compounds away from the oil is controlled by natural biogeochemical processes. Benzene, toluene, and *ortho-*, *meta*, and *para*-xylenes transported downgradient (*13*) are actively degraded by the *in situ* microbiota to the corresponding acid and phenol forms (*15*). Eganhouse *et al.* (*13*) have isolated and fractionated the bulk DOC in order to better characterize this component of transported carbon. While there are some compositional differences between DOC fractions in different geochemical zones, in general, DOC increases from 1.3 mg/L in background water samples to as high as 42 mg/L in water samples collected beneath the oil. The functional group composition (as determined by nuclear magnetic resonance) of the resin-isolated DOC beneath the oil and further downgradient indicates that it is oil-derived (*16*).

Experimental Methods.

PAH and other SVHCs were isolated from 4-8.0 L of water by pumping the sample sequentially through a 142-mm diameter glass-fiber filter (Whatman GF/F) with a

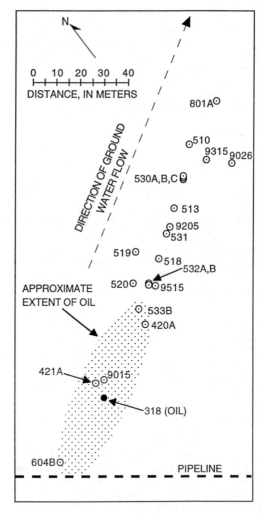

Figure 1. Location of study site at Bemidji, Minnesota, including locations of water and oil wells, oil body, and pipeline.

nominal pore size of 0.7 μm, and a 142-mm diameter C_{18} Empore disk (3M Corporation, St. Paul, Minnesota; mention of trade names is for identification purposes only and does not imply endorsement by the U.S. Geological Survey), contained in a single aluminum, stainless steel, and Teflon filter unit (Geotech, Inc.). Ground water was directly pumped through the filter unit using a Teflon and stainless steel submersible pump (Model SP-202, Fultz Pumps, Inc.). Prior to sampling, Empore disks were cleaned and conditioned in filter units by sequentially pumping a minimum of 100 mL each of pesticide-grade dichloromethane, methanol and water through the assembled unit. A modified version of the well purging protocol of Bennett *et al.* (*11*) was used prior to connection of the filter unit to the pump. After a minimum of 10 L plus 2 casing volumes of water had been removed from the well, the filter unit was connected to the pump and water was pumped through the filter and Empore disk at approximately 150 mL/min. After the desired volume of water had been pumped through the filter unit, the unit was disconnected and any residual water removed by vacuum. The GF/F and Empore disks were removed from the filter unit and separated. The Empore disks were stored in precleaned 40-mL amber vials, and the GF/F filters were wrapped in prebaked (450°C) aluminum foil and sealed in polyethylene bags. Samples were kept on ice in the field until they could be stored in the dark at -15°C.

PAH and other SVHCs were extracted from the Empore disks by Soxhlet extraction overnight using 350 mL of pesticide-grade dichloromethane. Approximately 30 mL of pesticide-grade methanol was added to each Soxhlet apparatus to remove residual water in the Empore disk. A method recovery sample and a method blank were coextracted with each set of 10 samples. Prior to extraction, an internal standard mixture of perdeuterated naphthalene, acenaphthene, phenanthrene, and pyrene was added to all samples. Each extract was dried over anhydrous sodium sulfate, concentrated to 0.5 mL in a 2-mL amber vial, and an injection internal standard of benzo[*e*]pyrene-d_{12} was added. Prior to analysis, all extracts were stored in the dark at -15°C.

Unsubstituted PAH were determined by capillary gas chromatography/mass spectrometry (GC/MS), using electron-impact ionization and selected-ion monitoring. Unsubstituted PAH concentrations were calculated using the isotopic dilution method. C_4-benzenes and alkyl-substituted naphthalenes were determined from full-scan, electron impact GC/MS, using a limited set of alkyl-substituted standards, with quantification by the internal standard method. The multiple isomeric alkyl-substituted naphthalenes were quantified from mass chromatograms as the sum of all isomers at each level (C_1-naphthalene, C_2-naphthalene, etc.). Total alkylnaphthalenes are the sum of all C_1- to C_4-substituted naphthalenes. Where authentic alkyl-substituted standards were unavailable, the closest eluting compound (one alkylation level lower) was used for quantitation.

This sample collection and analysis method has previously been described (*17*; Furlong *et al.*; in preparation). In reagent water samples fortified at concentrations of 1.23 μg component/L, recoveries ranged from 77 to 93% for 8 unsubstituted PAH, ranging from naphthalene to pyrene, with coefficients of variation of 1.3 to 2.8%. In a field sample fortified at 625 ng/L, matrix-corrected recoveries ranged from 76 to 121%, with the exception of anthracene, which was recovered at 42%. The efficient recovery of PAH from field samples is in contrast to the findings of Landrum *et al.* (*18*), who observed that PAHs were not quantitatively recovered from natural waters by solid-phase extraction (SPE) and used SPE cartridges to separate free PAH from DOC-bound PAH. Differences in DOC composition and the SPE isolation procedures could account for the differences between their results and ours. The percent differences in concentration observed for pairs of unfortified field duplicate samples ranged between 0.3 and 48% for all PAH and alkyl-PAH, across a concentration range of 258 to 84,900 ng/L, with the exception of total methylnaphthalene concentration in one sample, which was 113%. Our method was tested for unsubstituted PAH (*17*); we assume similar recoveries for C_4-benzenes and alkylnaphthalenes.

Total DOC samples were collected immediately after the Empore disk samples, using the same pump and tubing system. Each sample was filtered in-line through 0.45-μm silver membrane filters (Osmonics, Inc.). The filtrate was collected in precombusted glass bottles and immediately stored on ice until analysis. Total DOC concentrations were measured by the persulfate-wet oxidation method using an Oceanographic International Model 700 carbon analyzer.

Results and Discussion.

Hydrocarbon Concentrations In Ground Water. Well coordinates, screened interval depths, DOC, and individual PAH concentrations for these wells are contained in Table I and plotted in Figure 1. The wells sampled in this study are a subset of the entire well network at this site. The following criteria were used to select the wells used in this study. First, only water table wells that are downgradient of the oil body and whose screened interval midpoint elevations are above 420 m are included. Water samples collected from these wells should be representative of water in zones III, IV, and V of Baedecker *et al.* (*12*), the anoxic, suboxic, and oxic zones of the aquifer, respectively. Vertical variations in chemical concentration are not discussed, due to the range of screened interval lengths in the wells, that vary from 0.15 to 1.5m (Table I). With one exception, wells that were upgradient of the oil body or were in an area of surface remediation (*10*) were not sampled. Well 310, the background control used in this study, is listed in Table I, but is not plotted in Figure 1. PAH and DOC concentrations for samples collected from this well are listed in Table I, and are assumed to reflect ground water conditions before contact with oil-influenced portions of the aquifer. Naphthalene and phenanthrene were detected at very low concentrations in the water sample from well 310, and may reflect field conditions at that location during SPE isolation and processing. Samples from this well in 1993, collected as part of the original method development, did not contain detectable PAH or other SVHCs (*17*).

Four full-scan, total-ion current (TIC) chromatograms (Figures 2a-d) illustrate the changing compositions of both aromatic and aliphatic hydrocarbons in water samples collected at increasing distances downgradient from the oil body. Note that the vertical scales of each chromatogram in Figure 2 differ, reflecting sample extract dilution and aqueous concentration decreases downgradient. However, the qualitative changes in chromatogram shape suggest changing relative abundances of hydrocarbon components.

When compared to a chromatogram of the diluted oil itself (Figure 2e), the absence of *n*-alkanes in ground-water samples is striking. Even in ground-water samples collected immediately adjacent to the oil, *n*-alkanes were not detected. In contrast, Baedecker *et al.* (*19*) and Eganhouse *et al.* (*13*) detected measurable *n*-alkanes in both water and saturated-zone sediment samples from the anoxic and suboxic regions of the contaminant plume. The *n*-alkanes they detected were from 11 to 32 carbons in chain length, with no striking odd/even predominance. We observed this same pattern, typical of paraffinic crude oils (*20-21*), in an oil sample from well 318, collected in 1988 (Figure 2e).

While *n*-alkanes are absent in ground-water samples, a UCM "hump" is clearly present in TIC chromatograms from samples collected as far as 105 m downgradient from the center of the oil body (Figure 2a-c). This corresponds to approximately 71.5 m downgradient of the leading edge of the oil, estimated to extend just beyond well 533b in 1995 (Figure 1). The UCM hump is somewhat discernible in well 801A, 91.5 m downgradient of the leading edge (Figure 2d). C_4-benzenes and alkylnaphthalenes are a significant component of the of the TIC chromatograms beneath the oil in well 421a oil (Figure 2a), and they decrease downgradient. Over the same 71.4 m distance, the sum of C_4-benzenes, naphthalene, and total alkylnaphthalenes decreases over two orders of magnitude, from 517,000 ng/L in well 421A to 3,250 ng/L in well 510 (Table I).

Table I--Well locations, concentrations of polycyclic aromatic hydrocarbons (PAH) and dissolved organic carbon (DOC) in ground water from the spilled-oil site, Bemidji, Minnesota

Well	Distance from well 421 (m)	Center of screen (m)	Screened interval (m)	DOC (mg/L)	C_4-benzene isomers (ng/L)[a]	Naphthalene (ng/L)	Total alkyl naphthalenes (ng/L)[b]	Fluorene (ng/L)	Phenanthrene (ng/L)
310	-199.0	423.72	1.52	1.3	--[d]	120	--	--	14
604B	-35.2	421.60	0.15		1,550	1,060	143	--	--
421A	0.0	422.55	1.5e[c]	42.0	185,000	142,000	190,509	10,700	1,450
9015	3.8	421.66	0.15	32.8	66,900	274,000	222,120	4,040	1,700
420A	31.8	423.05	1.5e	34.4	83,700	103,000	240,800	2,390	1,230
533B	35.4	421.70	0.2e	24.3	38,000	128,000	28,249	1,440	--
520	43.5	422.66	1.52	21.9	21,500	2,390	34,680	1,300	316
532A	46.0	423.35	1.52	26.2	45,500	53,100	96,430	2,760	640
532B--Ave[e].	46.4	421.80	0.2e	33.0	84,900	64,500	74,263	1,510	423
9515C	46.8	421.89	0.54		--	123	--	13	10
519	56.3	423.36	1.52	13.8	--	490	--	1,050	105
518	57.3	423.39	1.52	29.2	--	2,450	--	1,860	159
531	67.8	422.95	1.5e	11.0	38,500	400	2,500	842	--
9205	71.0	422.41	0.152	14.8	35,000	401	8,256	1,050	--
513	78.3	423.15	1.52	10.4	8,040	395	890	903	--
530C	91.7	420.14	0.15	4.8	1,210	90	272	38	--
530B--Ave.	92.0	421.16	0.15	10.8	3,800	258	2,727	875	37
530A	92.3	421.66	1.52	5.6	2,190	212	1,570	412	27
9315	102.5	421.70	0.152	1.4	--	175	--	10	10
510	105.0	423.22	1.52	7.5	1,880	114	1,260	347	--
9026	106.3	422.18	0.15	4.4	427	89	36	88	--
801A	125.5	422.75	1.5e	5.8	113	77	--	42	--

Well	Distance from well 421 (m)	Center of screen (m)	Screened interval (m)	DOC (mg/L)	C_4-benzene isomers (µg/g)	Naphthalene (µg/g)	Total alkyl naphthalenes (µg/g)	Fluorene (µg/g)	Phenanthrene (µg/g)
Oil (Well 318)	6.8	423.4e	1.5e	ND[f]	2,060	699	8,300	179	209

[a]C_4-benzene isomers-sum of benzene isomers substituted with 4 carbons.
[b]total alkylnaphthalenes are the sum of C_1-, C_2-, C_3-, and C_4-substituted naphthalenes.
[c] The e indicates an estimated value. [d]The -- indicates compound not detected.
[e]Ave. indicates the average of duplicate determinations. [f]ND indicates not determined.

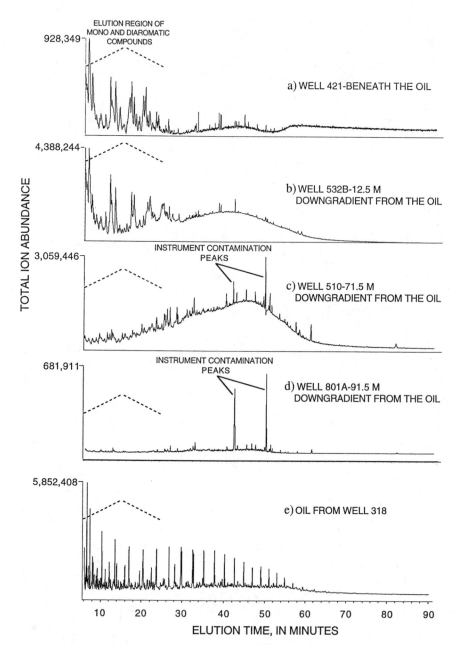

Figure 2. Total ion current chromatograms of water sample extracts from wells at Bemidji, Minnesota. Samples collected a) beneath the oil (0 m), b) 12.5 m, c) 71.5 m, and d) 91.5 m downgradient of the center of the oil body. e) total ion current chromatogram of a diluted oil sample collected from well 318 and analyzed under identical conditions.

The absence of n-alkanes, the persistence of the UCM, and the presence of significant aromatic components in our samples contrasts with the results of Eganhouse *et al.* (*13*), who found low concentrations of n-alkanes and a relatively small UCM hump in aqueous samples collected in 1987 from zones III and IV. These differences may have several causes. First, our results are from filtered water samples, in contrast to the whole-water samples collected by Eganhouse *et al.* (*13*). Although not extracted, all glass-fiber filters collected in our study contained visible, if very small, quantities of particles. Oil partitioned onto particles could contribute to the observed differences. Second, sample sizes and preservation techniques differed between the two studies. Our sample size was from 4.0 to 8.0 liters, and were frozen after field isolation. Eganhouse *et al.* (*8*) added solvent and mercuric chloride to their 250 mL samples after collection. Finally, an 8 year lag between these studies (*8*) also could account for these differences owing to changes in hydrologic or *in situ* microbiological characteristics over time .

While detectable PAH and alkylnaphthalene concentrations were measured over 100 m downgradient from the center of the oil body, individual PAH compositions varied. In general, the C_4-benzenes and total alkylnaphthalenes were present in greater abundance than naphthalene, fluorene, and phenanthrene (Table I). The greatest single-component concentration, 274,000 ng/L, was observed for naphthalene in a sample collected immediately adjacent to the oil body. Individual alkyl-PAH distributions also show downgradient changes. Dissolved-phase aromatic hydrocarbons were composed primarily of C_4-substituted benzenes and alkylated naphthalenes substituted with one to four alkyl carbons. In water samples adjacent to the oil, C_2-naphthalene and C_2-phenanthrene homologs predominate. Absolute concentrations decrease dramatically away from the oil body, and relative abundances of alkyl-substituted naphthalenes change immediately downgradient from the oil, where C_3-naphthalenes predominate. The alkylnaphthalene distribution in oil differed somewhat from water samples; C_1-naphthalenes were the dominant alkylnaphthalene component. The range of parent and alkylated aromatic hydrocarbons detected in oil in this study were also observed by Eganhouse *et al.* (*13*).

Effects of Biodegradation.   Petroleum and other complex mixtures of hydrocarbons can be degraded by both aerobic and anaerobic degradation processes, although the rates of aerobic degradation are faster and predominate in most field sites. The sequence of changes in petroleum composition resulting from aerobic biodegradation of petroleum are well established, and these compositions can be used to estimate the relative severity of degradation (*21*). At Bemidji, the bulk oil composition is dominated by n-alkanes. This is reflected in the total ion current chromatogram (Figure 2e), which indicates no significant biodegradation. In contrast, SVHCs extracted from water adjacent to the oil (Figure 2a) are compositionally similar to oils at Levels 3 to 4 in the biodegradation sequence of Volkman *et al.* (*21*), in which n-alkanes have been completely removed, but retain a significant aromatic hydrocarbon signature.

These compositional differences between the oil and oil-derived dissolved components may reflect biodegradation at the oil-water interface and subsequent water-oil partitioning of the biodegraded SVHCs into ground water. Typically, n-alkanes are the first petroleum component degraded and removed by either aerobic or aerobic biodegradation (*20-22*). The predominance of the UCM and absence of n-alkanes in our samples likely reflects this preferential removal. The highly aliphatic, hydrophobic DOC downgradient of the oil body (*10*) is the product of this degradation process. PAH can be aerobically and anaerobically degraded by naturally occurring microbial consortia (*23-27*), with degradation rates decreasing with increasing ring number (*28*). At the Bemidji site, PAH persist in ground water, both adjacent to the oil body and downgradient from it, reflecting relatively greater resistance to biodegradation (*22*) than the alkanes under anoxic conditions.

PAH Distributions in Ground Water. Hawley (29) calculated the expected aqueous concentrations of fluorene and phenanthrene for the Bemidji crude oil-ground water system at equilibrium and compared the calculated concentrations to our results from well 9015 (Table I). The screened interval of well 9105 is directly below the oil body, but at the time of sampling, ground water recovered from it did not contain observable quantities of free oil. Adjacent to the oil body, aqueous concentrations of phenanthrene and fluorene exceeded expected oil-water equilibrium concentrations, indicating that oil-derived DOC alters partitioning of PAH and other similar SVHC. The calculated equilibrium aqueous concentrations of fluorene and phenanthrene were 25% and 79%, respectively, of the observed concentrations for well 9015 (29). For comparison, we used the method described by Eganhouse *et al.* (14) to determine equilibrium aqueous concentrations of naphthalene, fluorene, and phenanthrene. This method differed from that of Hawley (29) in that aqueous equilibrium concentrations are calculated without making assumptions about oil composition. For both models, the effects of PAH sorbed onto aquifer solids were not considered. Using data from Mackay *et al.* (30) and a specific gravity of 0.85 for oil from well 318 (31), we calculated equilibrium aqueous naphthalene, fluorene and phenanthrene concentrations that were 27%, 64% and 72%, respectively, of the observed concentrations for well 9015. Our results confirm those of Hawley (29) and indicate that immediately beneath the oil, PAH concentrations are greater than oil-water equilibrium values.

The retardation coefficients for phenanthrene and fluorene also were estimated (29), and predicted transport distances were calculated based on a two-phase (aquifer-water) model. These calculated transport distances were compared to the distances that we observed phenanthrene and fluorene downgradient from the oil. The predictions underestimate the observed transport of fluorene and phenanthrene by 60% and 400%, respectively. Taken together, these results suggest that DOC or some component of the DOC acts to increase dissolved PAH concentrations in ground water above expected pure water equilibrium concentrations, enhancing transport.

In ground water downgradient from the oil body, DOC and dissolved PAH concentrations covary, although not linearly (Figure 3). An exponential curve model (designated by a straight line for each PAH) best fits these data, and shows that as DOC concentrations increase linearly, PAH concentrations increase exponentially. The correlation of each PAH to DOC is significant at the 99% confidence limit. The curves in the Figure 3 do not imply a causative relationship. The same processes that control the transport and attenuation of DOC, such as biodegradation, dilution, or partitioning onto aquifer solids, control the transport and attenuation of PAH. Ground water DOC concentrations, unlike PAH concentrations, may be supported by biodegradation of organic carbon sorbed onto aquifer material. However, the combination of support and attenuation processes must differ in magnitude or rate for DOC, compared to PAH, with PAH preferentially removed from solution.

In Figures 4a-d, the prevailing pattern of DOC and PAH concentrations downgradient from the oil body are plotted with respect to the main plume axis and the general direction of ground water flow (Figure 1). In Figure 4a, DOC concentrations decrease linearly with respect to distance downgradient from the leading edge of the oil body, and likely reflect losses due to dilution, degradation, and partitioning onto aquifer material. In contrast, naphthalene, fluorene, and phenanthrene concentrations (Figure 4b-d) decrease much more dramatically. The distribution of downgradient concentrations also vary between the three PAH. In one well, 9515C, all three PAH concentrations are anomalously low, while DOC concentrations follow the downgradient trends. We do not have an explanation for this anomaly, except that well 9515C was drilled and developed just prior to sample collection, and samples collected from it may reflect the drilling and development process.

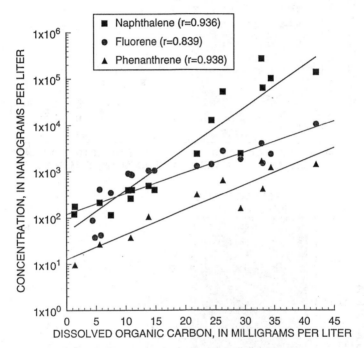

Figure 3. Property-property plot of dissolved organic carbon (DOC) versus naphthalene, fluorene, and phenanthrene for Empore extracts from the Bemidji study site. The three PAH are plotted on a logarithmic scale. DOC is plotted on a linear scale. The best-fit exponential curve is plotted for each PAH.

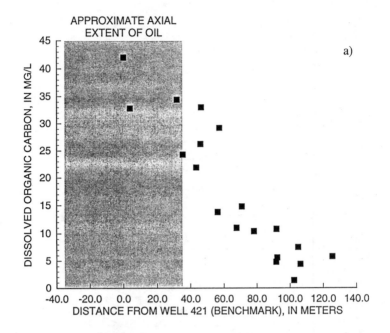

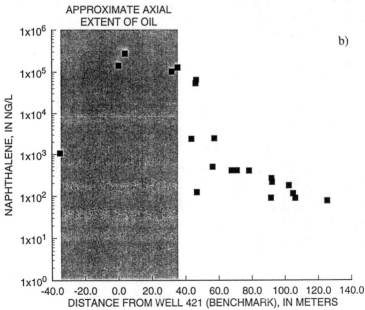

Figure 4. Axial distribution of a) dissolved organic carbon (DOC), b) naphthalene, c) fluorene, and d) phenanthrene along an axis parallel to general direction of ground water flow at the Bemidji study site and immediately downgradient from the leading edge of the oil body. The gray box indicates the approximate extent of the oil at the time of sample collection.

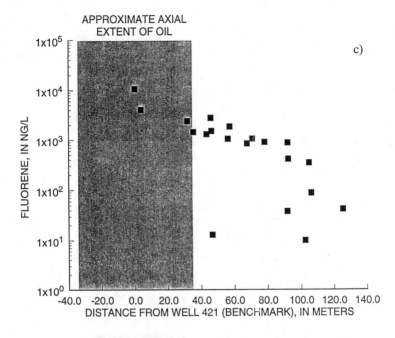

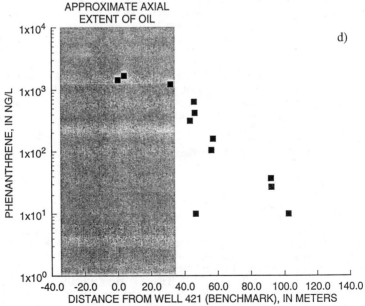

Figure 4. *Continued.*

Fluorene remained at elevated concentrations furthest downgradient to at least 60 m past the leading edge of the oil (Figure 4c). Across this distance, fluorene concentrations decrease from approximately 4,600 ng/L to approximately 600 ng/L, less than 1 order of magnitude. In contrast, naphthalene concentrations were variable but greatest closest to the oil body, and decreased by nearly three orders of magnitude at 25 meters downgradient (Figure 4b), from approximately 100,000 ng/L to 400 ng/L. Although detected less frequently than either fluorene or naphthalene, phenanthrene concentrations decreased exponentially from approximately 1,000 ng/L near the leading edge of the oil to approximately 30 ng/L at 55 meters downgradient of the leading edge of the oil (Figure 4d).

The differences in how naphthalene, fluorene, and phenanthrene concentrations are distributed downgradient reflect differences in the importance of volatilization, partitioning between aquifer solids and the mobile phase, and microbial degradation. Naphthalene is the most volatile and the most biodegradable of the three PAH (24, 28). The distribution of naphthalene concentrations downgradient is similar to the distributions of monoaromatic hydrocarbons reported by Eganhouse *et al.* over the same portion of the plume axis (14), including the magnitude of the step drop near 25 m. This likely reflects the importance of biodegradation and perhaps volatilization in controlling naphthalene concentrations. Aerobic and anaerobic half lives for naphthalene range from 0.5 to 28 days and 25 to 258 days, respectively (30). Unlike the monoaromatic hydrocarbons (14), naphthalene concentrations adjacent to the oil exceed those expected at equilibrium. However, as ground water is transported away from the oil body, microbial degradation controls the distribution of naphthalene and the monaromatic hydrocarbons.

In contrast, fluorene and phenanthrene distributions are less likely to be affected by biodegradation than naphthalene, since they are biodegraded less rapidly. The range of half lives of fluorene and phenanthrene overlap and range from 16 to 200 days under aerobic conditions and 64 to 200 days (estimated) under anaerobic conditions (30). Volatilization to the unsaturated zone is also less important, since fluorene and phenanthrene have much lower vapor pressures (0.715 and 0.113 Pascals, respectively; 30) than naphthalene (36.81 Pascals; 30). Partitioning between the DOC-containing ground water and the organic carbon and mineral matrix of the aquifer is the process most likely controlling fluorene and phenanthrene distributions downgradient. Although the downgradient distributions of these two PAH differ (Figure 4c and 4d), the differences cannot be ascribed to specific processes using only aqueous concentration data. Aquifer sediment PAH and organic carbon concentrations are needed to better model the behavior of fluorene and phenanthrene at this site, and to determine if the observed concentration distributions can be ascribed to differences in partitioning processes.

Conclusions.

In ground-water samples collected in 1995 downgradient from an oil spill, aqueous *n*-alkane concentrations are low. Aqueous hydrocarbon compositions are distinctly different from the parent oil with persistence of an unresolved complex mixture similar to moderately degraded petroleum. Directly beneath the oil, naphthalene, fluorene, and phenanthrene concentrations are at concentrations greater than would be predicted by equilibrium oil-water partitioning. Immediately downgradient from the oil, concentrations of each PAH decrease dramatically, with differing distributions for each PAH. For the same wells, exponential decreases in aqueous PAH concentrations are significantly correlated to linear decreases in DOC concentrations. The downgradient distribution of naphthalene is similar to that of previously reported monoaromatic hydrocarbons, and may similarly reflect the importance of biodegradation. Fluorene and phenanthrene concentrations persist further downgradient than naphthalene. The distributions of these two PAH may reflect partitioning processes between ground water and the aquifer. While the data presented are suggestive of the role of DOC in transport of persistent SVHCs, further work is needed. Determination of PAH and alkyl-PAH adsorbed onto aquifer solids, quantitative analysis of the concentration and distribution of individual alkyl naphthalenes, and estimation of aerobic and anaerobic biodegradation are required to better model the transport of PAH and by inference, SVHCs.

Acknowledgments.

We thank Tom Dorsey, Bennington College, for assistance in sample collection, and Galen Frohm, USGS-National Water Quality Laboratory, for help in sample extraction. Geoff Delin of the USGS-Minnesota District, and the Toxic Substances Hydrology Program, USGS, provided field support. Craig Markell and Rich Pieper, 3M Corporation, graciously provided the Empore disks used in this study. The comments of J. Leenheer and C. Rostad, National Research Program, USGS, and an anonymous reviewer substantially improved the quality of this manuscript.

Literature Cited

(1) Kan, A. T.; Tomson, M. B. *Environ. Toxicol. Chem.* **1990**, *9*, 253-263.
(2) Enfield, C.G.; Bengtsson, G. *Ground Water* **1988**, *26*, 64-70.
(3) Enfield, C.G.; Bengtsson, G.; Lindqvist, R. *Environ. Sci. Technol.* **1989**, *23*, 1278-1286.
(4) Amy, G.L.; Conklin, M.H.; Liu, H.; Cawein, C. In *Organic Substances and Sediments in Water. Volume 1--Humics and Soils*; Baker, R.A., Ed.; Lewis Publishers, Inc., Chelsea, MI; pp. 99-110.
(5) Liu, H.; Amy, G. *Environ. Sci. Technol.* **1993**, *27*, 1553-1562.
(6) McCarthy, J. F.; Zachara, J. M. *Environ Sci. Technol* **1989**, *23*, 496-502.
(7) Herbert, B. E.; Bertsch, P. M.; Novak, J. M. *Environ. Sci. Technol.* **1993**, *27*, 398-403.
(8) Johnson, W. P.; Amy, G. L. *Environ. Sci. Technol.* **1995**, *29*, 807-817.
(9) Magee, B. R.; Lion, L. W.; Lemley, A. T. *Environ. Sci. Technol.* **1991**, *25*, 323-351.
(10) Hult, M.F, *U.S. Geological Survey Water-Resources Investigations Report 84-4188*, 1984 pp. 1-15.
(11) Bennett, P.; Siegel, D. E.; Baedecker, M. J.; Hult, M. F. *Appl. Geochem.* **1993**, *8*, 529-549.
(12) Baedecker, M. J.; Cozzarelli, I.M.; Eganhouse, R.P.; Seigel, D.I.; Bennett, P.C. *Appl. Geochem.* **1993**, *8*, 569-586.
(13) Eganhouse, R. P.; Baedecker, M.J.; Cozzarelli, I.M.; Aiken, G.R.; Thorn, K.F.; Dorsey, T.F. *Appl. Geochem.* **1993**, *8*, 551-567.
(14) Eganhouse, R.P.; Dorsey, T.F.; Phinney, C.S.; Westcott, A.M. *Environ. Sci. Technol* **1996**, *30*, 3304-3312.
(15) Cozzarelli, I. M.; Baedecker, M. J.; Eganhouse, R. P.; Goerlitz, D. F. *Geochim. Cosmochim. Acta* **1994**, *58*, 863-877.
(16) Thorn, K.A.; Aiken, G.R. *U.S. Geological Survey Water-Resources Investigations Report 88-4220*, 1988, pp. 41-51.
(17) Furlong, E. T.; Koleis, J. C.; Gates, P. M. , Proceedings of the 43rd ASMS Conference on Mass Spectrometry and Allied Topics, May 21-26, 1995; p. 196.
(18) Landrum, P.F.; Nihart, S.R.; Eadie, B.J.; Gardner, W.S. *Environ. Sci. Technol.* **1984**, *18*, 187-192.
(19) Baedecker, M. J.; Eganhouse, R. P.; Lindsay, S. S.; U.S. *Geological Survey Water-Resources Investigations Report 84-4188*, 1984, p. 65-79.
(20) Tissot, B.P.; Welte, D.H. *Petroleum Formation and Occurence*, Second Revised and Enlarged Edition; Springer-Verlag: Berlin, 1984.

(21)   Volkman, J.K.; Alexander, R.; Kagi, R.I.; Rowland, S.J.; Sheppard, P.N. *Org. Geochem.* **1984,** *6,* 619-632.

(22)   Connan, J. In *Advances in Petroleum Geochemistry,.* J. Brooks and D. Welte, Eds. Academic Press, New York, NY, 1984, Vol. 1; pp. 299-335.

(23)   Madsen, E.L.; Sinclair, J.L.; Ghiorse, W.C. *Science* **1991,** *252,* 830-833.

(24)   Bauer, J.S.; Capone, D.G. *Appl. Environ. Microbiol.* **1985,** *50,* 81-90.

(25)   Madsen, E.L.; Mann, S.L.; Bilotta, S.E. *Environ. Toxicol. Chem.* **1996,** *15,* 1876-1882.

(26)   Madsen, E.L.; Thomas, C.T.; Wilson, M.S.; Sandoli, R.L.; Bilotta, S.E. *Environ. Sci. Technol.* **1996,** *30,* 2412-2416.

(27)   Schmitt, R.; Langguth, H.-R.; Puttmann, W.; Rohns, H.P.; Eckert, P.; Schubert, J. *Org. Geochem.* **1996,** *25,* 41-50.

(28)   Heitkamp, M.A.; Cerniglia, C.E. *Environ. Toxicol. Chem.* **1987,** *6,* 535-546.

(29)   Hawley, C.M. M.S. Thesis, University of Colorado, Boulder, CO 1996.

(30)   MacKay, D.; Shiu, W. Y.; Ma, K. C. *Illustrated Handbook of Physical-Chemcial Properties and Environmental Fate for Organic Chemicals. Volume II-Polynuclear Aromatic Hydrocarbons, Polychlorinated Dioxins, and Dibenzofurans;* Lewis Publishers: Chelsea, MI, 1992.

(31)   Landon, M.K.; Hult, M.F. *U.S. Geological Survey Water-Resources Investigations Report 91-4034,* 1991, pp. 641-645.

INDEXES

Author Index

Affiliation Index

Subject Index

Highlights from ACS Books

Desk Reference of Functional Polymers: Syntheses and Applications
Reza Arshady, Editor
832 pages, clothbound, ISBN 0–8412–3469–8

Chemical Engineering for Chemists
Richard G. Griskey
352 pages, clothbound, ISBN 0–8412–2215–0

Controlled Drug Delivery: Challenges and Strategies
Kinam Park, Editor
720 pages, clothbound, ISBN 0–8412–3470–1

Chemistry Today and Tomorrow: The Central, Useful, and Creative Science
Ronald Breslow
144 pages, paperbound, ISBN 0–8412–3460–4

Eilhard Mitscherlich: Prince of Prussian Chemistry
Hans-Werner Schutt
Co-published with the Chemical Heritage Foundation
256 pages, clothbound, ISBN 0–8412–3345–4

Chiral Separations: Applications and Technology
Satinder Ahuja, Editor
368 pages, clothbound, ISBN 0–8412–3407–8

Molecular Diversity and Combinatorial Chemistry: Libraries and Drug Discovery
Irwin M. Chaiken and Kim D. Janda, Editors
336 pages, clothbound, ISBN 0–8412–3450–7

A Lifetime of Synergy with Theory and Experiment
Andrew Streitwieser, Jr.
320 pages, clothbound, ISBN 0–8412–1836–6

Chemical Research Faculties, An International Directory
1,300 pages, clothbound, ISBN 0–8412–3301–2

For further information contact:
American Chemical Society
Customer Service and Sales
1155 Sixteenth Street, NW
Washington, DC 20036
Telephone 800–227–9919
202–776–8100 (outside U.S.)
The ACS Publications Catalog is available on the Internet at
http://pubs.acs.org/books

Bestsellers from ACS Books

The ACS Style Guide: A Manual for Authors and Editors
Edited by Janet S. Dodd
264 pp; clothbound ISBN 0–8412–0917–0; paperback ISBN 0–8412–0943–X

Writing the Laboratory Notebook
By Howard M. Kanare
145 pp; clothbound ISBN 0–8412–0906–5; paperback ISBN 0–8412–0933–2

Career Transitions for Chemists
By Dorothy P. Rodmann, Donald D. Bly, Frederick H. Owens, and Anne-Claire Anderson
240 pp; clothbound ISBN 0–8412–3052–8; paperback ISBN 0–8412–3038–2

Chemical Activities (student and teacher editions)
By Christie L. Borgford and Lee R. Summerlin
330 pp; spiralbound ISBN 0–8412–1417–4; teacher edition, ISBN 0–8412–1416–6

Chemical Demonstrations: A Sourcebook for Teachers, Volumes 1 and 2, Second Edition
Volume 1 by Lee R. Summerlin and James L. Ealy, Jr.
198 pp; spiralbound ISBN 0–8412–1481–6
Volume 2 by Lee R. Summerlin, Christie L. Borgford, and Julie B. Ealy
234 pp; spiralbound ISBN 0–8412–1535–9

From Caveman to Chemist
By Hugh W. Salzberg
300 pp; clothbound ISBN 0–8412–1786–6; paperback ISBN 0–8412–1787–4

The Internet: A Guide for Chemists
Edited by Steven M. Bachrach
360 pp; clothbound ISBN 0–8412–3223–7; paperback ISBN 0–8412–3224–5

Laboratory Waste Management: A Guidebook
ACS Task Force on Laboratory Waste Management
250 pp; clothbound ISBN 0–8412–2735–7; paperback ISBN 0–8412–2849–3

Reagent Chemicals, Eighth Edition
700 pp; clothbound ISBN 0–8412–2502–8

Good Laboratory Practice Standards: Applications for Field and Laboratory Studies
Edited by Willa Y. Garner, Maureen S. Barge, and James P. Ussary
571 pp; clothbound ISBN 0–8412–2192–8

For further information contact:
American Chemical Society
1155 Sixteenth Street, NW ♦ Washington, DC 20036
Telephone 800–227–9919 ♦ 202–776–8100 (outside U.S.)
The ACS Publications Catalog is available on the Internet at
http://pubs.acs.org/books

T
1 Month